AF577271

ECOLOGICAL CRISIS AND MANAGEMENT

Editor
Professor (Dr.) Arvind Kumar

2026
Daya Publishing House®
A Division of
Astral International Pvt. Ltd.
New Delhi – 110 002

First Published, 2006

Reprinted, 2026

ISBN: 978-81-7035-413-0 (HB)

Published by : **Daya Publishing House®**
A Division of
Astral International Pvt. Ltd.
– ISO 9001:2015 Certified Company –
4736/23, Ansari Road, Darya Ganj
New Delhi-110 002
Ph. 011-43549197, 8130496929
E-mail: info@astralint.com
Website: www.astralint.com

Printed at : Replika Press Pvt. Ltd.

Preface

Man is the real culprit for all the evils of today's ecological crisis. It is because of his progress and greed he damaged the fragile environment to an irreparable extent. Environmental issues have become more and more complex as a result of human beings unmindful exploitation of non-renewable natural resources. Population explosion, wrong development strategy and pollution are considered as series problems. Due to selfish motive of human beings nature has also reacted beyond the imagination of his supreme knowledge. Loss of food production, species extinction, global warming, changing sea-currents, occurrence of tsunami and ozone layer depletion are a few incidents of nature's anger towards the human attitude. Therefore, the need of the hour is to minimize the environmental threats by mass awakening right from grass root level to top ecoplanners as well as to develop management strategies so that we can lead a pollution free life. The main motto of this book is to trace some ways and means to improve the betterment of human beings without degrading natural resources.

Increasing poverty in developing and industrialized countries, global climate change, the loss of biodiversity, financial market crisis, rapid growth of megacities and growing unemployment due to the relocation of production abroad. These are just a few examples to illustrate the growing disparity between different world regions in social, economic and environmental terms. Phenomena of this kind and their roots in current globalization processes have been known and studied for decades. Now it is time to think about how to realize the concept of sustainable development with new aims and a new framework for the global economy. This event brought together leading representatives from science, policy and civil society in order to discuss strategies for reforming the world economy towards sustainability and for reducing injustice between world regions in economic, social and environmental terms. Keeping this view, the most recent articles of international level have been compiled to decipher the facts hidden hitherto to the students, researchers and academicians as well.

My special thanks and appreciation go to the scientists whose contributions have enriched this volume. I wish to express my sincere gratitude to Dr. P. C. Hembram, Hon'ble Vice Chancellor, S. K. M. University, Dumka who has been a source of constant inspiration. I am especially thankful to Professor (Dr.) B. N. Jha, Hon'ble Pro Vice-Chancellor, S. K. M. University, Dumka for his encouragement. I owe my special thanks to Professor M. C. Dash, Hon'ble Vice Chancellor of Sambalpur University, Professor

N. C. Datta of Calcutta University, Professor S. K. Konar of Kalyani University, Professor P. S. Murthy of Bangalore University, Professor P. Natarajan of Trivandrum University, Professor S. P. Roy of Bhagalpur University, Professor A. K. Quereshi of Bhopal University, Professor Tanmay Bhattacharya of Midnapore University, Professor P. C. Mishra of Sambalpur University, Professor B. D. Joshi of Hardwar University, Professor G. C. Pandey of Faizabad University, Professor K. C. Sharma of Ajmer University, Professor M. Raziuddin of Hazaribag University, Professor U. S. Bagde of Mumbai University, Dr. M. P. Sinha of Ranchi University, Professor Gurdeep Singh of I. S. M., Dhanbad, Dr. P. K. Goel of Karad, Dr. M. C. Varma and Shri T. Poddar of Bhagalpur University. I also acknowledge the incentives provided by one of my research scholars, Dr. Chandan Bohra, Head of Zoology, B. S. K. College, Barharwa. for helping me actively in bringing out this book.

I also express my deep sense of gratitude to my parents whose blessings have always prompted me to pursue academic activities deeply. I am also thankful to my sweet wife, Professor Kumari Bimla and my two lovely sons, Kumar Pallav Shivshankaran and Kumar Prasun Ramakrishnan whose natural smiles extended to me relief all through this tiresome endeavour.

Last but not the least, I am also thankful to Mr. Anil Mittal, Proprietor, Daya Publishing House, Delhi for taking keen interest in bringing out of this book. Finally, I will always remain a debtor to all my well-wishers for their blessings, without which this book would not have come into existence.

Dumka

Professor Arvind Kumar

Contents

	Preface	*v*
1.	**Vegetation Ecology Crisis of Macrophytes of Udhuwa Lake, Rajmahal (Jharkhand)** *Arvind Kumar and Chandan Bohra*	**1**
2.	**Environmental and Ecological Impact of Genetically Modified Organisms** *I. Sundar*	**10**
3.	**Influence of Thermal Stratification on Dissolved Oxygen in Subhas Sarobar, Kolkata** *N.R. Samal, D. Roy, A. Mazumdar and B. Bose*	**20**
4.	**Dust Accumulation Studies at Hyderabad** *Nirmala Babu Rao, I. Sobha Rani and Amena*	**32**
5.	**Water Quality Index for Ground Water Affected with Bicycle Manufacturing Industrial Wastes: An Environmental Quality Audit** *Vineeta Shukla, Sharda Abusaria, Monika Dhankhar and K.V. Sastry*	**39**
6.	**Changes in Sodium and Potassium Ratio in Pressmud with Influence of Earthworm *Eisenia foetida* (Savigny)** *Pramod Kumar and Shweta*	**46**
7.	**Status and Assessment of Noise Pollution Level at Rewa Town** *Bhupesh Kumar Mishra*	**50**
8.	**Age-related Changes in Biochemical Parameters in Long Bones of Indian Toad, *Bufo melanostictus*** *B.N. Andia and H.N. Behera*	**58**

9. **Distribution of *Salmonella* Antibodies in the Sera of Normal Individuals in Vivekanandha Hospital, Tiruchengode, Tamil Nadu** 72
S. Rajalakshmi, N. Kavitha and S. Uma

10. **Antimicrobial Activity of *Premna tomentosa* Willd. to Chosen Pathogenic Bacteria** 79
J. Matrtin Rathi and S. Gopalakrishnan

11. **Effect of Polyvinyl Pyrrolidone and Dimethyl Sulphoxide on Ethyl Cellulose Transdermal Patches of Verapamil Hydrochloride** 84
S.K. Sahoo, S. Chatterjee, D. Sahu, S.K. Mishra and B.B. Barik

12. **Effect of Scrotal Heating on the Reproductive Organs of the Laboratory Rat, *Rattus norwegicus*** 91
W. Vidyarani

13. **Magnesium in Scalp Hair and Fingernails in Relation to Different Parameters** 95
Rita Mehra and Meenu Juneja

14. **Heavy Metal Pollution in River Tunga, Bhadra and Tungabhadra at Kudali Near Shimoga, Karnataka** 100
G. Suresha, K. Ramadas and E.T. Puttaiah

15. **Chemical Impact on the Histological Studies on the Thyroid in the Freshwater Fish *Channa orientalis* (Sch.)** 104
S.V. Deshmukh and K.M. Kulkarni

16. **Thrombocytopenic Effect of Buprenorphine in Mice** 114
Dhriti Banerjee and Nirmal Kumar Sarkar

17. **Environmental Induced Changes in the Biomodal Gas Exchange and Haematology of Facultative Air-breathing Fish, *Mystus punctatus* (Jerdon)** 118
R. Devika, G.M. Natarajan, N. Parthi, P. Esther Joice, P. Palanisamy N. Bhuvaneshwari, G. Sasikala and S. Binu Kumari

18. **Investigation of Drinking Water Quality of Basavanhole Tank with Reference to Physico-chemical Characteristics** 122
J. Narayana, R. Purushothama, B.R. Kiran, K.P. Ravindra Kumar and E.T. Puttaiah

19. **Histological Alterations in Tadpoles of *Bufo Melanostictus*, Exposed to a Sublethal Concentration of Chromium** 127
D. Anusuya, D.J. Prakash and I. Christy

20. **Urban Development and Sound Level in Ichalkaranji City, Maharashtra** 131
C.T. Pawar and M.V. Joshi

21. **Studies on Limnological Characteristics of Guruvayanakere Pond Near Belthangady, S.K. District** 136
B.A. Kumara Hegde, G. Suresha, K. Ramadas and B. Yashovarma

22. **Diel Variation in Waterfowl During Winter at Sirpur Tank, Indore** **141**
Manjeet Malhotra, M.M. Prakash and K. Pawar

23. **Removal of Heavy Metals from Electroplating Industrial Effluent Using Plants–Phytoremediation** **152**
S. Anitha, V. Mahesh and C. Sheela Sasikumar

24. **Effects of DAP Toxicity on Certain Hematological Parameters of *Acridotheres tristis*** **157**
Anant Kumar, R.K. Yadav and B.P. Akela

25. **Comparison of Mosquito Fauna in Srivilliputhur Town and Krishnankovil Village, Tamil Nadu** **161**
K. Karuppasamy & T. Sooravan

26. **Biodegradation of Tannery Effluent Using *Pseudomonas aeruginosa* and *Bacillus subtilis*** **165**
P. Mythili, K.T.P.B. Ponneelan, R. Suchitra and M. Poonkothai

27. **Pesticide Disrupted Ovarian Physiology and Biochemistry of *Channa striata*** **169**
G.M. Natarajan, D. Kalavathi, N. Parthi, P. Esther Joice, P. Palanisamy, G. Sasikala and N. Bhuvaneshwari

28. **Effect of Metal Poisoning on Total Body Ascorbic Acid in *Sphaerodema rusticum* (Belostomatidae: Hemiptera)** **173**
S. Mumtazuddin and S. Ehteshamuddin

29. **Microbial Contamination in Drinking Water: Cause, Detection and Remedy** **178**
M.K. Bhutra and Ambica Soni

30. **Analysis and Seasonal Comparative Study of Amanishah Nallah and Neighbouring Ground Water Source in Sanganer Town, Jaipur** **187**
Dinesh Kumar, Hari Singh Shivran, Mahavir Prasad and R.V. Singh

31. **Status of Drinking Water Quality Awareness and its Impact on Student Health: A Study of Schools of Buldana District** **193**
S.V. Agarkar and B.S. Thombre

32. **Assessment of Pollution Soils Near Aluminium Industry (Industrial Estate) Warangal, Andhra Pradesh** **198**
B. Lalitha Kumari and M.A. Singara Charya

33. **Alteration in Oxygen Consumption in Freshwater Snail *Bellamya bengalensis* (Lamarck) During Pesticide Exposure** **206**
P.H. Rohankar & K.M. Kulkarni

34. **Estimating the Oxygen Requirement in a Mass Bathing Tank: A Case Study: Mahamaham 2004 (Southern Kumbhamela)** **210**
Ashutosh Das, V. Srihari, S. Madhan Babu and S. Ananthapadmanaban

35. A Report on Clinical Importance of Hypertension, Diabetes Mellitus and Heart Rate Associated with Acute Myocardial Infarction — 218
Rakesh K. Pandey, Arun K. Pandey, G.G. Potey, Arvind K. Pandey

36. Quenching of Diphenylamine by Benzoic Acid and Carbon Tetrachloride in Chloroform — 221
S. Bakkialalakshmi, B. Shanthi, D. Chandrakala and R. Santhi

37. Nutrient Status of Kanale Tank, Sagara Taluk with Reference to Diversity of Aquatic Macrophytes — 226
R. Purushothama, J. Narayana, B.R. Kiran and K. Harish Kumar

38. Antimicrobial Activity of *Syzygium aromaticum* (Clove) and *Millingtonia hortensis* (Indian Cork Tree) Against Human Pathogens — 231
A. Sujatha, B. Sandhya

39. Production and Biochemical Analysis of Bacteriocin (Nisin) from Immobilized and Non-immobilized *Lactococcus lactis* Subsp. Lactis — 236
D. Jegadeeshkumar, Suman Gulati, K. Revathi, K. Anbarasu, P. Vijayalakshmi and R. Kungumapriya

40. Species Distribution of the Family Verbenaceae in the Valley Districts of Manipur — 241
Y. Nanda Devi and P.K. Singh

41. Role of Zinc in Experimental Myocardial Infraction Influence of Abana: An Ayurvedic Formulation — 248
C. Sheela Sasikumar and C.S. Shymala Devi

42. Certain Herbal Ethno-medicines Used by the Karbi Tribe of Assam, India — 252
Sanjib Kumar Gogoi, Robindra Teron and Pratap J. Handique

43. Effect of Multifloral Honey on Blood Glucose Profile of Rabbits after Induced Lipidosis — 257
Shashikala, D. Belsare and B. Jayanti

44. Role of Soft Computing and Neural Network in Medical Analysis — 261
J. Justin Anand, M.S. Durairajan, J. Justin Suresh and P. Dhasarathan

45. *Symploca*: A New Report — 265
P.R. Sushama, Mary Esther Cynthia Johnson and P. Surekha Rani

46. Effect of Levels of Sodium Alginate on Chemical Quality of Dahi — 267
P.G. Chaudhari, D.K. Kamble, A.D. Kale and B.K. Pawar

47. Potential Difference Across the Cell of Skin around the Transmission Tower — 271
Vijay Kumar, Sachin Goyal, R.P. Vats and Tripti Johri

48. Antifertility Effect of Neem (*Azadirachta indica*) Bark in Male Albino Mice — 274
Anju Puri Chakravarty and Bir Hans

49.	**Palaeocurrent Analysis from Current Beddings and Pebbles of Tura Sandstone Group of Rocks Occurring in and Around Rangram Area, Meghalaya** *Mahananda Borah, Anima Gogoi and P.K Das*	278
50.	**Isolation and Characterization of Microorganisms Found in Gums and Adhesives from Post Offices of Madurai City** *G. Jayavarshni, A. Angelin, P. Mourin Rosee, M. Rajalakshmi, Mary Esther Rani*	285
51.	**Investigation of Physico-chemical Condition of Major Effluent Points of Nambul River, Manipur, India** *R.K Rajeshwari Devi*	289
52.	**Economic Appraisal of Biogas for Cooking and Electricity Generation: A Case Study** *Er. Sarbjit Singh Sooch, Er. Deepak Gupta and Indervir Singh*	294
53.	**Comparative Study of Community Biogas Plant Versus Family Size Biogas Plant: A Case Study** *Er. Sarbjit Singh Sooch, Er. Deepak Gupta and Sukhwinder Singh*	300
54.	**A Study of Arsenic Effect on Blood and Tissue Glucose Concentrations in a Fish Model** *Nimai Chandra Saha, Trilochan Midya and Nirmal Kumar Sarkar*	305
55.	**Season-induced Metabolic Alterations in *Abelmoschus esculentus* (L.) Moench (Okra)** *Supatra Sen and S. Mukherji*	310
56.	**Air Borne Fungal Spores at Sharavathi Reservoir, Karnataka, India** *K. Harish Kumar, B.R. Kiran, R. Purushotham, E.T. Puttaiah and S. Manjappa*	319
57.	**Hydrological Studies for Dams in Ungauged Small Catchments: A Case Study** *Er. Sarbjit Singh Sooch, Er. Deepak Gupta and N.K. Khullar*	324
58.	**Prediction of Nitrate Pollution of Groundwater: A Case Study** *Sarbjit Singh Sooch, Baljeet S. Kapoor, Bijay Singh and N.S. Grewal*	329
59.	**Design Flood by Frequency Analysis for Relatively Ungauged Small Watersheds: A Case Study** *Er. S.S. Sooch, N.K. Khullar, Er. Deepak Gupta*	337
60.	**Effectiveness of Three Botanicals Against Dermatophytes** *B.K. Dutta, I. Rahman and S. Karmakar*	347
	Index	353

[illegible]

56. [illegible] Spores [illegible] Karnataka, India 313

[illegible]

Index 353

Chapter 1
Vegetation Ecology Crisis of Macrophytes of Udhuwa Lake, Rajmahal (Jharkhand)

Arvind Kumar and Chandan Bohra

Environmental Science Research Unit, Post Graduate Department of Zoology, S.K.M. University, Dumka – 814 101, India

Introduction

Macrophytes are the most important constituents for the regulation of cycling of minerals and organic compounds in the natural aquatic system. One of the significant properties of the aquatic macrophytes is the capability of accumulating the heavy metal pollutants present in water bodies (Carpenter and Adam, 1977).

Floristic list of a particular area give reliable background information about species diversity in a community as each plant species has its own specific ecological amplitude and the same indicates the ecological nature of the habitat. It is one of the major anatomical characters of the plant community (Dansereau, 1960).

Life forms were studied extensively by Ellenberg and Mueller-Dombois (1967) and Mueller-Dombois and Ellenberg (1974). Besides the composition of the species, the life form composition in the plant community is found to be of special interest as the same may provide information of the response of community of particular environmental factors on the utilization of space and on the probable comparative relations within a community (Mueller-Dombois and Ellenberg, 1974).

Freshwater macrophytes play important role in aquatic ecosystem. They provide either directly or indirectly food, shelter and a variety of habitats for a large number of organisms, including wild fowl and economically important fish. Many aquatic plants like trapa, wild rice, lotus, typha, cyperus, bacopa, water hyacinth etc. are used as food, raw material for industrial processes, building materials, medicine and manure in agriculture. Macrophytes absorb dissolved minerals and enrich water with oxygen produced during photosynthesis. These properties are of benefit to man as they assist in the maintenance of clean water and the help in the recovery of polluted water. However, in disturbed water bodies rampant growth of aquatic plants may interfere with man's use of freshwater. They may obstruct water flow, navigation of water intakes, fish production and crops in aquatic environment on irrigated land, they may also create conditions favourable for pests, diseases and vectors affecting humans, animals and crop plants. They may also upset recreation persuits. Macrophytic communities are more productive per unit area than phytoplankton under comparable condition (Westlake, 1963), but it may be insignificant in deep water lakes (Gessner, 1959; Westlake, 1960 and Straskraba, 1963). The role of macrophytes in regulation of mineral nutrient concentrations in water and sediments of lakes and the relation of their appearance and biomass production in different physico-chemical conditions is very important. Under stress conditions the number of rare species is usually reduced. Thus, their presence in any community becomes important if we wish to determine the effect of pollution on the species distribution in an ecosystem. Under severe stress conditions, only a few species may survive and their frequency of occurrence may be very high. Studies on growth physiology of macrophytes especially on *Hydrilla* sp., *Myriophyllum* sp., *Elodea canadensis* and *Potamogeton* sp. will greatly contribute to the success of using these plants and others as bioindicators of aquatic pollution (Bhargava, 1988).

In Jharkhand, very little information is available till data on the studies pertaining to the freshwater ecosystem in relation to the ecology of macrophytes except the work of Kumar and Bohra (2002a and b) and Bohra and Kumar (2002). So, an attempt has been made in the present study to investigate the floristic distribution and life-form classification of macrophytes in Udhuwa lake.

Study Site

Udhuwa lake is located about 42 km from Sahibganj and 15 km. Barharwa, It is situated on the bank of the Ganges about 10 kms. South-east of Rajmahal. It is very old eutrophic lake. The total area of this lake is 5.65 km². The dimensions are found to be maximal in the rainy season. Average depth is about 1.5 meter. The morphology as well as the bathy matric characters of the lake is depicted in Table 1.1.

Table 1.1: Morphometry and Bathymetric Characters of Udhuwa Lake

Altitude (m)	150	Longitude	87°47' E
Latitude	24°52' N	Surface area (km²)	5.65
Maximum depth (m)	2.25	Minimum depth (m)	0.75
Mean depth (m)	1.50	Mean depth/maximum depth ratio	0.61
Basin Shape	Salter	Basin Slope	Gentle
Lake bottom	Silted	Shore-line (km)	2.52
Shore-line development ratio	1.25	Volume of lake (Cu. km)	0.0065
Surface area/volume ratio	869.23	Development of volume	2.01
Mean maximum temperature (°C)	31.15	Mean minimum temperature (°C)	20.78
Mean annual rainfall (mm)	113.58		

For the present investigation five different sites (Sites I, II, III, IV and V) were selected.

Materials and Methods

For studying the floristic composition, the five different sites were surveyed at monthly intervals throughout the year during the study period (2003). The macrophytes were classified after Rauknkiaer's life forms classification as modified by Ellenberg and Mueller-Dombois (1967) and Mueller-Dombois and Ellenberg (1974).

Results

A total number of 23 macrophytes were reported from different study sites. The total number of species recorded from each study site is depicted in Table 1.2. 18 species were recorded in site I, 2.2 species in site II, 18 species in site III, 17 species in site IV and 15 species in site V. Out of the total species, *Ceratophyllum demersum, Salvinia cuculata, Nymphaea* sp. and *Eichhornia crassipes* were found through out the different seasons in all the different sites. *Alternanthera sessilis* was found throughout the year only in site II and V. *Hydrilla verticillata* was found through out the year in all the study sites except in site II. *Naja* sp. was recorded only in site IV during the whole study period. The other remaining species were recorded occasionally. Among the occasionally distributed species, *Otellia alismoides, Ipomea aquatica, Eleocharis* sp. and *Rumek* sp. were found to be present only in site II. The group wise distribution of the 23 species have been depicted in Table 1.2. Out of the 23 species in Table 1.3, 5 species were recorded under the submerged group such as *Naja* sp., *Vallisnaria spiralis Hydrilla verticillata, Ottelia alismoides* and *Ceratophyllum demersum.* 11 species belonged to floating group and 7 species were recorded under the emergent group of the five study sites, the sites III, IV and V had 4 submerged species each while site II had 5 and site I had 3 submerged species each, sites I, III and IV recorded 9 floating species each while site II recorded 10 and site V recorded 6 floating species respectively. So far as the emergent species are concerned, sites III and V had 5 species site I had 6 species, site II had 7 species and site IV had 4 species.

Table 1.2: Group-wise Distribution of Macrophytes in the Five Study Sites

Sl.No.	*Name of Macrophytes*	*Site I*	*Site II*	*Site III*	*Site IV*	*Site V*	*Whole Lake*
1.	Submerged	3	5	4	4	4	5 (21.74%)
2.	Floating	9	10	9	9	6	11 (47.83%)
3.	Emergent	6	7	5	4	5	7 (30.43%)
	Total no. of Macrophytes	**18**	**22**	**18**	**17**	**15**	**23**

Out of the 23 species recorded one species belongs to the each family Poltedoriaceae, Salviniaceae, Marsileaceae, Poaceae, Gentianaceae, Lemnaceae, Onagraceae, Potamogitanaceae, Najadaceae, Ceratophyllaceae, Cyperaceae, Verbinaceae and Amranthaceae. Family Convulvulaceae, Polygonaceae and Nymphaceae contributed 2 species each while family Hydrocharitaceae had contributed 3 species.

The family Hydrocharitaceae contributed 13.04 per cent of the total species while the families Convulvulaceae, Polygonaceae and Nymphaceae contributed 8.70 per cent each of the total species. 14 families such as Pontedoriaceae, Salviniaceae, Marsileaceae, Poaceae, Gentianaceae, Lemnaceae, Onagraceae, Poaceae, Potamogitanaceae, Najadaceae, Ceratophyllaceae, Cyperaceae, Verbinaceae and Amranthaceae had equally contributed 4.34 per cent each (Table 1.4).

Table 1.3: Floristic Composition in the Study Sites

Sl.No.	*Name of Plant Species*	*Family*	*Jan*	*Feb*	*Mar*	*Apr*	*May*	*Jun*	*Jul*	*Aug*	*Sep*	*Oct*	*Nov*	*Dec*
1.	*Alternanthera sessilis* (Mart) Griseb	Amaranthaceae	+	+	+	+	+	+	+	+	+	+	+	+
2.	*Ceratophyllum demersum* Linn.	Ceratophyllaceae	+	+	+	+	+	+	+	+	+	+	+	+
3.	*Eichhornia crassipes* (Mart.) Solms	Pontendoriaceae	+	+	+	+	+	+	+	+	+	+	+	+
4.	*Eleocharis* sp.	Cyperaceae	+	+	+	–	–	–	–	+	+	+	+	+
5.	*Hydrilla verticillata* (Linn.f.) Royal	Hydrocharitaceae	+	+	+	+	+	+	+	+	+	+	+	+
6.	*Hygroryza aristata* (Retz) Nees	Poaceae	+	+	+	+	+	+	+	+	+	+	+	+
7.	*Ipomea aquatica* Forsk	Convulvulaceae	+	+	+	+	+	+	+	+	+	+	+	+
8.	*I. Cornea* Forsk	Convulvulaceae	+	+	+	–	–	–	–	+	+	+	+	+
9.	*Jussiaea adescendens* Hora	Onagraceae	+	+	+	+	+	+	+	+	+	+	+	+
10.	*Limnantherum cristatum*	Gentianaceae	–	–	+	+	+	+	+	+	+	+	+	+
11.	*Marsilea minuta* Linn.	Marsileaceae	+	+	+	+	–	–	–	–	–	+	+	+
12.	*Naja* sp. Allioni	Najadaceae	+	+	+	+	–	–	–	+	+	+	+	+
13.	*Nelumbium* sp. Gaertz	Nymphaceae	–	–	+	+	+	+	+	+	+	+	+	+
14.	*Nymphaea stellata* Wild	Nymphaceae	–	–	–	–	–	–	–	–	+	+	+	–
15.	*Ottelia alismoides* Pers.	Hydrocharitaceae	+	+	+	+	–	–	–	+	+	+	+	+
16.	*Phylla nodiflora*	Verbinaceae	+	+	+	–	–	–	–	–	+	+	+	+
17.	*Polygonum* sp.	Polygonaceae	+	+	+	–	–	–	+	+	+	+	+	+
18.	*Potamogeton indicus* Lin.	Potamogitanaceae	+	+	+	+	+	+	+	+	+	+	+	+
19.	*Rumux* sp.	Polygonaceae	+	+	+	+	–	–	–	+	+	+	+	+
20.	*Sacctiolepis*	Poaceae	+	+	+	+	+	+	–	–	+	+	+	+
21.	*Salvenia cuculata* Hoffim	Salviniaceae	+	+	+	–	–	–	–	–	+	+	+	+
22.	*Spirodela polyrhiza* Lin.	Lemnaceae	+	+	+	–	–	–	+	+	+	+	+	+
23.	*Vallisnaria spiralis* Lin.	Hydrocharitaceae	+	+	+	+	+	+	+	+	+	+	+	+

+: Denotes presence; –: Denotes absence.

Table 1.4: Family-wise Distribution of the Macrophytes

Sl.No.	Name of the Family	No. of Species	Percentage Compositions
1.	Hydrocharitaceae	3	13.04 per cent
2.	Convulvulaceae	2	8.70 per cent
3.	Polygonaceae	2	8.70 per cent
4.	Nymphaceae	2	8.70 per cent
5.	Pontedoriaceae	1	4.34 per cent
6.	Salviniaceae	1	4.34 per cent
7.	Marsileaceae	1	4.34 per cent
8.	Poacae	1	4.34 per cent
9.	Gentianaceae	1	4.34 per cent
10.	Lemnaceae	1	4.34 per cent
11.	Onagraceae	1	4.34 per cent
12.	Potamogitanaceae	1	4.34 per cent
13.	Najadaceae	1	4.34 per cent
14.	Ceratophyllaceae	1	4.34 per cent
15.	Cyperaceae	1	4.34 per cent
16.	Verbinaceae	1	4.34 per cent
17.	Amranthaceae	1	4.34 per cent

Life Forms

The microphytic species have been categorized into four major life forms *viz.*, Therophytes, Geophytes, hemicryptophytes and Errant vascular hydrophytes (Table 1.5). The therophytes included 8 species *viz.*, *Jussiaea adescendens*, *Sacciolepis intemipta*, *Potamogeton indicus*, *Eleocharis* sp., *Phylla nodiflora*, *Alternanthera sessilis*, *Rumex* sp. and *Polygonum* sp. The Geophytes had 2 species *viz.*, *Nelumbium* sp. and *Nymphaea* sp. The Hemicryptophytes comprised 6 species belonging to the floating and submerged communities *viz.*, *Ipomoea aquatica*, *I. Cornea*, *Hygroryza aristata*, *Limnan themum*, *cristatum*, *Spirodela polyrhiza* and *Hydrilla verticillata*. The submerged and free floating species like *Naja* sp., *Vallisnaria spiralis*, *Ottelia alismoides Ceratophyllum demersum*, *Eichhornia crassipes*, *Salvinia cuculata* and *Marsilea minuta* were included in Errant vascular Hydrophytes.

The per cent contribution of the different life forms have also been calculated. Therophytes have got the highest percentage of species (34.78 per cent) followed successively by Errant vascular hydrophytes (30.43 per cent), hemicryptophytes (26.09 per cent)and Geophytes (8.70 per cent). From a comparison with the normal biological spectrum of Raunkiaer (1934), it is evident that percentage of Therophytes is about two to three times higher than that in the normal biological spectrum. The percentage of Geophytes is about 1.5 times higher than that in the normal spectrum while the percentage of the Hemicryptophytes has been found almost equal to the normal biological spectrum (Table 1.6).

Table 1.5: Life Form Classification of Macrophytic Species in Udhuwa Lake

Life Form Group	*Name of Species*	*No. of Species*	*% Composition*
Therophytes (Th)	*Alternanthera sessilis*	8	34.78 per cent
	Eleocharis sp.		
	Jussiaea adescendens		
	Phylla nodiflora		
	Polygonum sp.		
	Potamogeton indicus		
	Rumex sp.		
	Sacciolepis intemipta		
Geophytes (G)	*Nelumbium* sp.	2	8.70 per cent
	Nymphaea sp.		
Hemicryptophytes (H)	*Hydrilla verticillata*	6	26.09 per cent
	Hygroryza aristata		
	Ipomoea aquatica		
	I. cornea		
	Limnanthemum cristatum		
	Spirodela polyrhiza		
Errant vascular (EVH)	*Hydrophytes ceratophyllum demersum*	7	30.43 per cent
	Eichhornia crassipes		
	Marsilea minuta		
	Naja sp.		
	Ottelia alismoides		
	Salvenia cuculata		
	Vallisnaria spiralis		

Table 1.6: Biological Spectrum of the Flora of Udhuwa Lake

Parameter	*Life Form*						
	Th	*G*	*H*	*Ch*	*Ph*	*EYH*	*Total*
Number of species	8	2	6	–	–	7	23
Life-form (%)	34.78%	8.70%	26.09%	–	–	30.43%	100
Raunkiaer's Normal spectrum (%)	13	6	26	9	46	–	100

Discussion

In the present investigation, altogether 23 macrophyte species were recorded and the plant species were grouped under different categories *viz.*, floating, submerged and emergent. Under the submerged group 5 species were recorded, 11 species in free floating and 7 species in the emergent. The present findings are in agreement with the findings of Kaul *et al.* (1978) in some water bodies of Kashmir where out of 43 species, 28 species were found belonging to the emergent, 7 species belonging to the rooted with floating leaves while 8 species were included in the submerged group. Handoo and Kaul (1982) reported 58 species in the wetlands of Kashmir, out of which 23 species were in the emergent group, 21 were the ground layer species, 7 species in the floating leaves and 7 species in the submerged. According to Shah and Abbas (1979), out of 28 species in the river Ganga at Bhagalpur, 22 were found to be emergents, 4 submerged and 2 floating species. Billore and Vyas (1982) recorded 3 floating, 7 submerged and 4 emergent species in the Pichhola lake, Udaipur. Devi (1993) reported 86 species in the Loktak lake, Manipur, out of which 73 species (all emergents) were recorded in Phumdi (floating mats) regions while 13 species were recorded in the non-phumdi (clear water) portion.

The non-phumdi species were divided into different categories *i.e.* 6 species in submerged, 4 species in the free floating and 3 species in the rooted with floating leaved species. Okram *et al.* (1996) reported 18 emergents 4 free–floating, 6 rooted with floating leaved and 6 submerged species in the Waithou lake, Maniupur. Devi and Sharma (1998) also studied the vegetation ecology of macrophytes of utrapat lake, Manipur.

In the present study no clear-cut zonations of floating, emergent and submerged species could be observed and as such all the communities are found accruing in intermized mats. Seshavatharam *et al.* (1982) observed the macrophytic species of emergent, submerged and floating groups occurring in intermixed mats in the Kolleru lake (Andhra Pradesh). Swindale and Curtis (1957) and Schmid (1965) also reported heterogeneous composition in the submerged vegetations of USA. Crowder *et al.* (1977) also reported heterogeneous composition of submerged vegetation in lake Opimicon (Cañada).

In the present investigation, out of23 species recorded, the family Hydrocharifaceae had contributed 3 species. The families Convulvulaceae Polygonaceae and Nymphaceae had contributed 2 species each while the families Pontedoriaceae, Salviniaceae, Marsileaseae, Gentianaceae, Lemnaceae, Onagraceae, Poaceae, Potamogitanaceae, Najadaceae, Cerattophylaceae, Cyperaceae, Verbinaceae and Amranthaceae had contributed 1 species each.

Out of four life forms, the Therophytes attained maximum percentage of 34.78 per cent followed by Errant Vascular hydrophytes (30.43 per cent), Hemicryptophytes (26.09 per cent) and Geophytes (8.70 per cent). Based on the percentage composition of the various life form classes macrophytes, the lake may be assigned to the Thero-Errant Vascular Hydrophytic type of phytoclimate.

The Udhuwa lake under the present investigation has got unique morphometric and bathymetric observations. Among the various parameters and occurrence of higher surface area, volume ratio, saucer shaped basin with silted bottom and gentle slope, higher ratio of mean depth to maximum depth and high figure of the shore line development ratio in the lake indicated the eutrophic nature of the lake. As already discussed the lake is rich in the number of macrophyte species belonging to the floating and emergent species. The occurrence of Thero-Errant vascular type of phytoclimate in the lake signifies that the lake water and sediments are very rich in nutrients. Thus, these observations do substantiate the fact that the eutrophication has set in at fast rate in the lake and necessary remedial measures to check the eutrophication in the lake has become indispensable to prevent the conversion of the lake into an aquatic desert.

References

Billore, D.K. and Vyas, L.N. 1982. Distribution and production of macrophytes in Pichhola lake, Udaipur (India). In: *Ecology and Management*, (Eds.) Gopal, B. *et al.* Int. Sci. Publ., Jaipur, p. 45–54.

Bohra, C. and Kumar, A. 2002. Bio-eradication of curse macrophytes (*Eichhoria crassipes*) from wetlands of Jharkhand: An applied approach. In: *Ecology of Polluted Water,* (Ed.) Kumar, A. Ashish Publ. House, New Delhi, p. 621–628.

Carpenter, S.R. and Adams, M.S. 1977. The macrophyte tissue nutrient pool of a hard water eutrophic lake: Implication for macrophyte harvesting. *Aquat. Bot.*, 3: 239–255.

Crowder, A.A., Bristow, J.M., Kind, M.R. and Vander, K.S. 1977. Distribution, seasonality and biomass of aquatic macrophytes in lake opinicon. *Naturalist Can.*, 104: 441–456.

Dansereau, P. 1960. The origin and growth of plant communities. In: *Growth in Living Systems*, Proc. of an Int. Symp. Held at Purdue Univ. Basic Books Inc., New York, p. 567–603.

Devi, N.B. 1993. Phytosociology, primary production and nutrient status of macrophytes of Loktak lake, Manipur. *Ph.D. Thesis*, Manipur University.

Devi, K.I. and Sharma, B.N. 1998. Vegetation ecology of macrophytes of Utrapat lake, Manipur India. *J. Freshwater Biol.*, 10: 1–9.

Ellenberg, H. and Mueller-Dombois, D. 1967. A key to Raunkiaer plant life forms with revised subdivisions. Ber. Geobot. Inst. Eth. Stiftg. Ruble. Zirich., 37: 56–73.

Gessner, F. 1959. Hydrobotanik: die physiologischen Grundlagen der flanzenverbreitung in Wasser. II. Stoffhaushalt VED Dentscher Verlag der Wissenschaften, Berlin.

Handoo, J.K. and Kaul, V. 1982. Phytosociological and standing crop studies in wetlands of Kashmir. In: *Wetlands Ecology and Management*, (Eds.) B. Gopal *et al.* Int. Sci. Publ., Jaipur, India, p.187–195.

Kaul, V., Trisal, C.L. and Handoo, J.K. 1978. Distribution and production of macrophytes in some water bodies of Kashmir. In: *Glimpses of Ecology,* (Eds.) Singh, J.S. and Gopal, B. Int. Sci. Publ., Jaipur, p. 313–334.

Kumar, A. and Bohra, B. 2002a. Impact of environmental stress on the growth behaviour of water hyacinth, *Eichhornia crassipes* (Maets) with special reference to removal of pollutants. In: *Ecology and Ethology of Aquatic Biota*, (Eds.) Kumar, A. Daya Publ. House, Delhi, p. 345–353.

Kumar, A. and Bohra, C. 2002b. Periodicity and biomass potentials of macrophytes in polluted aquatic environment of Jharkhand. In: *Ecology of Polluted Waters*, (Ed.) Kumar, A. Ashish Publ. House, New Delhi.

Mueller-Dombois, D. and Ellenberg, H. 1974. *Aims and Methods of Vegetations Ecology*. John Wiley and Sons.

Okram, I.D., Sharma, B.M. and Singh, E.J. 1996. Ecology study of Waithou lake, Manipur Morphometry and qualitative analysis of microphytic vegetation. *J. Freshwater Biol.*, 8: 177–189.

Raunkiaer, C. 1934. *The Life Forms of Plants and Statistical Plant Geography*. Clarendon Press, Oxford.

Schmid, W.D. 1965. Distribution of aquatic vegetation as measured by line Intercept with SCUBA. *Ecology*, 46: 816–823.

Seshavatharam, V., Datt, B.S.M. and Venu, P. 1982. An ecological study of the vegetation of the Kollenu lake. *Bull. Bot. Surv., India*, 24: 7–75.

Shah, J.D. and Abbas, S.G. 1979. Seasonal variation in frequency, density, biomass and rate of production of some aquatic macrophytes of the river Ganges at Bhagalpur (Bihar). *Trop. Ecol.*, 20: 127–134.

Straskarba, M. 1963. The share of the littoral region in the productivity of two ponds in Southern Bohemia. Rozpr. Css. Aead Ved. (Mat. Prirod. Ved.), 73: 1–63.

Swindale, D.N. and Curtis, T.T. 1957. Phytosociology of the larger submerged plants in Wisconsin lake. *Ecology*, 38: 397–407.

Westlake, D.F. 1960. Water weed and water management. *Instn. Publ. J.*, 59: 148–160.

Westlake, D.F. 1963. Comparison of plant productivity. *Biol. Rev.*, 38: 385–425.

Chapter 2
Environmental and Ecological Impact of Genetically Modified Organisms

I. Sundar

Lecturer in Economics, Directorate of Distance Education, Annamalai University, India

ABSTRACT

Genetically modified organisms are the result of gene manipulation. This involves artificially inserting a gene from one organism into another producing a change in a plant or animals' biological characteristics through means other than conventional breeding programmes. Organisms that contain an artificially inserted gene are known as transgenic. This paper analyses the implication of genetically modified organisms on ecology and environment. The implication of such GMOs is the subject of widespread debate as to the safety and efficacy of the new products and the ethical and socio-economic issues surrounding their development and use. This paper concludes with some policy measures to overcome the negative environmental and ecological impacts of genetically modified organisms.

Introduction

There is a growing public concern about the risks and benefits of genetically modified organisms. In addressing the risks posed by the cultivation of plants in the environment five environmentally related safety issues are often raised. These include the following: the first one is gene transfer meaning the movement of genes from a crop through out crossing with wild relatives to form new hybrid plants. The second one is weediness meaning the tendency of a plant to spread beyond the field where first planted and to establish itself as a week species. The third one is trait effects meaning effects of traits that are potentially harmful to non-target organisms. The fourth one is genetic and phenotypic variability meaning the tendency of a plant to exhibit unexpected characteristic. The last one is genetic material from pathogens such as the risk of genetic recombinations following mixed virus infections. The application of modern biotechnology to agriculture, particularly the development of genetically modified foods and other living modified organisms are subject to widespread public debate as to the safety and efficiency of the new products and the ethical and socio economic issues surrounding their development and use. This paper makes a comprehensive attempt to analyze the impacts of genetically modified organisms including transgenic plants and transgenic animals.

Transgenic Plants

Genetic engineering is becoming a useful tool in the improvement of plants but concern has been expressed about the potential environmental risks of releasing genetically modified (GM) organisms into the environment. Attention has focused on pollen dispersal as a major issue in the risk assessment of transgenic crop plants (De Marchis, Bellucci and Arcioni, 2003).

Genetic engineering is becoming a useful tool in the improvement of plants and plant-based raw materials. Varieties with value-added traits are developed for nonfood use in industrial and medical production, and different production lines must be kept separate. For good management practices, knowledge of relevant gene flow parameters is required (Ritala, Nuutila, Aikasalo, Kauppinen and Tammisola, 2002).

The process of introgression between a transgenic crop modified for better agronomic characters and a wild relative could lead potentially to increased weediness and adaptation to the environment of the wild species. However, the formation of hybrid and hybrid progeny could be associated with functional imbalance and low fitness, which reduces the risk of gene escape and establishment of the wild species in the field (Gueritaine, Sester, Eber, Chevre and Darmency, 2002).

Alien transgene escape from genetically engineered rice to non–transgenic varieties or close wild relatives including weedy rice may lead to unpredictable ecological risks. However, for transgene escape to occur three conditions need to be met:

1. Spatially, transgenic rice and its non-transgenic counterparts or wild relatives should have sympatric distributions;
2. Temporally, the flowering time of transgenic rice and the non-transgenic varieties or wild relatives should overlap; and
3. Biologically, transgenic rice and its wild relative species should have such a sufficiently close relationship that their interspecific hybrids can have normal generative reproduction (Lu, Song, and Chen, 2003).

Gene flow and introgression from cultivated to wild plant populations have important evolutionary and ecological consequences and require detailed investigations for risk assessments of transgene escape into natural ecosystems. Sugar beets *Beta vulgaris* ssp. vulgaris are of particular concern because:

1. They are cross-compatible with their wild relatives the sea beet, *B. vulgaris* ssp. maritima;
2. Crop-to-wild gene flow is likely to occur via weedy lineages resulting from hybridization events and locally infesting fields. Thus it is an environmental problem (Arnaud, Viard, Delescluse and Cuguen, 2003).

Pollen can function as a vehicle to disseminate introduced, genetically engineered genes throughout a plant population or into a related species. The measurement of the risk of inadvertent dispersal of transgenes must include the assessment of accidental dispersion of pollen. Factors to be considered include the rate of pollen spread, the maximal dispersion distance of pollen, and the spatial dynamics of pollen movement within seed production fields; none of which are known for alfalfa (*Medicago sativa* L.), an insect-pollinated crop species (Snow, Uthus and Culley, 2001).

Potential risks of gene escape from transgenic crops through pollen and seed dispersal are being actively discussed and have slowed down full utilization of gene technology in crop improvement. To ban the trans gene flow, barren zones and 'terminator' technology were developed as GMO risk management technologies in transgenic crops. Unfortunately, the technologies have not protected reliably the transgene migration to wild relatives (Driessen, Pohl and Bartsch, 2001).

Weed beets pose a serious problem for sugar beet *Beta vulgaris* crops. Traditionally, the only efficient method of weed control has been manual removal, but the introduction of transgenic herbicide-tolerant sugar beets may provide an alternative solution because non-tolerant weed beets can be destroyed by herbicide. The appearance of transgenic weed beets is possible but can best be retarded if the transgenic for herbicide tolerance is incorporated into the tetraploid pollinator breeding line (Desplanque, Hautekeete and Van Dijk, 2002).

The spread of plant populations into new habitats and their establishment there is an important object of ecological research, particularly for those plant species consisting of cultivated, weedy and wild forms. One of these species is *Beta vulgaris* L., which is taxonomically divided into the subspecies vulgaris (sugar beet, Swiss chard, red beet, some weedy beet forms) and maritima (sea beet). In northeastern Europe, the distribution range of the different Beta subspecies is not well known. The fact that sea beet populations can now be found sympatrically to flowering sugar beets and weed beets in coastal agricultural areas offers new possibilities of gene flow. This is important fundamental knowledge for the risk assessment of gene flow from transgenic sugar beet (Ritala, Nuutila, Aikasalo, Kauppinen and Tammisola, 2002).

One element of the current public debate about genetically modified crops is that gene flow from transgenic cultivars into surrounding weed populations will lead to more problematic weeds, particularly for traits such as herbicide resistance. Evolutionary biologists can inform this debate by providing accurate estimates of gene flow potential and subsequent ecological performance of resulting hybrids (Meagher, Belanger and Day, 2003). In this context it is significant to note that the potential of transferring herbicide resistance from transgenic rice (*Oryza sativa* L.) varieties to sexually compatible weeds is of paramount importance for development of effective weed control strategies (Zhang, Linscombe and Oard, 2003).

Development of plant genetic engineering has led to the deployment of transgenic crops and, simultaneously, to the need for a thorough assessment of the risks associated with their environmental release. A study investigated the occurrence of gene flow from transgenic rice to non-transgenic rice plants under agronomic conditions using a herbicide resistance gene as a tracer marker (Kuvshinov, Koivu, Kanerva and Pehu, 2001). Weed species are known to evolve rapidly with their associated crops. A better understanding of the mechanisms and rates of weed evolution could aid in limiting or at least anticipating this process. Spontaneous hybridization between crops and related weed species can transfer crop genes coding for fitness-enhancing traits to wild populations, but little is known about how easily this takes place in various weed-crop complexes (Messeguer, Fogher, Guiderdoni, Marfa, Catala, Baldi and Mele, 2001).

Different methods are used to study the incidence of genetically modified oilseed rape on honeybees, that depend on the type of transgene and on the transformation induced in the plant.

According to a study two examples are chosen to present risk assessment procedures. One deals with oilseed rape resistant to pest insects (by expressing protease inhibitors PI), the other concerns oilseed rape tolerant to an herbicide (glufosinate). It is unlikely that tolerance to herbicide has toxic effects, but a pleitropic effect of the transgene could effect the value and attractiveness of the plants. So, pollen and nectar production, flower size and density are evaluated and experiments are carried out in the field to study the foraging behaviour in particular the ability of honeybees to cross-visit transgenic and traditional oilseed rape or oilseed rape and weedy relatives is observed. Both types of studies show how complementary methods and collaborations between teams have been designed to study various aspects of the impact of genetic engineered oilseed rape on honeybees (Arnand, Skinner and Peaden, 2000).

Gene flow and introgression from cultivated plants may have important consequences for the conservation of wild plant populations. Cultivated beets (sugar beet, red beet and Swiss chard: *Beta vulgaris* ssp. vulgaris) are of particular concern because they are cross-compatible with the wild taxon, sea beet (B.vs. ssp. maritima). Cultivated beet seed production areas are sometimes adjacent to sea beet populations; the numbers of flowering individuals in the former typically outnumber those in the populations of the and latter. In such situations, gene flow from cultivated beets has the potential to alter the genetic composition of the nearby wild populations (Pierre and Pham-Delegue, 2000).

According to a study high rates of hybridization and introgression have been reported between the cultivated sunflower and its wild progenitor (both *Helianthus annuus*), raising concerns that neutral or favorable transgenes might escape and persist in wild *H. annuus* populations. However, little consideration has been given to the possibility that other wild sunflower species may hybridize with the cultivated sunflower (Bartsch, Lehnen, Clegg, Pohl-Orf, Schuphan and Elistrand, 1999).

Gene flow from crops to wild related species has been recently under focus in risk-assessment studies of the ecological consequences of growing transgenic crops. However, experimental studies addressing this question are usually temporally or spatially limited. Indirect population-structure approaches can provide more global estimates of gene flow, but their assumptions appear inappropriate in an agricultural context. In an attempt to prevent negative effects on the release of transgenic crops, new method has been pretended to estimate the quantity of genes migrating from crops to populations of related wild plants by way of pollen dispersal. This method provides an average estimate at a landscape level. Its originality is based on the measure, of the inverse gene flow, *i.e.* gene flow from the wild plants to the crop. Such gene flow results in an observed level of impurities from wild plants in crop seeds. This level of impurity is usually known by the seed producers and, in any case, its measure is easier than a direct screen of wild populations because crop seeds are abundant and their genetic profile is known. By assuming that wild and cultivated plants have a similar individual pollen dispersal function, one infers the level of pollen-mediated gene flow from a crop to the surrounding wild populations from this observed level of impurity (Lavigne, Klein and Couvet, 2002).

An increasing number of genetically engineered cultivars of several crops is being experimentally released into the environment. In future, crops with new transgenic traits will probably play an important role in agricultural practice. The long-term effect of transgenes on community ecology will depend on the distribution and establishment of transgenic plants in the wild, on the sexual transfer of their new genes to the environment and on the potential ecological impact of the transgenic trait (Wang, Chen, Reboud and Darmency, 1997).

Gene how is a key concern associated with the contamination of seed multiplication fields and the use of transgenic crops. The release of herbicide-resistant germplasms and the use of male-sterile varieties make foxtail millet (*Setaria italica*) an appropriate material to investigate this concern. Pollen dispersal from pollen donor sources and gene flow in fertile and male-sterile varieties of foxtail millet were investigated in experiments in China and France (Rieseberg, Kim and Seiler, 1999).

With the development of transgenic crops, concern has been expressed regarding the possible escape of genetically-engineered genes via hybridization with wild relatives. This is a potential hazard for sunflowers because wild sunflowers occur as weeds in fields where cultivated sunflowers are grown and hybridization between them has been reported. In order to quantify the potential for gene escape a study has been conducted: Here two experimental stand of sunflower cultivars were planted at two sites with different rainfall and altitude profiles. Populations of wild plants were planted at different distances from each cultivar stand. An allele homozygous in the cultivar (6Pgd-3-a), but absent in the wild populations, was used as a molecular marker to document the incidence and rate of gene escape from the cultivar into the wild populations of sunflowers. Three-thousand achenes were surveyed to determine the amount of gene flow from the cultivated to the wild populations. The marginal wild populations (3 m from the cultivar) showed the highest percentage (27 per cent) of gene flow. Gene flow was found to decrease with distance; however, gene flow occurred up to distances of 1000 m from the source population. These data suggest that physical distance alone will be unlikely to prevent gene flow between cultivated and wild populations of sunflowers (Bartsch and PohlOrf, 1996).

Genetically modified plants containing selectable markers offer a unique opportunity for pollination biologists to investigate some of the major, but intractable questions about paternity distributions and their causes. Here, a method is reported that uses transgenic plants to enable the quantification of the outcrossed fertilizations that result from a single pollinator visit. Gene flow mediated by worker bumblebees (*Bombus terrestris*) was studied among plants of oil seed rape (*Brassica napus* L. cv. Westar) where transgenic paternity in seeds of a non-transgenic plant was manifested as herbicide resistance. Overall, 91 per cent of the resistant seeds resulted from the first four flowers that were visited after the bumblebee left the transgenic plant, and none was found beyond the 14th successively visited newer.

Transgenic Animals

The gene transfer technique, transgenesis, has permitted the transfer of genes from one organism to another to create new lineages of organisms with improvement in traits important to aquaculture. Genetically modified organisms (GMOs), therefore, hold promise for producing genetic improvements, such as enhanced growth rate, increased production and efficiency, disease resistance and expanded ecological ranges. The basic procedure to generate transgenic fish for aquaculture includes:

1. Design and construction of transgenic DNA;
2. Transfer of the gene construct into fish germ cells;
3. Screening for transgenic fish;
4. Determination of transgene expression and phenotype;
5. Study of inheritance; and
6. Selection of stable lines of transgenics.

GMOs offer economic benefits, but also pose environmental threats. Optimising the mix of benefits and risks is of fundamental importance. The potential economic benefits of transgenic technology to aquaculture are obvious. Transgenic fish production has the goal of producing food for human consumption; thus the design of genetic constructs must take into consideration the potential risks to consumer health, as well as marketing strategies and product acceptance in the market (Levy, Marins and Sanchez, 2000).

Transgenic technology is developing rapidly; however, consumers and environmentalists remain wary of its safety for use in agriculture. Research is needed to ensure the safe use of transgenic technology and thus increase consumer confidence. This goal is best accomplished by using a thorough, unbiased examination of risks associated with agricultural biotechnology (Muir and Howard, 2002).

Any release of transgenic organisms into nature is a concern because ecological relationships between genetically engineered organisms and other organisms (including their wild-type conspecifics) are unknown. One model has predicted that, for a wide range of parameter values, transgenes could spread in populations despite high juvenile viability costs if transgenes also have sufficiently high positive effects on other fitness components. Sensitivity analyses has indicated that transgene effects on age at sexual maturity should have the greatest impact on transgene frequency, followed by juvenile viability, mating advantage, female fecundity, and male fertility, with changes in adult viability, resulting in the least impact (Muir and Howard, 2001).

Organisms modified by the techniques of modern biotechnology may differ significantly from normal organisms or organisms modified by other methods. Before transgenic organisms are introduced into the environment, the potential environmental effects should be assessed. In general, modification of ecologically important traits in undomesticated species presents the greatest environmental risk. Transgenic livestock probably pose low risk to the environment. Transgenic fish and live virus-based vaccines pose greater risks and present challenging questions for environmental risk assessment (Bruggemann, 1993).

In recent years, there has been widespread concern about the ecological and genetic effects of genetically modified organisms. In salmon and other fishes, transgenic growth hormone genes have been shown to have large effects on size and various traits related to fitness (Hedrick, 2001).

As per the observation of a study growth hormone (GH) gene transgenesis has allowed the production of salmon with an inherently increased growth potential, on average two to threefold higher compared with daily specific growth rates observed in normal, non-transgenic fish. This difference quickly results in animals of very different sizes at age, and is associated with specific morphological effects and enhanced appetites in transgenic animals. However, less is known of the feeding and antipredator behaviour of GH-transgenic fish, information that can help with predictions of potential ecological consequences of release or escape of transgenic fish into the wild (Sundstrom, Devlin, Johnsson and Biagi, 2003).

The ability to genetically engineer arthropods using recombinant DNA meopens new opportunities for improving pest management programs but also creates new responsibilities, including evaluation of the potential risks of releasing transgenic arthropods into the environment. It is now becoming easier to transform diverse species of arthropods by a variety of recombinant DNA methods. Useful genes and genetic regulatory elements are being identified for pest arthropods, but less effort is being expended to identify genes that could improve the efficacy of beneficial arthropods (Hoy, 2000).

Genetic engineering is not comparable to evolution or any other natural event, or a system of breeding. This is because, with genetic engineering, single genes are separated, from their original genotype, and added elsewhere, instead of natural cross-breeding which involves the whole genome.

Past breeding programmes have achieved a marked increase in productivity, brit at the price of an increased risk for a number of diseases. This is despite the fact that breeding systems should be for the protection of animal welfare, whereas past breeding systems have had the singular objective of increasing productivity. Given this consideration, the research for resistance genes does nor benefit the health of animals has the objective of reducing the economic disadvantages by reducing the incidence of animal disease instead of improving animals husbandry and appropriate animal breeding. The use of genetic engineering in animal breeding reduces the genetic diversity considerably. The conclusions are that the current objectives of breeding systems do not have the necessary objectives to match the current ethical requirements (Idel, 1998).

Environmental risk assessment of chemicals depends on the production of toxicity data for surrogate species of mammals, birds, and fish and on making comparisons between these and estimated or predicted environmental concentrations of the chemicals. It could be noted that of biomarker assays and strategies that might be used as alternatives, that is, to replace, reduce, or refine currently used ecotoxicity tests that cause suffering to vertebrates. Here a biomarker is a biologic response to an environmental chemical at the individual level or below which demonstrates a departure from normal status. Of immediate interest and relevance are nondestructive assays that provide a measure of toxic effect in vertebrate species and that can be used in both laboratory and parallel field studies. A major shortcoming of this approach is that such assays are currently only available for a limited number of chemicals, primarily when the mode of action is known. Nondestructive assays can be performed on blood, skin, excreta, and eggs of birds, fish, reptiles, and amphibians. An interesting recent development is the use of vertebrate cell cultures, including transgenic cell lines that have been developed specifically for toxicity testing. The ultimate concern in ecotoxicology is the effects of chemicals at the level of populations and above. Current risk assessment practices do not address this problem. The development of biomarker strategies could be part of a movement toward 14 more ecologic end points in the safety evaluation of chemicals, which would effect a reduction in animal tests that cause suffering (Walker, 1998).

Recombinant DNA and gene transfer technology now allow the transfer, inheritance and expression of specific DNA or gene sequences into fish. Preliminary results on the performance of the resulting transgenic fish have been quite dramatic in some cases, especially when growth hormone genes are transferred. Utilization of high performance transgenic fish has the potential to greatly increase aquaculture production in developing countries and increase the income of poor farmers. Growth of some transgenic fish has been increased more than to fold in laboratory conditions. Response appears to be greatest in unimproved fish, which in most cases would benefit developing countries the most. The potential increase in production and production efficiency from successful transgenic fish application could relieve pressure on habitat destruction for food production, relieve pressure on overfished natural stocks and discourage introduction of exotic species. Application of transgenic fish in aquaculture has just begun and could expand within a few years. However, prior to commercialization of transgenic fish, public education, environmental risks and food safety issues should be addressed. Genetically improved fish generated by recombinant DNA technology probably do not pose any greater risk to the environment than fish genetically improved through traditional selective breeding, but environmental risk data is lacking to verify this hypothesis. Environmental risk data will be needed in a case-by-case basis until more is known concerning the aquaculture potential and ecological risk of transgenic fish. Research institutions need to address the lack of environmental risk data to help ensure that any future application of transgenic fish in developing (and developed) countries be done in an environmentally and socio-economically sound manner. Socio-economic study is lacking for detailed cost-benefit analysis, and policy research is needed for proper application or regulation of transgenic fish in these countries (Dunham, 1999).

Conclusion

It could be seen clearly from the above discussion that the characteristics of genetically modified organisms have risk of inter-breeding with closely related wild of cultivated species. There are growing public concerns about the risks and benefits of genetically modified organisms in the environment. Most important impact of genetically modified organisms on bio–diversity including their target direct impact on non-target species. Some of the ecological effects of genetically modified organisms in include gene transfer in terms of movement of genes from a crop through in terms of movement of genes from a crop through out crossing with wild relatives to form new hybrid varieties, weediness is the tendency of a plant planted and to establish itself as a week species trait effects in terms of effects of traits that are potentially harmful to non-target organisms, genetic and phenotypic variability is the tendency of a plant to exhibit unexpected characteristics and expression of genetic material from pathogens, such as the risk of genetic recombination following mixed virus infections. In order to overcome the negative environmental and ecological impacts of genetically modified organisms the following policy suggestions can be considered.

1. There is a need for greater science based understanding of the risks and benefits in the application of biotechnology in agriculture and the environment.
2. Data and research capacity in the use of geographic information systems that could be used to model and evaluate the likely behaviour of genetically modified organisms in different environments and assess the risks and benefits associate with particular traits in those environments.
3. Efforts should be made to develop geographically dispersed network of research centres, located throughout the world's major agro-ecosystems and across the centre of diversity of the world's major food crops.
4. The greater understanding of the environmental risks and benefits posed by gene technology may lead to the better design of future genetically modified organisms.
5. There is a need go for further ecological research and developing agreed standard and protocols to enable the continuing monitoring of the behaviour of genetically modified organisms after their experimental and commercial release into the environment.
6. There is a need to set up effective monitoring systems to detect gene transfer and research to assess its ecological impacts. Most of the present research has been undertaken in developed nations. Little has been undertaken in developing nations which are the centres of diversity of the most of the world's major food crops.
7. Ecological research should require additional support by national governments and international agencies in their efforts to develop methodologies and undertake participatory field studies on the environmental impact of genetically modified organisms. These assessments should be undertaken using participatory approaches so as to involve local communities in the evaluation of the risks and benefits of new technologies.
8. Additional data could be supplemented in risks assessment of genetically modified organisms so as to inform future decisions on the choice of appropriate technology in addressing specific problems including the development and management of genetically modified crops for agricultural purposes.

9. Research capabilities should be enhanced in the application of gene technologies to crops livestock, forestry and fisheries and associated socio-economic policy and management expertise.
10. Efforts should be made to use new discoveries from functional genomics to identify useful genes within species and understanding how better to regulate them to control useful traits. This approach will place more emphasis on the control of genes already existing within species particularly those that require gene movement amongst distantly related species.

References

Amand, P.C.S., Skinner, D.Z. and Peaden, R.N. 2000. Risk of alfalfa transgene dissemination and scale-dependent effects. *Theor. Appl. Genet.*, 101: 107–114.

Arias, D.M. and Rieseberg, L.H. 1994. Gene flow between cultivated and wild sunflowers. *Theor. Appl. Genet.*, 89: 655–660.

Amaud, J.F., Viard, F., Delescluse, M. and Cuguen, J. 2003. Evidence for gene flow via seed dispersal from crop to wild relatives in Beta vulgaris (Chenopodiaceae): Consequences for the release of genetically modified crop species with weedy lineages. *Proceedings of the Royal Society of London Series B–Biological Sciences*, 270: 1565–1571.

Bartsch, D. and Pohl-Orf, M. 1996. Ecological aspects of transgenic sugar beet: Transfer and expression herbicide resistance in hybrids with wild beets. *Euphytica*, 91: 55–58.

Bartsch, D., Lehnen, M., Clegg, J., Pohl-Orf, M., Schuphan, I. and Ellstrand, N.C. 1999. Impact of gene flow from cultivated beet on genetic diversity of wild sea beet populations. *Mol Ecol.*, 8: 1733–1741.

Bruggemann, E.P. 1993. Environmental safety issues for genetically modified animals. *J. Anim. Sci.*, 71: 47–50.

De Marchis, F., Bellucci, M. and Arcioni, S. 2003. Measuring gene flow from two birdsfoot trefoil (Lotus corniculatus) field trials using transgenes as tracer markers. *Mol Ecol.*, 12: 1681–1685.

Desplanque, B., Hautekeete, N. and Van Dijk, H. 2002. Transgenic weed beets: possible, probable, avoidable? *J. Appl. Ecol.*, 39: 561–571.

Driessen, S., Pohl, M. and Bartsch, D. 2001. RAPD-PCR analysis of the genetic origin of sea beet (*Beta vulgaris* ssp. maritima) at Germany's Baltic Sea coast. *Basic Appl. Ecol.*, 2: 341–349.

Dunham, R.A. 1999. Utilization of transgenic fish in developing countries: Potential benefits and risks. *J. World Aquacult. Soc.*, 30: 1–11.

Gueritaine, G., Sester, M., Eber, F., Chevre, A. M. and Darmency, H. 2002. Fitness of backcross six of hybrids between transgenic oilseed rape (*Brassica napus*) and wild radish (*Raphanus raphanistrum*). *Mol Ecol.*, 11: 1419–1426.

Hedrick, P.W. 2001. Invasion of transgenes from salmon or other genetically modified organisms into natural populations. *Can. J. Fish Aquat. Sci.*, 58: 841–844.

Hoy, M.A. 2000. Transgenic arthropods for pest management programs: Risks and realities. *Exp. Appl. Acarol.*, 24: 463–495.

Idel, A. 1998. Genetically engineered animals: A critique. *Tierarztl Umsch*, 53: 83–87.

Kuvshinov, V., Koivu, K., Kanerva, A. and Pehu, E. 2001. Molecular control of trans gene escape from genetically modified plants. *Plant Sci.*, 160: 517–522.

Lavigne, C., Klein, E. K. and Couvet, D. 2002. Using seed purity data to estimate an average pollen mediated gene flow from crops to wild relatives. *Theor. Appl. Genet.*, 104: 139–145.

Levy, J.A., Marins, L.F. and Sanchez, A. 2000. Gene transfer technology in aquaculture. *Hydrobiologia*, 420: 91–94.

Lu, B.R., Song, Z.P. and Chen, J.K. 2003. Can transgenic rice cause ecological risks through transgene escape? *Progress in Natural Science*, 13: 17–24.

Meagher, T.R., Belanger, F.C. and Day, P.R. 2003. Using empirical data to model transgene dispersal. *Philosophical Transactions of the Royal Society of London Series B–Biological Sciences*, 358: 1157–1162.

Messeguer, J., Fogher, C., Guiderdoni, E., Marfa, V., Catala, M.M., Baldi, G. and Mele, E. 2001. Field assessments of gene flow from transgenic to cultivated rice (*Oryza sativa* L.) using a herbicide resistance gene as tracer marker. *Theor. Appl. Genet.*, 103: 1151–1159.

Muir, W.M. and Howard, R.D. 2001. Fitness components and ecological risk of transgenic release: A model using Japanese medaka (*Oryzias latipes*). *Am. Nat.*, 158: 1–16.

Muir, W.M. and Howard, R.D. 2002. Assessment of possible ecological risks and hazards of transgenic fish with implications for other sexually reproducing organisms. *Transgenic Res.*, 11: 101–114.

Pierre, J. and Pham-Delegue, M.H. 2000. How to study the impact of genetically modified colza on bees? *OCL–OI Corps Gras Lipids*, 7: 341–344.

Rieseberg, L.H., Kim, M.J. and Seiler, G.J. 1999. Introgression between the cultivated sunflower and a sympatric wild relative, Helianthus petiolaris (Asteraceae). *Int. J. Plant Sci.*, 160: 102–108.

Ritala, A., Nuutila, A. M., Aikasalo, R., Kauppinen, V. and Tammisola, J. 2002. Measuring gene flow in the cultivation of transgenic barley. *Crop Sci.*, 42: 278–285.

Snow, A.A., Uthus, K.L. and Culley, T.M. 2001. Fitness of hybrids between weedy and cultivated radish: Implications for weed evolution. *Ecol. Appl.*, 11: 934–943.

Sundstrom, L.F., Devlin, R.H., Johnsson, J.I. and Biagi, C.A. 2003. Vertical position reflects increased feeding motivation in growth hormone transgenic coho salmon (*Oncorhynchus kisutch*). *Ethology*, 109: 701–712.

Walker, C.H. 1998. Biomarker strategies to evaluate the environmental effects of chemicals. *Environ. Hlth. Perspect.*, 106: 613–620.

Wang, T.Y., Chen, H.B., Reboud, X. and Darmency, H. 1997. Pollen-mediated gene flow in an autogamous crop: Foxtail millet (*Setaria italica*). *Plant Breed.*, 116: 579–583.

Zhang, N.Y., Linscombe, S. and Oard, J. 2003. Out-crossing frequency and genetic analysis of hybrids between transgenic glufosinate herbicide-resistant rice and the weed, red rice. *Euphytica*, 130: 35–45.

Chapter 3

Influence of Thermal Stratification on Dissolved Oxygen in Subhas Sarobar, Kolkata

N.R. Samal, D. Roy**, #A. Mazumdar*** and B. Bose*****

**Senior Research Fellow, **Lecturer, ***Joint Director, ****Retired Professor and Ex-Director School of Water Resources Engineering, Jadavpur University (JU), Kolkata – 700 032, West Bengal, India*

ABSTRACT

Thermal stratification is a physical phenomenon, which occurs in many lakes and/or reservoirs, and refers to the formation of distinct temperature layers within a waterbody. While thermal stratification is a physical phenomenon, it has profound effects on physical, chemical and biological processes within a lake. The effects are largely due to the formidable mixing constraints imposed by thermal stratification. As a result of it, the exchange of oxygen-rich surface water in the epilimnion into the bottom lying hypolimnion is hindered and such the waterbody can no longer sustain any higher life-forms and the lake bottom becomes biologically dead. In the present paper, an attempt has been made to develop a correlation between temperature and its associated density and the effect of thermal stratification on dissolved oxygen in the Subhas Sarobar, a National Lake in East Kolkata.

Keywords: Thermal stratification, Dissolved Oxygen, Subhas Sarobar, Hypolimnion.

Introduction

The thermal stratification in lakes and reservoirs is an important fundamental regulator of the biological and chemical processes occurring in these systems, which in turn, govern lake production

---

Corresponding Author.

and decomposition processes (Wetzel, 1983). Measurements of thermal stratification cycles have been an integral part of basic limnological studies since the pioneering work of Forel (Hutchinson, 1957). The very dynamic nature of Lake Stratification, due to its dependence on air/water heat exchange and on wind induced turbulent mixing, has not always been conceived. The interaction of heating, cooling, and mixing processes and their opposing effects were qualitatively described by Birge (Ford, *et al.* 1980). In the present study, the paramount aspects of the hydrodynamics of Subhas Sarobar is analyzed keeping in view the influence of three largely dominant external agents, like the exchange of thermal energy with the surrounding environment (in particular with the atmosphere), the mechanical energy transferred by wind to the watermass and the kinetic and potential energy produced by the inflows into the lake/waterbody. All together these three agents clearly concern the energy inter-relationships between the waterbody and the different surrounding environments, even if some other forms of energy is neglected as insignificant. The presence of a surface layer of "warm water" (epilimnion), which is separated from the relatively cold water mass (hypolimnion) by a steep thermal gradient, is one of the most important physical aspects of lakes in certain periods of their yearly cycles. The thermocline represents a temperature and density barrier which consistently limits any exchanges between epilimnetic and hypolimnetic watermasses: the physical characteristics of these two layers are so different, particularly at the height of thermal stratification, that they may even be considered completely isolated. As a result of thermal stratification it is quite evident that a very close relationship exists between temperature and oxygen concentration distributions along the water column in a lake: both are very useful for understanding lake hydrodynamics, and especially for evaluating the depth of mixed layers in the lake (Barbanti *et al.*, 1989). The objective of the study described herein is to study lake temperature stratification at short enough time intervals at three sampling points in the mid-reach of the National Lake, Subhas Sarobar during the hot period in the climate of Kolkata (*i.e.* premonsoon) and attempt is also made to develop a correlation between temperature and its associated density and the effect of thermal stratification on dissolved oxygen.

Study Area

The Subhas Sarobar (Figure 3.1) covers an area of 39.6 ha, of which the water area is about 6.04 ha. Two islands, one small and the other big, constitute an excellent habitat for diverse species of life. The mean length and width of the Lake are 0.584 km and 0.326 km respectively while the maximum depth and the mean depth of the lake are 10.36 m and 9.6 m respectively. The water-body is given on rent to the Department of Fisheries of Government of West Bengal. A part of the water-body is used as a public swimming pool. The open space in both the lakes is an oasis in the space-starved congested city. The lake complex is used for sports including water-sports, recreation and cultural activities. Presently, water of the Sarobar is intensively used for bathing, washing, cleaning and community purposes by a large number of people who live in the slum area adjacent to the Sarobar. The lake is under the administrative control of Kolkata Improvement Trust.

Materials and Methods

In any water quality study, sampling is crucial with regards to mode of collection (*i.e.* time, location, depth of water, etc.). The observation sites are selected considering the maximum water depth along the mid-reach (Stations I, II and III) and considering the less human intervention of water of the Lake, to study the thermal stratification. The water temperature and Dissolved Oxygen (DO) have, however, been measured using a WTW DO meter along the depth at an interval of 0.5 m below the water surface. The concentration of density and Biochemical Oxygen Demand (BOD) has been estimated at the same depth in the water analysis laboratory of School of Water Resources Engineering

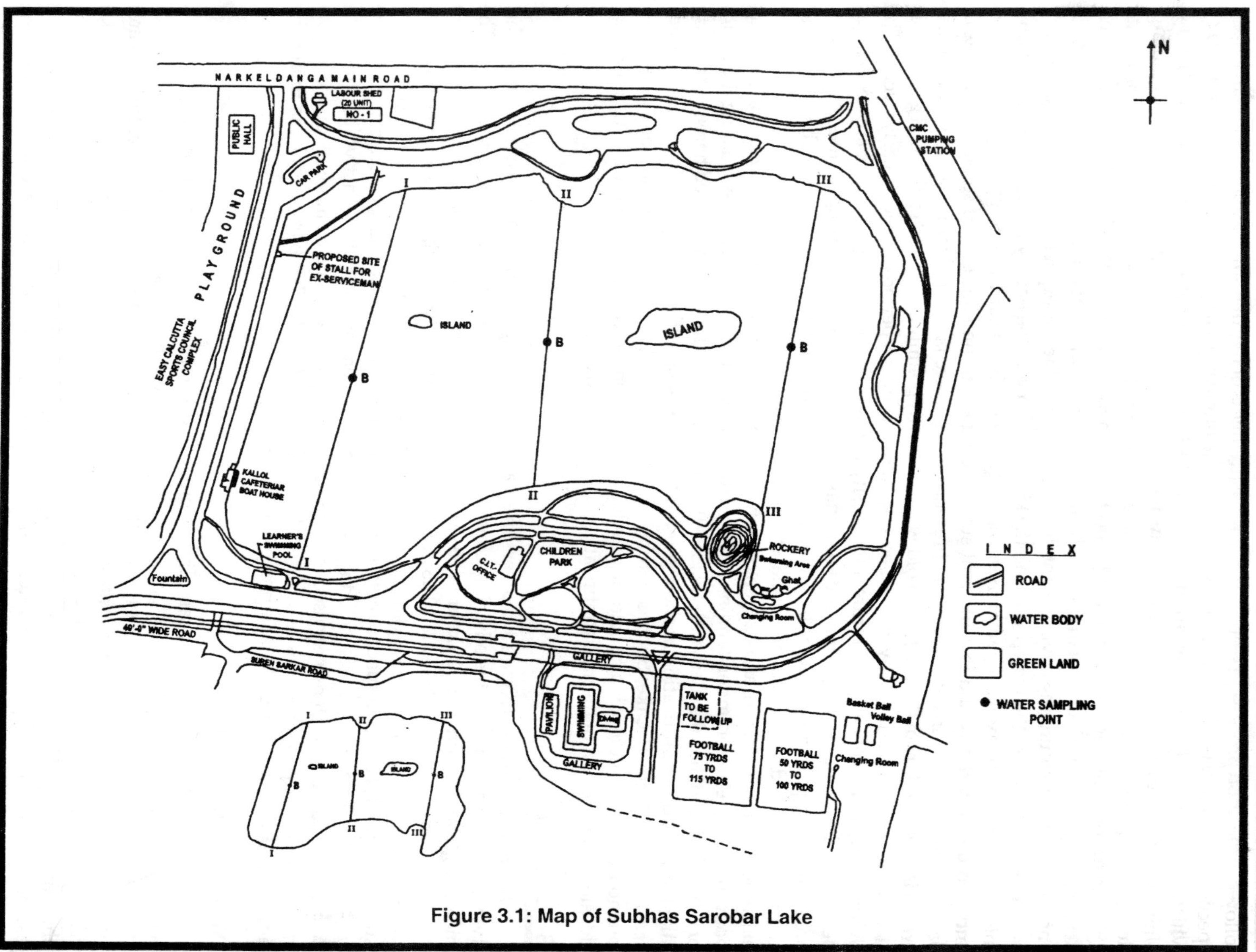

Figure 3.1: Map of Subhas Sarobar Lake

following the standard method of APHA (1989). The water samples have been collected from these specific sites on 28-05-2003 by a water sampler having capacity one litre separately for each site, which is locally made and specially designed by the School Water Resources Engineering, Jadavpur University. The samples after collection have been stored in the deep freeze to avoid any microbial decomposition.

Results and Discussion

The vertical distribution of temperature and the transport of pollutants from surface to the bottom in lakes is important to study for maintaining a better water quality and its surrounding ecosystem. In the present investigation, the vertical profile of the temperature, dissolved oxygen (DO), per cent DO and the Biochemical Oxygen Demand (BOD) along the depth at stations IB, IIB, IIIB for the Subhas Sarobar is given in Figures 3.8–3.10. These data show the development of the thermal stratification and dissolved oxygen stratification in the lake. The pattern of thermal stratification is observed to be clear and indicates the position of thermocline from 3.8 m to 5.5 m at station IB (maximum depth: 8.5 m), 3.45 m to 5.0 m at station IIB (maximum depth: 7.5 m) and 3.1 m to 4.75 m at station IIIB (maximum depth: 6.0 m). The development of an appreciable shear at the epilimnion level thus makes possible a relative deepening of the thermocline towards the bottom over a time scale (observation period: 9.50 a.m. to 1.20 p.m.). It is well understood that as the rate of solar radiation increases, the kinematics rate of pollutant transport along the vertical direction also increases. The thermocline thus form a thermal barrier, which is responsible for inhibiting oxygen exchange between the upper and lower waters and also works to limit thermal gain by the lower waters, thus maintaining lower temperatures at depth. The surface water temperature and bottom water temperature is found to be maximum (34.6°C) at station IIB and minimum (21.5°C) at stations IB and IIB. The average air temperature and wind speed above the lake water surface is observed to be 36.6°C and 1.88 m/s for the Subhas Sarobar. The water temperature and the density at all the stations are related along the depth through a linear regression,

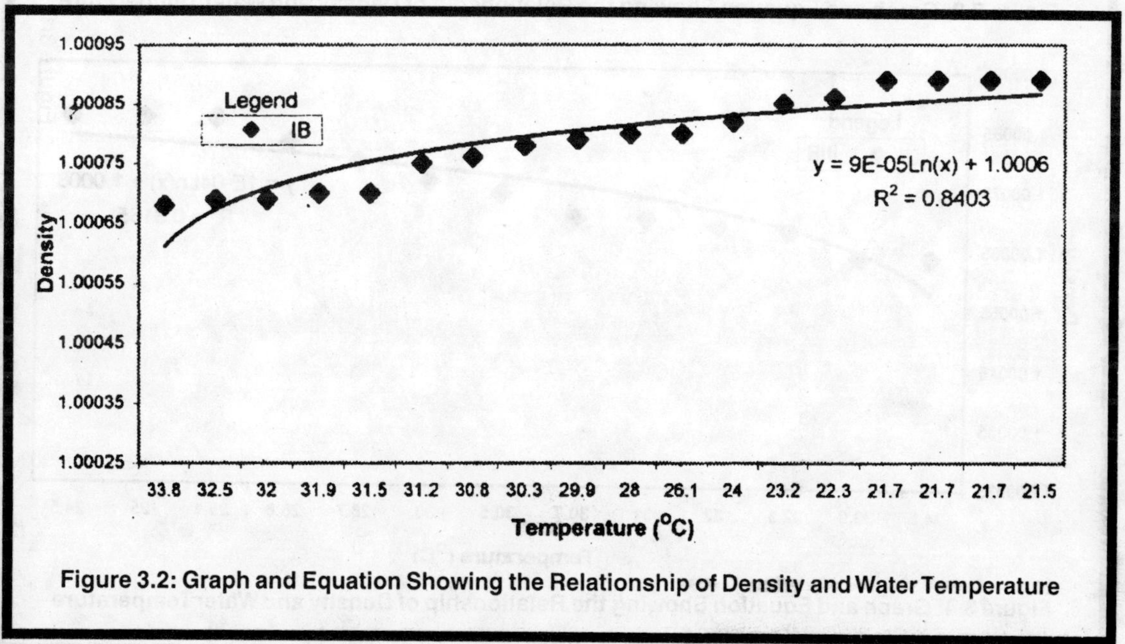

Figure 3.2: Graph and Equation Showing the Relationship of Density and Water Temperature

expressed in the Figures 3.2–3.4. Again the temperature exhibits a good correlation with depth at all the stations for the lake, which is detailed in Tables 3.1–3.3.

Hypolimnetic Dissolved Oxygen

The dissolved oxygen concentration remain constant in the hypolimnion zone in the lake and however, follows the same trend of variation with temperature and opposite trend of variation with Biochemical Oxygen Demand (BOD). The hypolimentic dissolved oxygen maintain a zero slope from

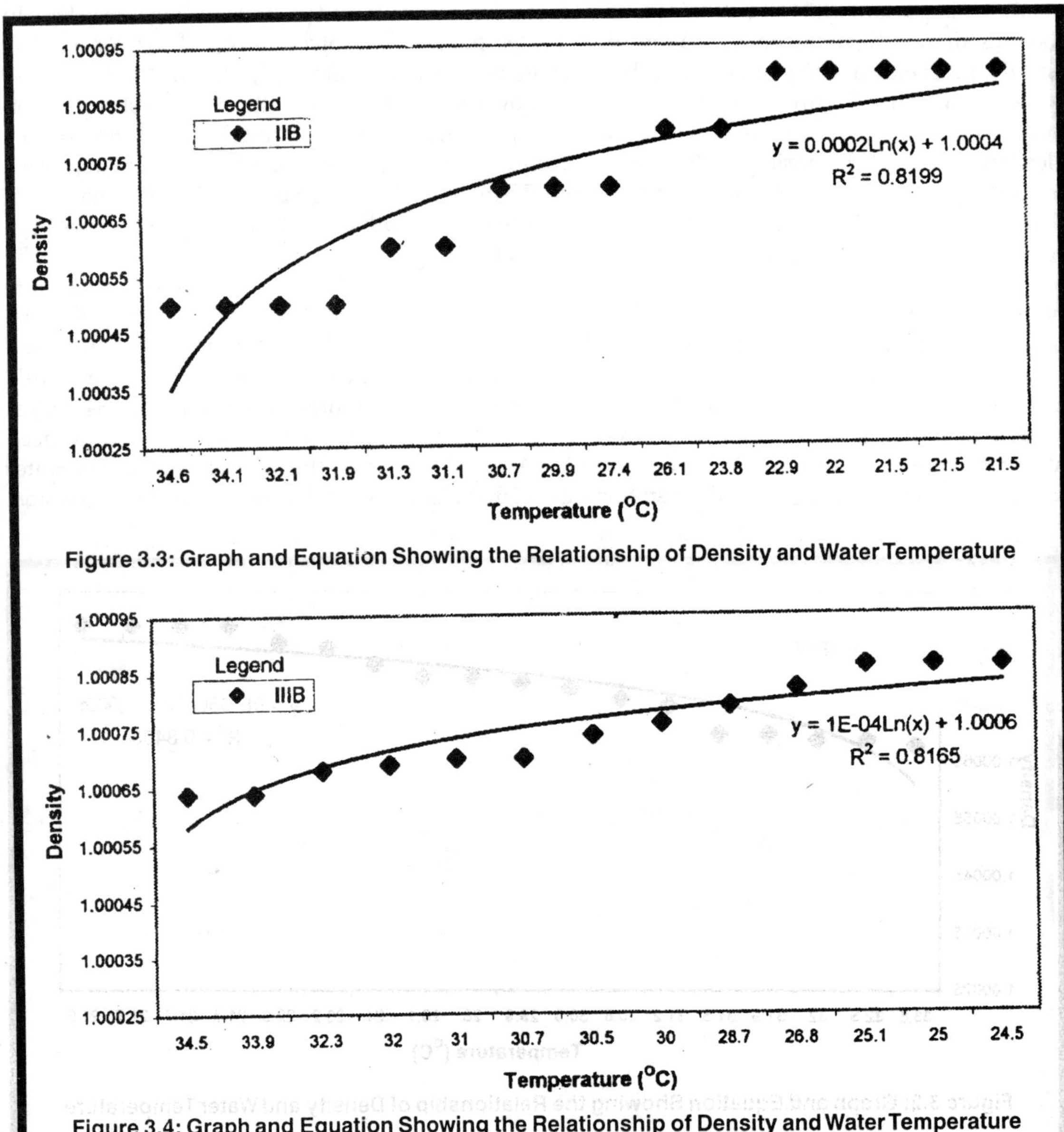

Figure 3.3: Graph and Equation Showing the Relationship of Density and Water Temperature

Figure 3.4: Graph and Equation Showing the Relationship of Density and Water Temperature

3.5 m to the bottom at the station IB, from 3.4 m to the bottom at station IIB and from 2.5 m to the bottom at station IIIB in the lake. The abrupt change in the dissolved oxygen content in the epilimnion is due to the diffusion of wind intensity over the lake water surface. The dissolved oxygen content in the surface water and bottom water is found to be maximum (6.67 mg/l) at station IB and minimum (0.48 mg/l) at station IIIB. Low value of dissolved oxygen near the sediment-water interface develops the hypoxic condition (DO < 5 mg/l) and it gradually extended upward throughout the summer. As a

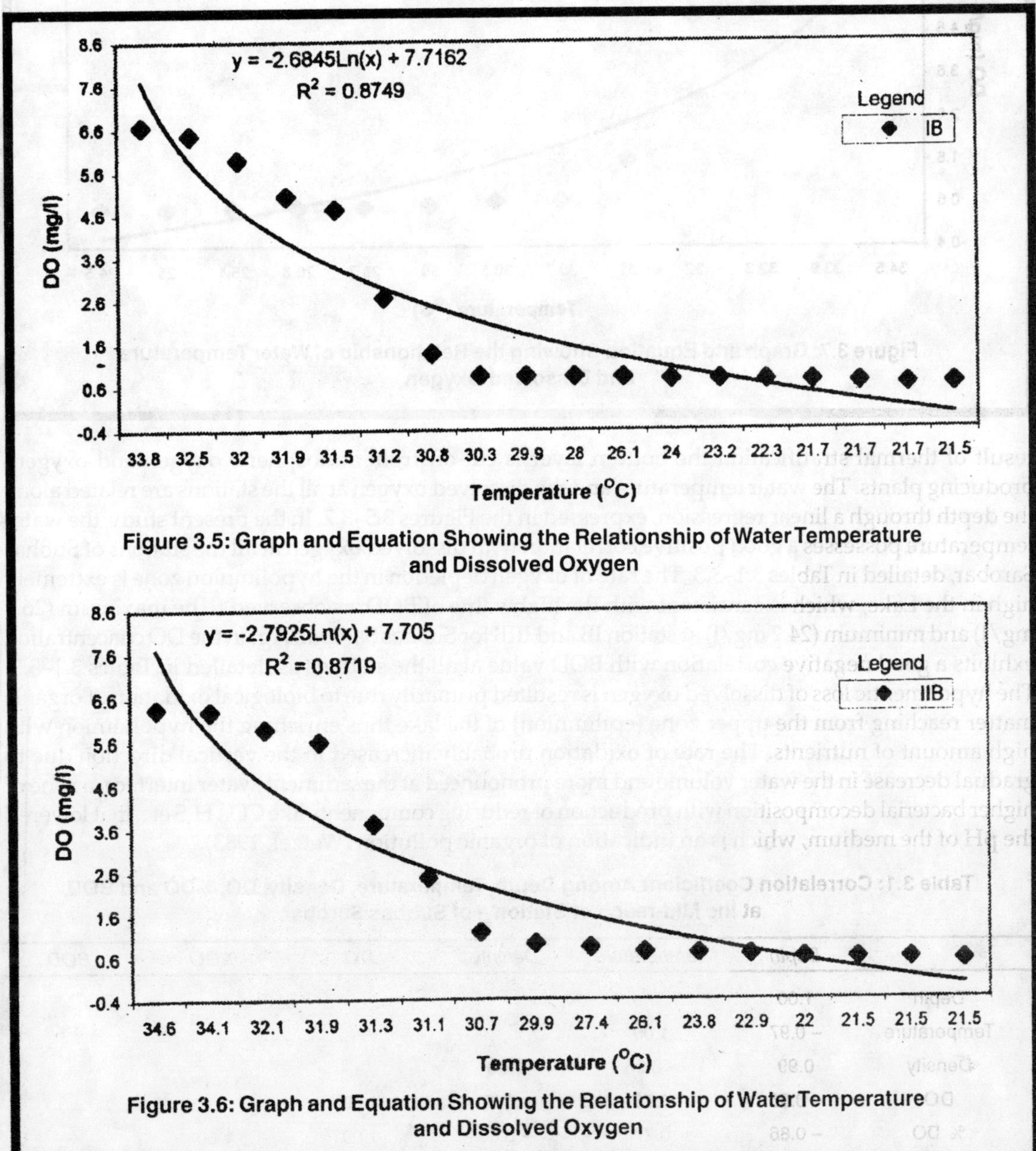

Figure 3.5: Graph and Equation Showing the Relationship of Water Temperature and Dissolved Oxygen

Figure 3.6: Graph and Equation Showing the Relationship of Water Temperature and Dissolved Oxygen

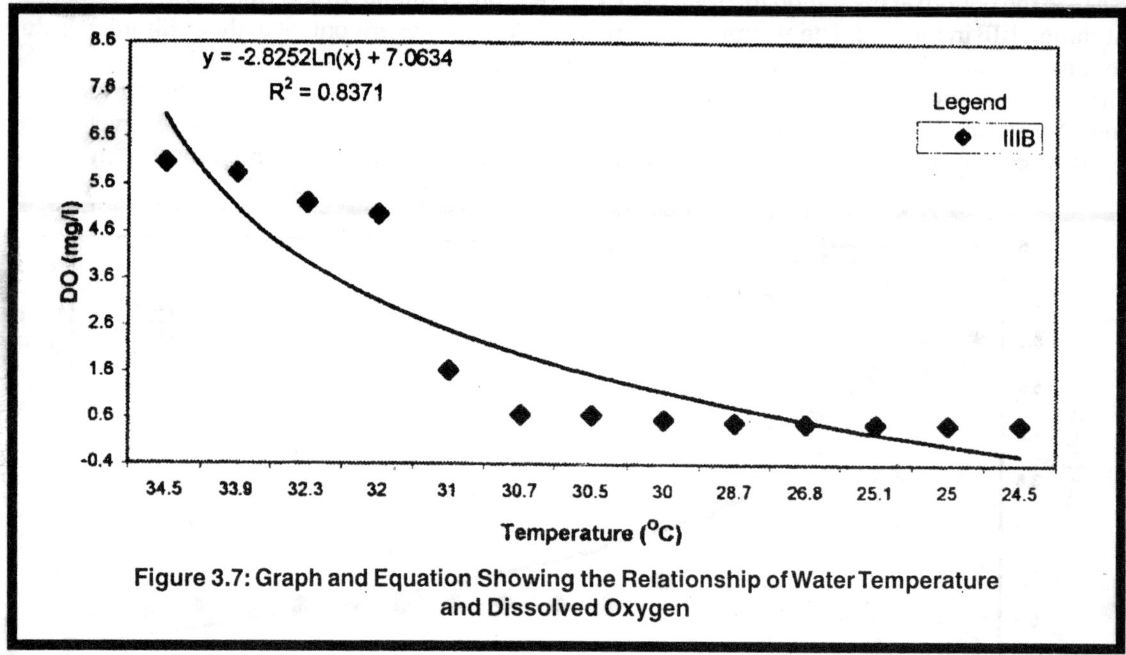

Figure 3.7: Graph and Equation Showing the Relationship of Water Temperature and Dissolved Oxygen

result of thermal stratification the bottom layer is cut-off from atmospheric oxygen and oxygen-producing plants. The water temperature and the dissolved oxygen at all the stations are related along the depth through a linear regression, expressed in the Figures 3.5–3.7. In the present study, the water temperature possesses a good positive correlation with dissolved oxygen at all the stations of Subhas Sarobar, detailed in Tables 3.1–3.3. The rate of oxygen depletion in the hypolimnion zone is extremely high in the Lake, which is consistent with the high value of BOD, as observed to be maximum (26.1 mg/l) and minimum (24.2 mg/l) at station IB and IIIB for Subhas Sarobar. Again the DO concentration exhibits a good negative correlation with BOD value at all the stations as detailed in Tables 3.1–3.3. The hypolimentic loss of dissolved oxygen is resulted primarily due to biological oxidation of organic matter reaching from the upper zone (epilimnion) of the lake thus enriching the hypolimnion with high amount of nutrients. The rate of oxidation probably increased in the vertical direction due to gradual decrease in the water volume and more pronounced at the sediment-water interface, a zone of higher bacterial decomposition with production of reducing components like CO_2, H_2S etc. that lowered the pH of the medium, which is an indication of organic pollution (Wetzel, 1983).

Table 3.1: Correlation Coefficient Among Depth, Temperature, Density, DO, %DO and BOD at the Mid-reach of Station I of Subhas Sarobar

	Depth	*Temperature*	*Density*	*DO*	*%DO*	*BOD*
Depth	1.00					
Temperature	– 0.97	1.00				
Density	0.99	– 0.96	1.00			
DO	– 0.86	0.76	– 0.87	1.00		
% DO	– 0.86	0.77	– 0.88	1.00	1.00	
BOD	0.98	– 0.95	0.96	– 0.75	– 0.76	1.00

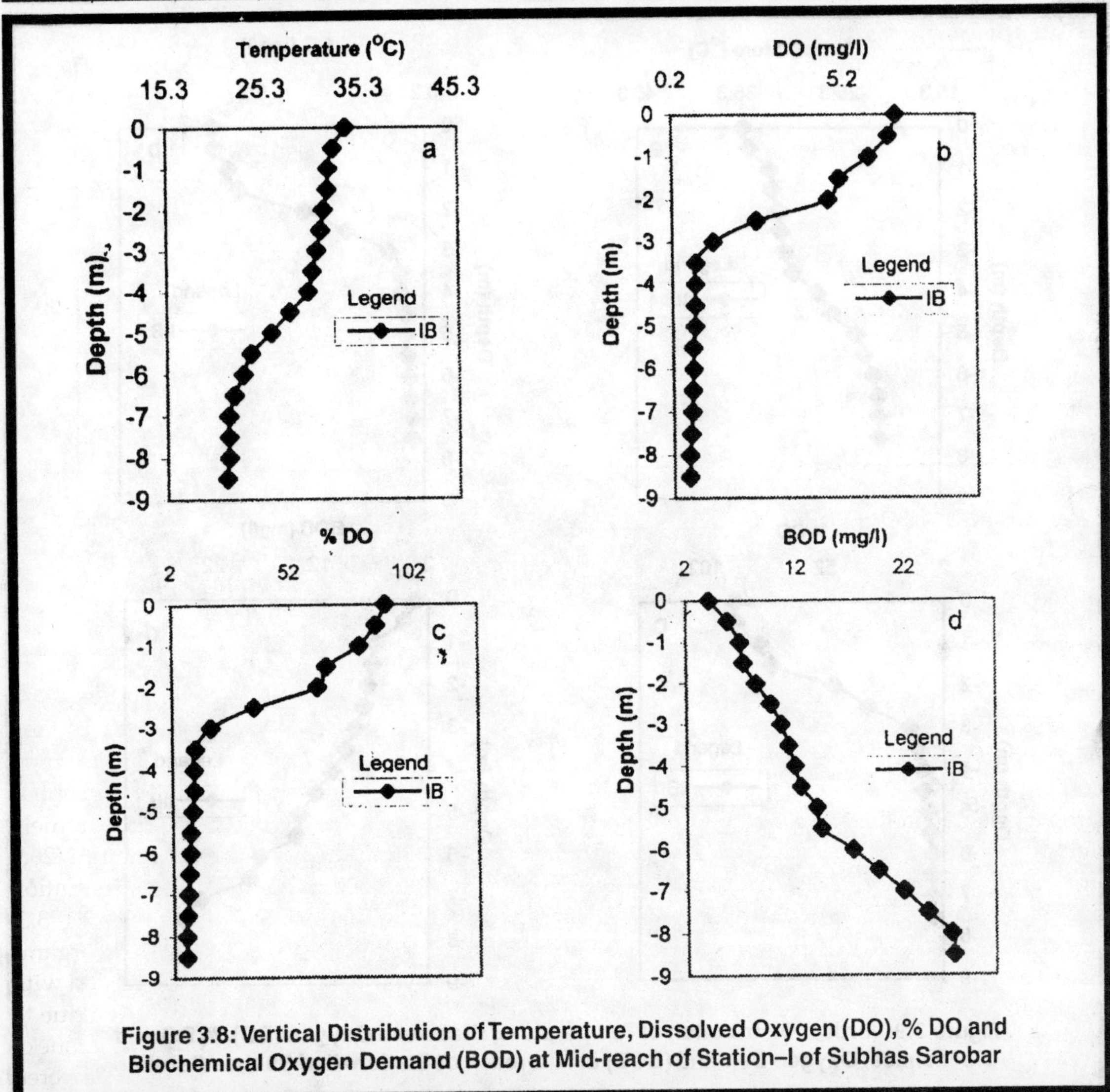

Figure 3.8: Vertical Distribution of Temperature, Dissolved Oxygen (DO), % DO and Biochemical Oxygen Demand (BOD) at Mid-reach of Station–I of Subhas Sarobar

Table 3.2: Correlation Coefficient Among Depth, Temperature, Density, DO, %DO and BOD at the Mid-reach of Station II of Subhas Sarobar

	Depth	*Temperature*	*Density*	*DO*	*%DO*	*BOD*
Depth	1.00					
Temperature	– 0.98	1.00				
Density	0.67	– 0.97	1.00			
DO	– 0.88	0.83	– 0.90	1.00		
%DO	– 0.89	0.83	– 0.90	1.00	1.00	
BOD	0.97	– 0.94	0.90	– 0.76	– 0.76	1.00

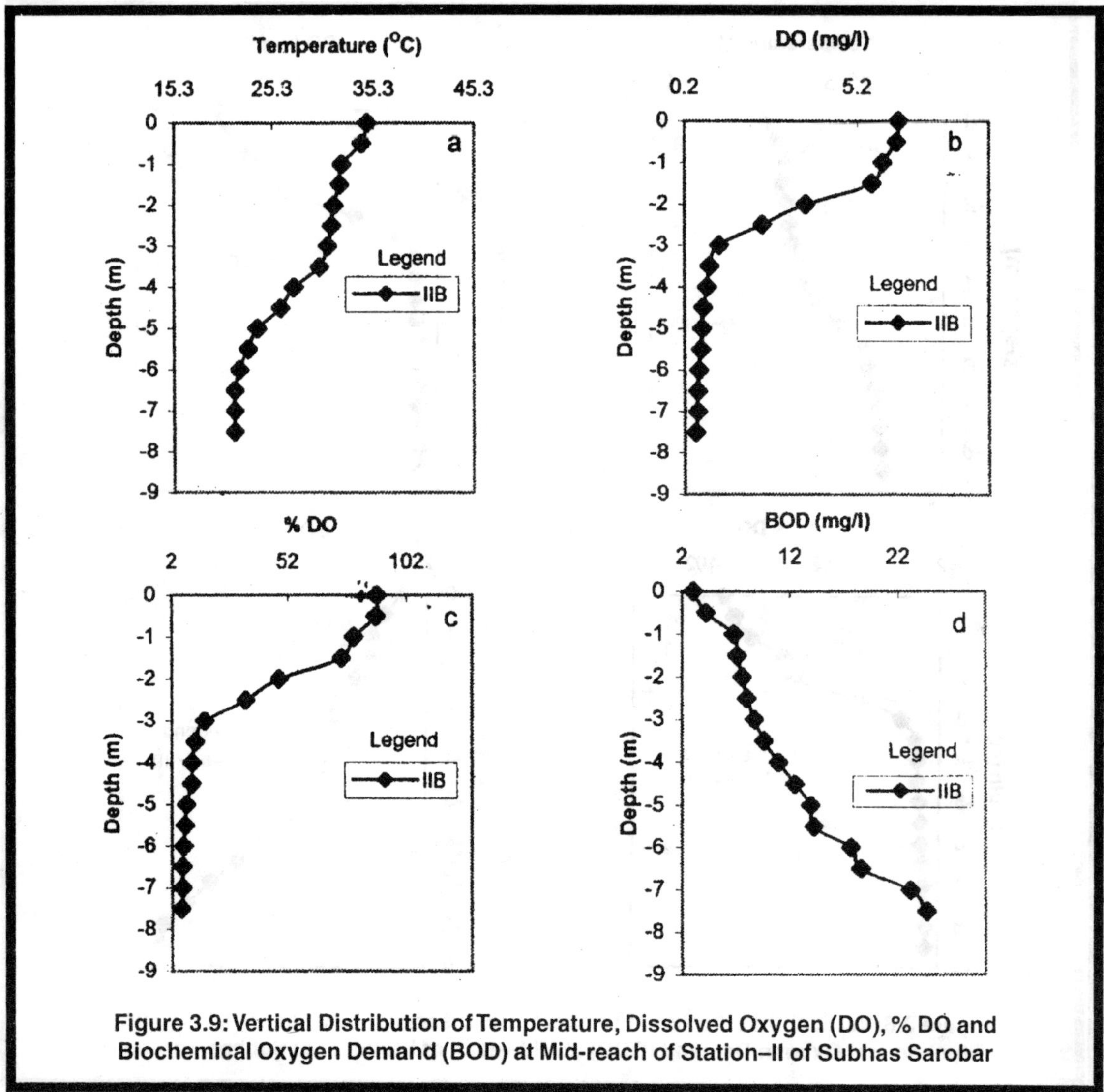

Figure 3.9: Vertical Distribution of Temperature, Dissolved Oxygen (DO), % DO and Biochemical Oxygen Demand (BOD) at Mid-reach of Station–II of Subhas Sarobar

Table 3.3: Correlation Coefficient Among Depth, Temperature, Density, DO, %DO and BOD at the Mid-reach of Station III of Subhas Sarobar

	Depth	*Temperature*	*Density*	*DO*	*%DO*	*BOD*
Depth	1.00					
Temperature	– 0.98	1.00				
Density	0.98	– 0.99	1.00			
DO	– 0.86	0.79	– 0.80	1.00		
% DO	– 0.86	0.80	– 0.80	1.00	1.00	
BOD	0.98	– 0.97	0.97	– 0.80	– 0.80	1.00

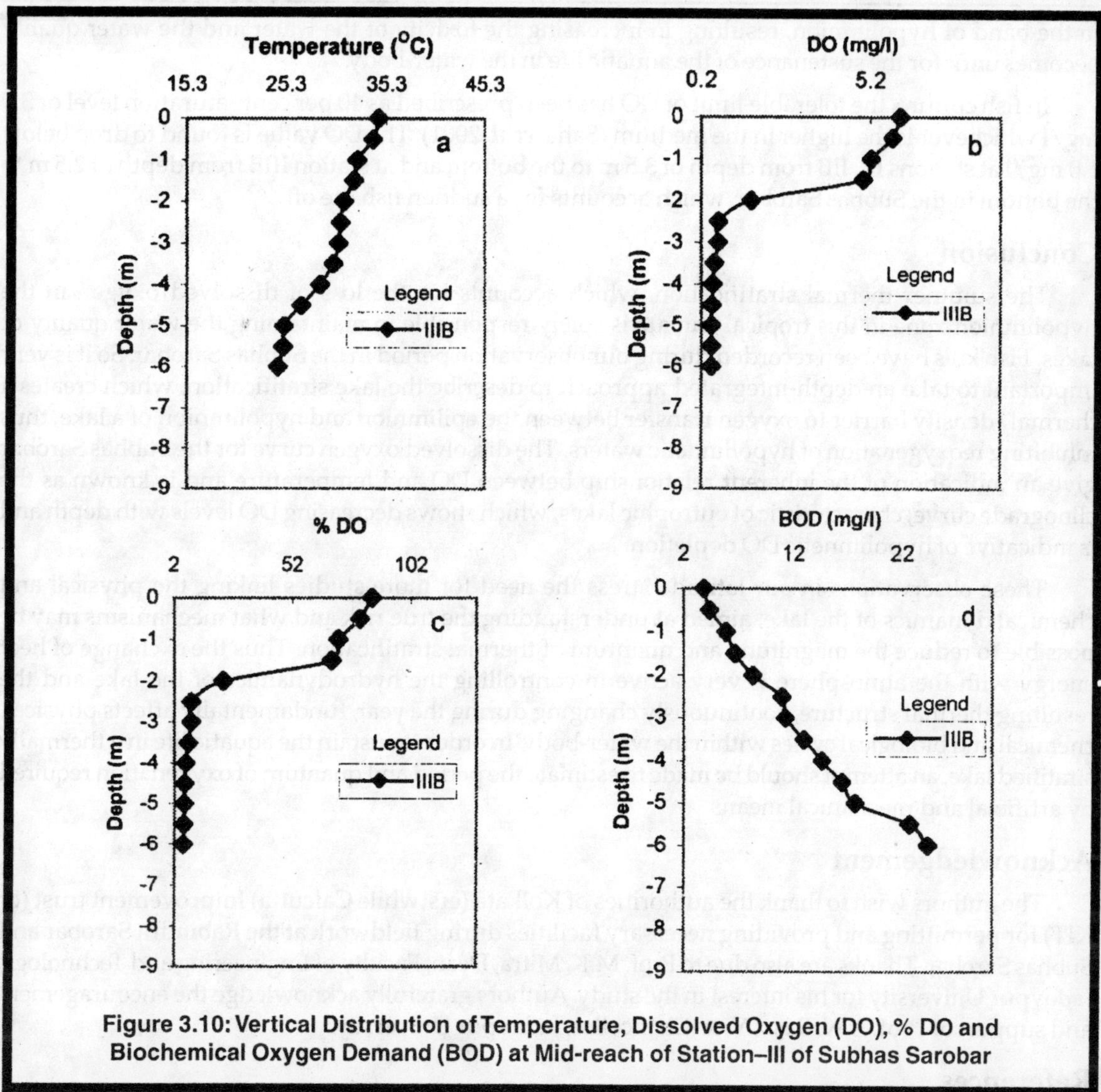

Figure 3.10: Vertical Distribution of Temperature, Dissolved Oxygen (DO), % DO and Biochemical Oxygen Demand (BOD) at Mid-reach of Station–III of Subhas Sarobar

The percentage of oxygen saturation (per cent DO) at the surface is found to be maximum (91.8 per cent) at station IB and minimum (85.2 per cent) at station IIIB for the Subhas Sarobar. Thus the water-body remained under-saturated with dissolved oxygen during our observation period indicating that the oxygen will diffuse into water in proportion to the oxygen deficit. Diffusion oxygen losses and gain can also be quite substantial, especially at windy sites when DO values are substantially different from 100 per cent saturation. The spatial and temporal variation in the hypolimnion thickness indicates a warning for the aquatic species to look for oxygenated rich environment. As observed from the temperature and oxygen profile (Figures 3.8–3.10), the rate of oxygen consumption is rapid in the thermocline zone due to high temperature gradient and contribute to the development of minimum DO and is continued through out the summer and fall and persisted until thickening of the mixed layer destroyed it in early winter. Low value of DO can not enhance other chemical oxidation processes in the band of hypolimnion, resulting in increasing the toxicity of the water and the water quality becomes unfit for the sustenance of the aquatic life in the water-body.

In fish culture, the tolerable limit of DO has been prescribed as 40 per cent saturation level or 3.0 mg/l whichever is the higher in the medium (Saha, *et al.* 2001). The DO value is found to drop below 1.0 mg/l at stations IB, IIB from depth of 3.5 m to the bottom and at station IIIB from depth of 2.5 m to the bottom in the Subhas Sarobar, which accounts for a sudden fish die off.

Conclusion

The summer thermal stratification, which accounts for the loss of dissolved oxygen in the hypolimnion zone in this tropical climate is solely responsible in maintaining the water quality of lakes. Fish kills have been recorded during our observation period in the Subhas Sarobar. So it is very important to take an depth-integrated approach to describe the lake stratification, which creates a thermal/density barrier to oxygen transfer between the epilimnion and hypolimnion of a lake, thus inhibiting reoxygenation of hypolimnetic waters. The dissolved oxygen curve for the Subhas Sarobar give an indication of the inherent relationship between DO and temperature and is known as the clinograde curve, characteristic of eutrophic lakes, which shows decreasing DO levels with depth and is indicative of hypolimnetic DO depletion.

These observations in our latitude stress the need for more studies linking the physical and chemical dynamics of the lake, aimed at understanding the true risk and what mechanisms may be possible to reduce the magnitude and quantum of thermal stratification. Thus the exchange of heat energy with the atmosphere is very active in controlling the hydrodynamics of the lake and the resulting thermal structure, continuously changing during the year, fundamentally affects physical, chemical and biological cycles within the water-body. In order to sustain the aquatic life in a thermally stratified lake, an attempt should be made to estimate the period and quantum of oxygenation required by artificial and mechanical means.

Acknowledgement

The authors wish to thank the authorities of Kolkata (erstwhile Calcutta) Improvement trust (or KIT) for permitting and providing necessary facilities during fieldwork at the Rabindra Sarobar and Subhas Sarobar. Thanks are also due to Prof. M.K. Mitra, Dean, Faculty of Engineering and Technology, Jadavpur University for his interest in the study. Authors gratefully acknowledge the encouragement and support of Prof. A.N. Basu, Vice-Chancellor, Jadavpur University.

References

APHA, AWWA, WPCF 1989. *Standard Methods for the Examination of Water and Wastewater*, 7th Edn., Washington, D.C.

Barbanti, L., W. Ambrosetti and A. Rolla 1981. Dipendenza della temperature dell'acqua superficiale da temperature atmosferica e radiazione solare nei laghi italiani, Mem. Ist. Ital. Idrobiol., 39: 65–98.

Barbanti, L. and W. Ambrosetti 1989. The Physical Limnology of Lago Maggiore: A review, Mem. Ist. Ital. Idrobiol., 46: 41–68.

Bonnet, M.P., Poulin, M. and Devaux, J. 2000. Numerical Modeling of thermal stratification in a lake reservoir: Methodolgy and case study. *J. Aquatic Sciences., Basel*, 62: 105–124.

Ford, D.E., and Stefan, H. 1980. Stratification variability in three morphometrically different lakes under identical meteorological forcing. *Water Resources Bulletin, AWRA,* 16(2): 243.

Hutchinson, G.E. 1957. *A Treatise on Limnology*, John Wiley & Sons, New York, Vol. 1: 1015.

Saha, T., Ghosh. P.B., Mazumdar, S.S. and Bandyopadhya, T.S. 2001. Extent of Hygienic Level of Dissolved Oxygen in Subhas Sarobar, Kolkata. *J. Public Health Engineers, India*, 1: 33–37.

Wetzel, R.G. 1983. The oxygen content in freshwater. *Limnology*, M. Brown, pp. 172.

Chapter 4

Dust Accumulation Studies at Hyderabad

Nirmala Babu Rao, I. Sobha Rani and Amena

Department of Botany, Osmania University College For Women, Koti, Hyderabad

ABSTRACT

Dust accumulation studies show that there was more dust accumulation at Railway Station and MC EME Mess samples than Indira Park, Women's college and Saroor Nagar. A continuous accumulation of dust particles from 1st to 30th day was noticed with an increase in size and weight. There was more accumulation of living aggregates on the slides in summer season than in winter and rainy seasons. Rains cleared the atmosphere and the load of dust aggregates were less in the air. Electrical conductivity was less in rainy season than in winter and summer seasons. This type of studies help to monitor the dust pollution levels of a place.

Keywords: *Dust, Pollution, Aero-allergies, Aggregates*

Introduction

Due to dust accumulation air loses its freshness and a continuous encounter with dust causes damage to lungs, bronchus, eyes and skin often beyond repair (Misra, 1978). The dust fall in Chembur suburbs in Mumbai was found to be 25 tons/month/km^3 which was close to that in New York or Tokyo (Sunavala, 1978). Kolkata air becomes more loaded with dust in the winter. Chennai and Hyderabad seem to be in a better position having particulate matter below 150 μg/m^3 (Seth, 1976). Due to non-availability of information on dust pollution in Hyderabad the present investigation was taken up to estimate the quantum of dust fall, concentration of living and non-living aerosols, size of the particulate matter, with their pH and conductivity at different localities exhibiting Urban, Suburban, Rural, and protected environment in and around Hyderabad.

Materials and Methods

Dust is collected by glass plate method where 10 cm^2 plates were employed for the measurement of the particulate matter. Glass plates were washed, dried and weighed and then were coated with grease and again weighed. Glass plates were exposed over a height of 30 feet al Secunderabad Railway Station, MC EME Mess, Saroor Nagar, Women's College and Indira Park. The glass plates were exposed horizontally by placing them on a tripod stand. Nine glass plates were exposed at each place and the first set of 3 slides were removed after 24 hours, next 3 after 15 days and the last 3 on 30th day of exposure to air in each season. The experiment was similarly performed at all the localities. The slides were brought back to the laboratory and weighed to know the amount of dust accumulated per unit area per unit time (Chaphekar *et al.*, 1980). pH was found with the pH meter and the conductivity with the help of conductivity meter.

Result and Discussion

Weight of dust particles per unit area (g/cm^2) were presented in Table 4.1. Greater amount of dust has accumulated on the slides kept over Secunderabad Railway Station. There was more smoke at the Railway Station resulting in greater dust accumulation over a given period of time per unit area. Lowest quantity of dust was accumulated on Women's college and Indira Park slides, which were regarded as protected areas having a plant cover of several types of trees. Saroor Nagar provided a rural atmosphere with less dust accumulation. Further the barrier of trees in the College Campus and Indira Park, Saroor Nagar intercepted the dust particles carried by air. There was approximately twenty times increase in the dust accumulation values from I to 30 days interval in all the localities similar to the observations of Chaphekar *et al.*, 1980. But the quantum increase was not the same as observed by Kayastha and Kumar (1978) and Chaphekar *et al.* (1980) at Kanpur and Mumbai conditions respectively. Hyderabad is also more populated and polluted area as far as dust pollution is concerned. In general the accumulation of dust particles in rainy season was less than that was observed in summer and winter season. However, the rate of increase was higher initially followed by a steep fall, except in the rainy season.

Table 4.1: Weight of the Dust Particles per Unit Area (g/m^2) Collected from the Air Sample from the Height of 30 Feet above the Ground Level

Locality	*Dust Accumulated in*								
	Winter			*Summer*			*Rains*		
	1 day	*15 days*	*30 days*	*1 day*	*15 days*	*30 days*	*1 day*	*15 days*	*30 days*
Railway Station	0.23	3.10	5.10	0.27	3.40	5.40	0.25	0.50	3.06
(Urban) Rate	–	2.87	1.93	–	3.13	1.00	–	0.25	2.56
MC EME Mess	0.13	1.83	2.59	0.15	2.00	2.90	0.00	0.30	1.86
(Sub Urban) Rate	–	1.70	0.76	–	1.85	0.90	–	0.30	1.56
Saroor Nagar	0.08	0.92	1.21	0.10	1.10	1.41	0.00	0.25	1.16
(Rural) Rate	–	0.84	0.29	–	1.00	0.31	–	0.25	0.91
Women's College	0.05	0.52	0.99	0.08	0.80	1.06	0.02	0.08	0.86
(Protected) Rate	–	0.47	0.47	–	0.72	0.26	–	0.06	0.78
Indira Park	0.10	1.00	1.95	0.12	1.22	2.03	0.11	0.18	1.40
(Protected) Rate	–	0.90	0.90	–	1.10	0.81	–	0.07	1.22

Table 9.2: List of Trees and Shrubs from Twin Cities of Hyderabad

Railway Station (Urban)	MC EME Mess (Suburban)	Saroor Nagar (Rural)	Women's College (Protected)	Indira Park (Protected)
Annona squamosa, Linn.	*Annona squamosa,* Linn.	*Annona squamosa,* Linn.	*Annona squamosa,* Linn.	*Annona squamosa,* Linn.
Citrus limonum, Linn.	*Citrus limonum,* Linn.	*Citrus limonum,* Linn.	*Citrus limonum,* Linn.	*Citrus limonum,* Linn.
Zizyphus jujuba, Lamk.	*Zizyphus jujuba,* Lamk.	*Zizyphus jujuba,* Lamk.	*Zizyphus jujuba,* Lamk.	*Zizyphus jujuba,* Lamk.
Ipomea carnea, Jacq.	*Ipomea carnea,* Jacq.	*Ipomea carnea,* Jacq.	*Ipomea carnea,* Jacq.	*Ipomea carnea,* Jacq.
Lantana camera, Linn.	*Lantana camera,* Linn.	*Lantana camera,* Linn.	*Lantana camera,* Linn.	*Lantana camera,* Linn.
Cassia occidentalis, Linn.	*Cassia occidentalis,* Linn.	*Cassia occidentalis,* Linn.	*Cassia occidentalis,* Linn.	*Cassia occidentalis,* Linn.
Hyptis, Jack.	*Hyptis,* Jack.	*Hyptis,* Jack.	*Hyptis,* Jack.	*Hyptis,* Jack.
Bauhinia variegata, Linn.	*Bauhinia variegata,* Linn.	*Bauhinia variegata,* Linn.	*Bauhinia variegata,* Linn.	*Bauhinia variegata,* Linn.
Tamarindus indica, Linn.	*Tamarindus indica,* Linn.	*Tamarindus indica,* Linn.	*Tamarindus indica,* Linn.	*Tamarindus indica,* Linn.
Albizzia lebbek, Benth.	*Albizzia lebbek,* Benth.	*Albizzia lebbek,* Benth.	*Albizzia lebbek,* Benth.	*Albizzia lebbek,* Benth.
Delonix regia, Ref.	*Delonix regia,* Ref.	*Delonix regia,* Ref.	*Delonix regia,* Ref.	*Delonix legia,* Ref.
Saraca indica, Linn.	*Saraca indica,* Linn.	*Saraca indica,* Linn.	*Saraca indica,* Linn.	*Saraca indica,* Linn.
Ficus religiosa, Linn.	*Ficus religiosa,* Linn.	*Ficus religiosa,* Linn.	*Ficus religiosa,* Linn.	*Ficus religiosa,* Linn.
Syzigium jambolanum, D.C.	*Syzigium jambolanum,* D.C.	*Syzigium jambolanum,* D.C.	*Syzigium jambolanum,* D.C.	*Syzigium jambolanum,* D.C.
Mangifera indica, Linn.	*Mangifera indica,* Linn.	*Mangifera indica,* Linn.	*Mangifera indica,* Linn.	*Mangifera indica,* Linn.
Melia azedarach, Linn.	*Melia azedarach,* Linn.	*Melia azedarach,* Linn.	*Melia azedarach,* Linn.	*Melia azedarach,* Linn.
	Calotropis gigantea, R.B.r.	*Calotropis gigantea,* R.B.r.	*Calotropis gigantea,* R.B.r.	*Calotropis gigantea,* R.B.r.
	Hibiscus rosa sinensis, Linn.	*Hibiscus rosa sinensis,* Linn.	*Hibiscus rosa sinensis,* Linn.	*Hibiscus rosa sinensis,* Linn.
	Hibiscus mutabilis, Linn.	*Hibiscus mutabilis,* Linn.	*Hibiscus mutabilis,* Linn.	*Hibiscus mutabilis,* Linn.
	Murraya koenogii, Spr.	*Murraya koenogii,* Spr.	*Murraya koenogii,* Spr.	*Murraya koenogii,* Spr.
	Pongamia glabra, Vent.	*Pongamia glabra,* Vent.	*Pongamia glabra,* Vent.	*Pongamia glabra,* Vent.
	Ficus benghlansis, Linn.	*Ficus benghlansis,* Linn.	*Ficus benghlansis,* Linn.	*Ficus benghlansis,* Linn.
	Artocarpus integrifolia, Linn.	*Artocarpus integrifolia,* Linn.	*Artocarpus integrifolia,* Linn.	*Artocarpus integrifolia,* Linn.
	Cocos nucifera, Linn.	*Cocos nucifera,* Linn.	*Cocos nucifera,* Linn.	*Cocos nucifera,* Linn.
	Borassus, Linn.	*Borassus,* Linn.	*Borassus,* Linn.	*Borassus,* Linn.
	Rosa, Linn.	*Rosa,* Linn.	*Rosa,* Linn.	*Rosa,* Linn.
		Nerium odorum, Soland	*Nerium odorum,* Soland	*Nerium odorum,* Soland
			Polyalthia longifolia H k.f &T.	*Polyalthia longifolia* H k.f &T.
			Jatropha pandurifolla, Linn.	*Jatropha pandurifolla,* Linn.

Contd...

Table 9.2–Contd...

Railway Station (Urban)	MC EME Mess (Sub Urban)	Saroor Nagar (Rural)	Women's College (Protected)	Indira Park (Protected)
			Bombax malabaricum, D.C.	*Bombax malabaricum*, D.C.
			Aegle marmelos, Corr.	*Aegle marmelos*, Corr.
			Caesalpinia	*Caesalpinia*
			Pulcherrina, Sw.	*Pulcherrina*, Sw.
			Ceasalpinia bonducella, Flem.	*Ceasalpinia bonducella*, Flem.
			Cassia fistula, Linn.	*Cassia fistula*, Linn.
			Cassia glauca, Lam.	*Cassia glauca*, Lam.
			Acacia arabica, Wild.	*Acacia arabica*, Wild.
			Eucalyptus globules, Labill.	*Eucalyptus globules*, Labill.
			Adhatoda vasica, Ness.	*Adhatoda vasica*, Ness.
			Clerodendron inerme, Gaertn.	*Clerodendron inerme*, Gaertn.
			Minusops, Linn.	*Minusops*, Linn.
			Spathlobus, Hassk.	*Spathlobus*, Hassk.
			Guazuma tomentosa, L.	*Guazuma tomentosa*, L.
			Boerhaavia diffusa, L.	*Boerhaavia diffusa*, L.
			Tinospora cordifolia, L.	*Tinospora cordifolia*, L.
			Thespesia populenea	*Thespesia populenea*
			Sida acuta	*Sida acuta*
			Sida rhombifolia	*Sida rhombifolia*
			Argemone maxicana	*Argemone maxicana*
			Malva sylvestris	*Malva sylvestris*
			Thuja occidentalis	*Thuja occidentalis*
			Arucaria spp.	*Arucaria* spp.
			Pongamia glabra	*Pongamia glabra*
			Santalum album	*Santalum album*
			Millingtonia hortensis	*Millingtonia hortensis*
			Sida acuta	*Sida acuta*
			Corchorus acutangularis	*Corchorus acutangularis*
			Klenohaavia hospata	*Klenohaavia hospata*
			Abutilon indicum	*Abutilon indicum*
			Commelina benghalensis	*Commelina benghalensis*
			Cleome viscose	*Cleome viscose*
			Gynandropsis gynandra	*Gynandropsis gynandra*
			Tribulus terrestris	*Tribulus terrestris*

The list of trees and shrubs in different localities is given in Table 4.2. When compared with Mumbai 25 tons/month/km^3 and Chennai and Hyderabad had less particulate matter (Seth, 1976). However, a minimum of dust particles (10 tons/month/km^3) in air acts as condensing nuclei for water vapour, which gives rise to rain and this confines to the primary standards of the dust fall (EPA, 1971). However, continuous dust monitoring is to be done to check pollution levels of environment.

Number and the percentage composition of living and non-living matter accumulated during the study period were given in Table 4.3. It increased during 1 to 30 days exposure. Up to 19 to 50 per cent increase in living matter was observed between 15 to 30 days of exposure in summer than in winter and rainy season. This is due to the quantitative reduction in the dust aggregates in the air due to rain. In 30 days samples due to polarity repulsion reduction in the number of particulate matter was observed. The settlement of the particulate matter was higher up to 15 days and declines thereafter in the slides kept for 30 days.

Table 4.3: Percentage Composition of Living Matter in the Air Dust Samples Collected Over a Height of 30 Feet from Different Localities of Twin Cities of Hyderabad

Localities		*Accumulation on Slides Exposed for 1 Day*		*Accumulation on Slides Exposed for 15 Days*		*Accumulation on Slides Exposed for 30 Days*	
		Living	*Non-living*	*Living*	*Non-living*	*Living*	*Non-living*
Railway	W	15.9	94.1	17.2	82.8	17.6	82.4
Station	S	16.9	83.1	19.2	80.8	33.9	66.1
(Urban)	R	5.5	94.5	15.0	85.0	9.7	90.3
MC EME	W	8.1	91.9	17.1	82.9	24.0	76.0
Mess	S	8.0	92.0	19.2	80.8	41.0	59.0
(Suburban)	R	–	–	15.2	84.9	5.9	94.1
Saroor	W	2.4	97.6	24.8	75.2	36.5	63.5
Nagar	S	11.1	88.9	55.1	44.9	44.8	55.2
(Rural)	R	–	–	5.8	94.2	2.7	97.3
Women's	W	1.3	98.7	7.3	92.7	6.6	93.7
College	S	9.2	90.8	41.5	58.5	42.3	57.7
(Protected)	R	1.5	98.5	3.3	96.7	3.4	96.6
Indira	W	1.6	98.4	7.6	92.4	19.3	80.8
Park	S	15.2	84.8	48.4	51.6	40.0	60.0
(Protected)	R	1.2	98.8	6.7	93.3	5.9	94.1

W = Winter; S = Summer; R = Rainy.

Size of the particulate matter settled per unit area on the glass plate per unit time were given in the Table 4.4. It was observed that there was no uniform pattern in the number of particles collected on the slides. An increase in the size of the particles was observed in the samples accumulated over a period of 30 days due to aggregation of dust particles resulting in the increase of their size. Increase in the size of the aggregates has made the number compact and less.

Table 4.4: Size in Microns of Aggregate Particles Collected from the Air Samples Over a Height of 30 Feet in Different Localities of Twin Cities of Hyderabad

Localities	Winter			Summer			Rains		
	1 day	15 days	30 days	1 day	15 days	30 days	1 day	15 days	30 days
Railway Station (Urban)	9.7	21.3	48.2	10.7	13.2	49.4	9.1	9.4	80.6
MC EME Mess (Suburban)	7.7	17.3	23.3	8.7	9.1	9.8	0.0	25.1	90.6
Saroor Nagar (Rural)	6.6	14.5	20.5	7.45	7.4	22.6	0.0	10.1	69.3
Women's College (Protected)	5.5	9.4	16.7	6.4	6.7	17.8	3.2	6.4	88.2
Indira Park (Protected)	7.2	15.3	30.9	8.2	8.3	33.1	7.6	8.7	19.2

Table 4.5 shows electrical conductivity measured in micro ohms / per cm. Electrical conductivity (EC) was higher in summer and winter than in rainy season. EC was very much reduced in rainy season probably due to variation in the types of particulate matter that settle in various seasons.

Table 4.5: Electrical Conductivity of Dust Pollutants in Different Areas of Twin Cities of Hyderabad Placed 30 Feet Above the Ground

Localities	1 day			5 days			30 days		
	Winter	Summer	Rains	Winter	Summer	Rains	Winter	Summer	Rains
Railway Station (Urban)	0.7 x 10 70	0.75 x 10 75	9.5 x 10 50	1.1 x 10 110	1.25 x 10 125	0.6 x 10 60	1.5 x 10 150	1.6 x 10 160	1.25 x 10 125
MC EME Mess (Suburban)	0.15 x 10 15	0.2 x 10 20	0 0	0.5 x 10 50	0.6 x 10 60	0.55 x 10 55	1.0 x 10 100	0.05 x 10 105	0.9 x 10 90
Saroor Nagar (Rural)	0.25 x 10 25	0.35 x 10 35	0 0	0.65 x 10 65	0.75 x 10 75	0.45 x 10 45	0.7 x 10 70	0.8 x 10 80	0.65 x 10 65
Women's College (Protected)	0.1 x 10 10	0.15 x 10 15	0.05 x 10 5	0.4 x 10 40	0.5 x 10 50	0.4 x 10 40	0.5 x 10 50	0.60 x 10 60	0.55 x 10 55
Indira Park (Protected)	0.6 x 10 60	0.65 x 10 65	0.4 x 10 40	1.05 x 10 105	1.15 x 10 115	0.3 x 10 30	1.2 x 10 120	1.25 x 10 125	1.15 x 10 115

Table 4.6: pH of Dust Pollutants in Urban, Suburban, Rural and Protected Areas at Different Place in Twin Cities of Hyderabad Placed at 30 Feet Above the Ground

Localities	1 Day			15 Days			30 Days			Average
	Winter	Rains	Summer	Winter	Rains	Summer	Winter	Rains	Summer	
Railway Station (Urban)	7.0	7.5	7.0	7.5	8.5	7.5	9.0	8.5	6.5	7.6
MC EME Mess (Suburban)	5.5	6.0	7.0	5.5	6.5	6.5	6.5	7.0	6.0	6.2
Saroor Nagar (Rural)	6.0	6.5	7.0	6.0	7.0	6.0	6.5	7.0	6.0	6.3
Women's College (Protected)	6.5	7.0	7.0	7.0.	8.0	7.0	7.5	7.5	7.0	7.1
Indira Park (Protected)	5.5	6.0	7.0	5.5	6.5	7.0	6.5	6.5	6.5	6.3

Table 4.6 represented the data on pH of the dust particles. Slightly acidic range in the air dust samples were observed at and near Indira Park, Saroor Nagar and MC EME Mess due to small scale industries *viz.*, Iron melters, hand made paper mills, cement dust etc. At Saroor Nagar which has chemical industries also show acidity of air dust samples. Due to coal and wood burning at the MC EME Mess there was accumulation of smoke and the samples showed acidic pH. Women's College dust environment was neutral.

References

Chaphekar, S.B., Boralkar, D.B. and R.P. Shetye (1980). Effects of industrial air pollutants on plants. *Final Reports of the U.G.C. Sponsored Research Project* No. F. 23–170/75 (SR. 11), pp. 83.

E.P.A. (1977). National air quality standard. *J. Air Poll. Control Assoc.*, 21: 149.

Kayastha, S.C., and V.K. Kumar (1978). Air pollution problem in Kanpur city. *Indian J. Air Poll. Control,* 1: 38–42.

Misra, R.C. (1978). Hazards of dust pollution. *Indian J. Air Poll. Control,* 1: 32–34.

Seth, G.K. (1976). Know your environment. *Science Reporter,* 13: 7–11.

Sunavala, P.D. (1977). Chemical processing and engineering [Misra, R.C. (1978) Hazards of dust pollution]. *Indian J. Air Poll. Control,* 1: 32–34.

Chapter 5

Water Quality Index for Ground Water Affected with Bicycle Manufacturing Industrial Wastes: An Environmental Quality Audit

Vineeta Shukla, Sharda Abusaria**, Monika Dhankhar* and K.V. Sastry**

**Department of Biosciences, Maharshi Dayanand University, Rohtak – 124 001, Haryana*
***Lecturer, Department of Zoology, A.I.J.H.M. College, Rohtak*

ABSTRACT

The present study is intended to calculate water quality index for ground water at Sonepat in Haryana State of India from various parameters, *viz.*, pH, dissolved oxygen, total hardness, chlorides, total alkalinity, biological oxygen demand, sulphates, iron, chromium, cyanide and MPN coliforms to ascertain the quality of water for human. consumption and to assess the impact of bicycle manufacturing industrial wastes on ground water quality. The high values of water quality indices for ground water samples from various sources located in the industrial premises and vicinity of the industry indicate the influence of industrial wastes on ground water quality and make it unsuitable for drinking purposes. The trend of variation in water quality index values collected around and inside the industry was irregular while for control samples, water quality index values were well below the permissible limit throughout the study period and did not vary widely. The importance of water quality index and pollution control has also been discussed and emphasized.

Keywords: Ground water, Bicycle manufacturing industrial wastes, Water quality index, Pollution control.

Introduction

Environment is our biological capital. It makes up a huge, enormously complex living machine on which our total productivity depends. Water is one of the basic natural resources needed for survival of life. Nobel laureate A. Szent-Gyorgy, physician and biochemist, called water the "matrix of life" Many organisms can live without air, but none without water. The use of this under ground fluid for agriculture and drinking purposes has a long history. But ground water is safe only as long as it is free from disease causing organisms and chemicals that are deleterious to human health. Unfortunately, in recent years industrial growth has created severe environmental pollution, and degradation has not spared even ground water aquifer system. A number of places in the country face serious ground water contamination problem due to indiscriminate dumping of toxic chemicals by polluting industries. Water soluble chemicals are not retained by soil matrix. they percolate through porous medium and enter the ground water. Environmental protection and mankind's survival are closely linked. Ground water once polluted cannot be treated, therefore, the effect is long lasting and problem is of vital concern. Hence, there is need for protection and management of ground water quality. Accurate and timely information on water quality is essential to frame a sound public policy and to implement the water quality improvement programs.

Water quality index is the most efficient tool of environmental audit process to convey the overall water quality of a water resource. It is a rating of water quality parameters by a single numerical expression reflecting the complete influence of water quality parameters. Unfortunately in India water quality index determination has been much neglected. Few attempts have been made to calculate water quality for river waters. Studies on ground water quality index are scanty.

Sonepat in Haryana State is known for bicycle manufacturing industry whose product is available globally. But the industrial effluents and sludge generated as byproduct may cause great environmental stress locally. Spent waters of different operations such as electroplating and heat treatment are normally stored inside the industrial premises before discharge. For the past more than forty five years, surrounding land has been receiving untreated/partially treated effluents of the industry. Sludge is being dumped within the premises and in low lying areas located near the industry. The dumping and holding of such chemical wastes are likely to affect the ground water quality in these areas. So far, no detailed investigation on the effect of these industrial wastes on ground water quality has been carried out. The present study has, therefore, been undertaken to calculate the water quality index for ground water in and around the industry to measure the extent of damage to water quality.

Material and Methods

In the industry and surrounding regions ground water is the only source of drinking water and is being tapped by tube wells. Four sampling sites were selected for the present investigation. Details of these sites are given in Table 5.1. In the formulation of water quality index, the importance of various parameters depends upon the intended use of water. In this study eleven water quality parameters of ground water are studied from the point of view of suitability of water from different sources at Sonepat for human consumption. For this purpose monthly samples of ground water from selected sites were collected and analysed as per Standard Methods during March, 1999 to February, 2000.

Based on the ground water analysis data, water quality index was calculated for each sample by a simple method. The 'standards' (permissible values) of various parameters for drinking water, recommended by the ISI are used for comparison. For dissolved oxygen and BOD, European Economic Community (EEC) Standard and United Public Health Service (UPHS) have been considered respectively.

Table 5.1: Particulars of Ground Water Sampling Sites

Reference Number	*Name and Location*	*Description*	*Age (Yrs)*	*Depth (M)*
GW1	Tubewell No. 3 in bicycle industry premises in contaminated area	20M from effluents bodies	40	115 up to May, 1999. June onwards 120. Closed in August
GW2	Tubewell No. 7 in bicycle industry premises in contaminated area	Earlier 30M, now 1000M from sludge dumping area and effluents bodies	8	105
GW3	Sujan Singh Tubewell in Agricultural field contaminated with industrial effluents	Near cyanide sludge dumping area of industry and about 1000M from effluents flow channel	8	75
GW4	Public Health Tube well in unpolluted area	Beyond the influence of industrial wastes and it served as control ground water	5	83

Calculation of Subindex (SI)

To calculate the Water Quality Index, first of all subindex (SI) i for the ith parameter was calculated which was the product of the quality rating (qi) and the unit weight (wi) of the ith parameter *i.e.*

$$(S)\,i = qi\,wi \qquad ....(1)$$

Calculation of Quality Rating (qi)

The quality rating, qi for the ith water quality parameter was obtained from the equation:

$$qi = 100\,(Vi/Si) \qquad ....(2)$$

where,

Vi: Value of the ith parameter at a given sampling station; and

Si: Recommended standard for the ith parameter.

Equation (2) ensures that qi = 0 when pollutant (the ith parameter) is absent in the water, while qi = 100 if the value of this parameter is just equal to its permissible value for drinking water.

Quality ratings for pH, dissolved oxygen and MPN coliforms required special handling.

The permissible range of pH for drinking water is 6.5 to 8.5. Therefore, the quality rating qpH was calculated from the relation:

$$qpH = 100[(V_{pH} \sim 6.5)/(8.5-6.5)] \qquad ....(3)$$

where,

V_{pH}: Value of pH; and

Symbol "~" means simply the numerical differences between V_{pH} and 6.5 ignoring the algebraic sign.

The situation is slightly complicated in the case of dissolved oxygen, since, in contrast to other pollutants, the quality of water is enhanced if it contains more dissolved oxygen. Therefore quality rating qDO was calculated from the relation.

$$qDO = 100\left[\frac{14.6}{14.6-5.0} - \frac{V_{DO}}{14.6-5.0}\right] \qquad ....(4)$$

where,

V_{DO}: Value of dissolved oxygen.

14.6 is the solubility of oxygen (mg/l) in distilled water at 0°C; and 5.0 is the standard (mg/l) for drinking water.

In case of MPN coliforms, the permissible value for the drinking water is 1/100 ml, while their actual number in untreated water may be several thousands/100 ml. Therefore to get a convenient number for quality rating qMPN was calculated from the relation

$$qMPN = 100\left[\frac{Log_{10}V_{MPN}{}^{+1}}{Log_{10}S_{MPN}{}^{+1}}\right] \qquad ....(5)$$

where,

V_{MPN}: MPN count/100 ml; and

S_{MPN}: MPN standard for drinking water.

Calculation of Unit Weight (wi)

$$wi = k/Si \qquad ....(6)$$

where,

wi: Unit weight;

Si: Standard for the ith parameter (i = 1, 2..............11); and

k: Constant of proportionality which was determined from the condition

$$\sum_{i=1}^{i=11} wi = k\sum_{i=1}^{i=11}\left(\frac{1}{si}\right) = 1 \qquad ....(7)$$

Final Value of Water Quality Index

Finally the water quality index was obtained by aggregating these eleven subindices (SI) linearly

$$W.Q.I. = \sum_{i=1}^{i=11}(qiwi) \qquad ....(8)$$

Results and Discussion

Unit weights assigned to water quality parameters and a sample calculation of overall water quality index from the subindices for ground water samples from different sources have been shown in Table 5.2. Water quality indices for one year study period are depicted in Table 5.3. On extensive comparison with control sample to document the impact of industrial pollution on ground, water quality present study reveals an alarming feature that in 100 per cent samples from affected area quality of ground water is much altered with respected to natural and local background indicating pollution status of ground water and is not fit for human consumption at any time period of the year due to non-conforming water quality index with drinking water limit (100). A critical study of Table 5.2 indicates that for GW1, GW2 and DW3 samples the main dominating contribution to water quality index comes from the existence of un-natural toxic chemicals–chromium and cyanide–which are used in bicycle manufacturing industry. It is further evidenced by the fact that the more harmful a given pollutant is, the smaller is its permissible value for drinking water. So the, unit weights for various water quality parameters are assumed to be inversely proportional to the recommended standards for the corresponding parameters. Maximum value of the unit weight for water quality parameters in the present study was in the order of chromium and cyanide > iron > MPN coliforms > BOD and dissolved oxygen> pH > total alkalinity and sulphates > chlorides> total hardness, consequently due to high values of chromium and cyanide subindices WQI increased drastically. It may thus be, interred that higher water quality indices for ground water samples from different sources–in and around the bicycle manufacturing industry–may be due to leaching of chemicals from industrial wastes through soil into ground water. Further, WQI for GW1, GW2 and GW3 samples varied depending upon the location of sampling site and var-ation was sizeable. Very high value of WQI for GW1 indicates the influence of proximity of effluents tanks. There was a wide difference in the values of WQI between GW1 and other two samples coinciding with levels of pollutants. Amongst above three samples, GW1 water exhibited a maximum value of WQI in the month of April which is about 39 times greater than the limit for drinking water. However, GW2 and GW3 samples also exhibited deteriorated status of water quality. On the other hand chromium and cyanide were not the constituents of WQI for GW4 sample (control) coinciding with their nil quality rating values. WQI for GW4 was quite less than other three samples and remained below the limit for drinking water. The variability in WQI among different ground water samples was measured statistically by calculating the coefficient of variation. It confirnled well marked variation in WQI for ground water samples from different sources (Table 5.3).

Table 5.2: Water Quality Index of Ground Water Samples (March)

Water Quality Parameters	*Drinking Water Standard (Si)*	*Unit Weight (wt)*	*Parameter Sub-index (qiwi)*			
			GW1	*GW2*	*GW3*	*GW4*
pH	6.5–8.5	0.00262	0.1179	0.1048	0.1179	0.1048
Dissolved oxygen	5.0 mg/l	0.00446	0.3836	0.3791	0.4371	0.4371
Total hardness	300.0 mg/l	0.00007	0.0104	0.0238	0.0149	0.0141
Chlorides	250.0 mg/l	0.00008	0.0028	0.0214	0.0071	0.0053
Total alkalinity	200.0 mg/l	0.00011	0.0157	0.0092	0.0109	0.0144
BOD	5.0 mg/l	0.00446	1.6484	0.8028	0.0347	0.2382
Sulphates	200.0 mg/l	0.00011	0.0037	0.0049	0.0037	0.0028
Iron	0.3 mg/l	0.7430	0.2194	3.4178	3.2194	12.4764
Total chromium	0.05 mg/l	0.44580	3584.2320	53.4960	53.4960	0.0000
Cyanide	0.05 mg/l	0.44580	71.3280	53.4960	53.4960	0.0000
MPN coliforms	1/100 mg/l	0.02229	0.0000	0.0000	0.0000	0.0000
Water quality index			3660.9619	111.7528	111.8377	3.2931

Table 5.3: Monthly Variation in Water Quality Index of Ground Water Samples

Months	*Water Quality Index (WQI)*			
	GW1	*GW2*	*GW3*	*GW4*
Mar	3360.96	111.75	111.84	3.29
Apr	3911.12	111.02	120.73	3.43
May	3633.90	114.32	142.60	3.07
Jun	3635.88	110.46	123.77	4.28
Jul	3654.14*	110.69	148.53	9.48
Aug		119.45	113.44	7.43
Sep		132.22	131.19	3.18
Oct		141.64	139.60	3.19
Nov		129.55	122.18	3.74
Dec		119.80	130.35	3.42
Jan		129.87	121.90	3.73
Feb		138.14	130.91	3.42
Average	3639.20	122.41	128.09	4.30

* GWI (Tube well No.3) was closed by the industry from August, 1999 CV (%) = 182.64.

Another striking feature emerging from the present investigation is that, different sources do not synchronize in monthly WQI variation. This may be attributed to the fact that an index significantly increases or decreases when some of the critical parameters exceed or become less than the specified permissible value. Over a period of about half a century enormous quantity of pollutants has been added to the soil which may act as a bank of pollutants. Thus, chemical characteristics of percolating effluents, and nature and status of soil may be the possible reasons for irregular variations in WQI for affected sample? However, WQI for control samples indicates uniformly good water quality, reflecting pollution free status of ground water of area contaminated, without any irregular variation. But if the bicycle industry is allowed to dump its wastes without proper treatment the possibility of spread of pollution of ground water in this area also cannot be ruled out.

It is clearly brought out by the findings of this study that the ground water environment of bicycle manufacturing industrial area has already become grossly polluted. Operational phase of the industry established beyond doubt the potential threat to ground water. It is unfortunate that groundwater, a gift of nature, turns out to be a man-made curse. Good quality ground water, as of now is a rare and scarce commodity in the area. Therefore, there is an urgent need for strict enforcement of Environment (Protection) Act, 1986. Violation of these rules can be disastrous and can have far reaching consequences. Present study also clearly confirms the utility of WQI as a proper indicator for conveying the idea of the overall quality of water with a single number. This type of information is helpful for public to understand the quality of drinking water and may have an administrative use in the field of water quality management.

References

APHA 1985. *Standard Methods for the Examination of Water and Wastewater*. American Public Health Association, Washington, D.C.

Horton, R.K. 1965. An index number system for rating water quality. *J. Water Poll. Cant. Fed.*, 37(3): 300–306.

ISI 1991. *Indian Standard Specification for Drinking Water* (IS: 10500), Indian Standards Institution, New Delhi.

Jeevan Rao, K. and Shantaram, M.V. 1995. Ground water pollution from refuse dumps at Hyderabad. *Indian J. Environ. Hlth.*, 37(3): 197–204.

Mariappan, P., Vasudevan, T. and Yegnaraman, V. (1999). Surveillance of ground water quality in Thiruppathur Block of Sivagangai district. *Ind. J. Environ. Prot.*, 19(4): 250–254.

Singh, A.P. and Ghosh, S.K. 1999. Water quality index for river Yamuna. *Poll. Res.*, 18(4): 435–439.

Singh, K.P. and Parwana, B.K. 1999. Ground water pollution due to industrial wastewater in Punjab State and strategies for its control. *Ind. J. Environ. Prot.*, 19(4): 241–244.

Tiwari, T.N., Das, S.C. and Hose, P.K. 1986. Water quality index for the river Jhelum in Kashmir and its seasonal variation. *Poll. Res.*, 5(1): 1–5.

Venkata Mohan, S., Rama Mohan, S. and Jayaraman Reddy, S. 1998. Determination of water quality index (GWQI) of ground water: An application. *J. Environ. Poll.*, 5(3): 219–223.

Chapter 6

Changes in Sodium and Potassium Ratio in Pressmud with Influence of Earthworm *Eisenia foetida* (Savigny)

Pramod Kumar and Shweta

Vermiculture Unit, Department of Zoology, D.S. College, Aligarh – 202 001, UP, India

ABSTRACT

Different organic wastes (Pressmud, Bagasse, Buffalo dung, Cow dung, Kitchen waste and Trash) were subjected individually and in combination for vermicomposting by using the earthworm *Eisenia foetida*. The sodium and potassium concentration before and after the conversion were recorded. It was found that Bagasse and kitchen waste (1 : 1) were best substrate to increase the percentage of both sodium and potassium.

Introduction

In recent year, vermicomposting technique has been accepted as suitable tool for its efficiency in recycling a wide range of organic waste into useful product. In India, manurial resources and organic wastes are abundantly available for their bioconversion into vermicompost with the employment of one or another species of earthworm. The only alternate to boost up food production in the face of these shortages, is through processing and utilization of agro-industrial residues available in plenty at the processing sites. Sugar milling industry is the second largest industry in India with a network of 298 sugar factories (Anon, 1981).

During the manufacture of sugar, bagasse molasses and pressmud are available in the form of residues (Paturau, 1969). Bagasse is used mainly as fuel in the factory. Molasses serve as a raw material for the production of alcohol.

Pressmud is sold to the farmers for use as manure at a nominal price (Rao, 1981) and (Humbert, 1980) have critically reviewed sugarcane as source of energy. In; sugar factory after milling of canes the raw juice is processed by adding specific quantity of triple super phosphate solution (Patil, 1979), then heated and further processed either by carbonation or sulphitation process.

Pressmud is a soft, spongy, amorphous dark brown to brownish white material which contains sugar, fibre, coagulated colloids, including cane wax, albuminoids inorganic salts and soil particles. On the basis of 3 per cent of sugarcane milled, the quantity of pressmud produced during the year 1979–80 amounted to 1.171 million tones (Anon, 1981).

The composition of pressmud in vaccum filter and filter press (Paturau, 1969) showing wide variation: sucrose 2.1–7.3 per cent, moisture 78.2–60.4 per cent; wax 1.6–1.8 per cent, fat 1.6–1.8 per cent, nitrogen 0.4–0.7 per cent, P_2O_5 0.4–0.7 per cent, K_2O 0.02–0.01 per cent respectively.

Agarwal (1951) while critically reviewing the distribution of nutrient in the byproduct of sugar industry observed that over 15 per cent of major plant nutrients in sugarcane find their way to the pressmud cake with the highest concentration of phosphorus. Da Giloria *et al.*, (1972) analysed twenty two composite sample of pressmud in Brazil and observed that N, P, Ca and Fe were the main source of nutrients. Vimal and Kale (1982) reported 1.2 per cent N, 2.4 per cent P_2O_5 and 0.5–1.5 per cent K_2O in pressmud and observed that it acts as a rich source of phosphorus.

The present investigation carried out to record the changes in sodium and potassium during bioconversion of pressmud with influence of earthworm.

Materials and Method

The pressmud were collected from the Kishan Sahkari Chini Mill Ltd. Satha factory located at 15 km. away from Aligarh (U.P.), India and experiment area coagulated in the vicinity of vermiculture Unit, D.S. College Aligarh, U.P. for determining the relative efficiency of earthworm species *viz. Eisenia foetida* were used to under following treatments T_1, T_2, T_3, T_4, T_5, T_6, T_7, T_8, T_9 and T_{10} by using Windrows method.

Treatment	Substrate
T_1	Pressmud
T_2	Pressmud + Bagasse (1 : 1)
T_3	Pressmud + Buffalo dung (1 : 1)
T_4	Pressmud + cow dung (1 : 1)
T_5	Pressmud + kitchen waste (1 : 1)
T_6	Bagasse
T_7	Bagasse + Buffalo dung (1 : 1)
T_8	Bagasse + cow dung (1 : 1)
T_9	Bagasse + kitchen waste (1 : 1)
T_{10}	Pressmud + Trash (1 : 1)

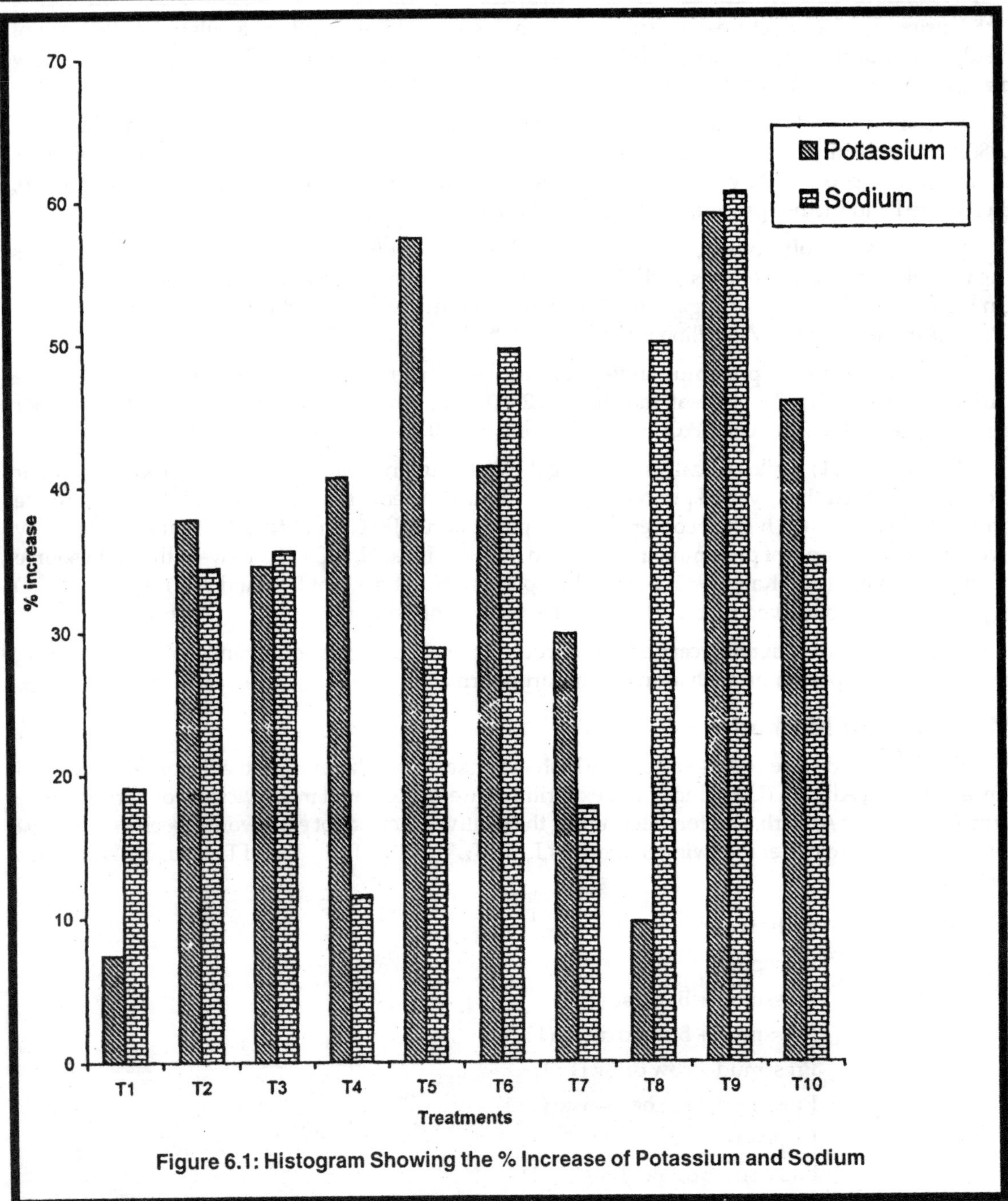

Figure 6.1: Histogram Showing the % Increase of Potassium and Sodium

Bed were covered with a thick layer of mulching. The beds were monitored constantly throughout the experimentation. Their physico-chemical analysis were carried out under control condition by using standard method.

Results and Discussion

Maximum percentage of increase of potassium were recorded in T_9 and minimum in T_1, while maximum percentage of increase of sodium were recorded in T_9 and minimum in T_4.

Yaduvanshi and Yadav (1991) recorded fillering and harvest of the crop increased significantly with direct residual or cumulative application of pressmud based vermicompost. Singh *et al.*, (1986) recorded 0.86 per cent N_2 0.28 per cent P_2O_5 and 0.51 per cent K_2O in pressmud cake. Sahi and Rai (1978) recorded 0.57 per cent nitrogen in cow dung, 0.42 per cent in carbonation pressmud, 0.5 per cent in pressmud + cow dung, 0.96 per cent in sulphitation pressmud and cow dung 0.36 per cent P_2O_5 in cow dung, 0.45 per cent in carbonation pressmud, 0.39 per cent in carbonation pressmud + cow dung, 0.74 per cent in sulphitation pressmud and 0.59 per cent sulphitation pressmud + cow dung, 0.57 per cent K_2O in cow dung, 0.67 per cent in carbonation pressmud, 0.59 per cent in carbonation pressmud + cow dung, while 7.9 per cent organic carbon in cow dung, 6.2 per cent in carbonation pressmud, 6.8 per cent in carbonation pressmud + cow dung and 11.3 per cent in sulphitation pressmud.

Finally, it can be concluded that bagasse + kitchen waste were best substrate to increase the potassium and sodium content of vermicompost.

Acknowledgement

Authors are thankful to University Grants Commission, New Delhi for providing the financial support.

References

Agarwal, R.N. 1951. Sugarcane Agriculture in Queensland. *Indian Sug. J.*, 1(6): 219–224.

Anonymous, 1981. Statements showing the cane crushed and sugar produced in India. *Indian Sug.*, 31(1): 61–65.

Da Gloria, N.A., A.G. Santa Ana and H. Monteiro 1972. Composition of sugar factory and distillary wastes during the cane harvest. *Int. Sug. J.*, 76(901): 30.

Humbert, R.P. 1980. The growing of sugarcane for energy. Paper presented at 17th ISSCT Congress. Manila, Phillipines. *Sugar J.*, 43(1): 19–22.

Patil, S.P. 1979. Some studies on pressmud cake, its residual effect on yield and quality of sugarcane and soil chemical properties. *M.Sc. (Agri.) Thesis*, MPKV, Rahuri (Maharashtra).

Paturau, J.M. 1969. *Byproducts of the Sugarcane Industry: Introduction to their Industrial Utilization*. Elseveir Publishing Co. Barking; Essex England, U.K.

Rao, P.J.M. 1981. National fuel alcohol plans sugarcane and cassava, the potential energy crops to produce fuel alcohol. *Indian Sug.*, 31(1): 11–35.

Sahi, B.P. and Rai, Y. 1978. Studies on better utilization of sulphitation pressmud Proc. 42nd Ann. Conv. Sugar Technol. Assoc. India, Kanpur, G: 105–114.

Singh, K.D.N., Prasad, C.R. and Singh, Y.P. 1986. Comparative study of pyritus and sulphitation pressmud on soil properties, yield and quality of sugarcane in calcareous saline sodic soil of Bihar. *J. Indian Soc. Soil Sci.*, 34: 152–154.

Vimal, O.P. and S.P. Kale 1982. Utilization ofpressmud status of information. *Indian Sug.*, (10): 457–61.

Yaduvanshi, N.P.S. and Yadav, D.V. 1991. Effect of combined application of sulphitation, pressmud and fertilizer nitrogen on the economy and recovery of added nitrogen in sugarcane. *Pert. News*, 36(4): 49–51.

Chapter 7

Status and Assessment of Noise Pollution Level at Rewa Town

Bhupesh Kumar Mishra

School of Environmental Biology, A.P.S. University, Rewa, M.P.

ABSTRACT

Rewa is a important town in M.P. and is a district head quarters for the Rewa district. The ambient noise level were measured at selected zone 5 (five) specific location (Silence and traffic, Residential, Commercial Industrial zone) at Rewa town to assess the content of noise pollution in huge place. Minimum, maximum L_{10}, L_{50}, L_{90}, L_{eq}, L_{np} and TNI noise level were computed. The status of noise pollution at Rewa varied from 44.25 dB (A) to 103.55 dB (A). Other parameters like–L_{10}, L_{50}, L_{90}, L_{eq}, L_{np} and TNI were also computed for that hour during day time. None of the places recorded noise level within the prescribed limits by the C.P.C.B. Vehicular traffic with air horn was found to be the main reason for these high noise levels. Strict enforcement of low to be use of air horns inside the town and any type of horns was the silence zone, proper maintenance of vehicles, laying of proper roads and their maintenance, planting of trees etc. can be under taken to control this menace in Rewa town.

Keywords: *Decibel (dB), Transport Noise Index (TNI), Equivalent Continuous Noise Level (L_{eq}), Noise Pollution Level (L_{np}).*

Introduction

Rewa is an improving city. It is the old capital of the Vindhya Pradesh and Rewa State. It is improving industrial and improved educational center of Madhya Pradesh. Rewa is a Vainwe of concrit. Rewa town is situated between longitude 24°–19′ and 82°–5′east with an average height of 318.5 meter from sea level. The town is a part of Vindhya plateaue of inked by Vindhya rision in heart if Kaimore ranges in south. The geographical area of Rewa is 628.555 sq.km. and this is 17 per cent of the whole M.P. and has population more than.

A useful definition of noise is "objectionable sound" just now objectionable obtrusive or irritating a sound is depends on its intensity and character. The character of a sound is defined by the frequencies it contains such as base and treble its duration and now suddenly the sound level rises. The level of annoyance will depend on whether the noise is predictable and also on many individual factors relating to the circumstances and activities those affected.

Sound can be defend as atmospheric air born vibration perceptible to the ear. Noise is usually unwanted or undesired sound. Consequently a particular sound can be noise to one person and not to others, are hear at one time and not at other times. Noise is a form of pollution, because it can cause hearing impairment and psychological stress and noise is responsible to hearing loss, high blood pressure, sleepless and harmful physiological and phycological effects. In the first place, Eurasia noise can damage the ear depending upon the intensity and duration of noise, the person expressed to it may suffer from temporary or permanent noise induced hearing loss. The temporary hearing loss is called temporary threshold shift (TTS) (Bugliurell *et al.*, 1976).

A number of studies are available on noise pollution from different part of India (Pandey, 1992 and Sastri *et al.*, 1996; Ravichandran *et al.*, 1997 a and b) and the main course of high noise is attributed to vehicle and its hooting. No study was available for Rewa town. In the Rewa municipality area may be situated TAMAS, BTL Cattha Factory and many more small scale industries are located in Rewa. In the present investigation an attempt has made to assess the noise level at five zone (different locations such as industrial; Silence, redundancies and traffic zone area as per ministry of environmental and forest guidelines. The six locations where survey was under taken are:

1. Sirmour, Chack (traffic zone)
2. Bus stand and Jaustambh (Indu. and Comm. zone)
3. Sanjay Gandhi and Mahatma Gandhi hospital (Sitinco zone)
4. Tamus and Gupta steel industry (Industrial area)
5. Different residential areas (Included localities)
6. Different commercial areas (total main market).

This investigation time is 4 months and in season October 2001 to January 2002. Their investigation is present to the noise level report on winter season.

Materials and Methods

A preliminary survey of noise levels in different areas at Rewa town were conducted for a period for October 2001 to January 2002. The measurement of noise levels were carried out between phe Period 8 am to 8 pm (only one site 7 am +8 pm with the help of a "Precision sound level meter" type 2232 manufactured by the Firmbruel and Kajer of Denmark. A world leader in the field of acoustic instruments. It is an easy to use precision class instrument with digital display. Measuring maximum root mean square (RMS) sound pressure level in dB (A) '(A' weighted decibel). The type 2232–instrument used for present survey has built in "A"–weighting network and "fast" and "slow" detector characteristic.

During each hourly interval sound pressure level (SPI) values has been measured for period on minute's. From the measured value, the parameters L_{10}, L_{50}, L_{90}, L_{eg}, L_{np} and TNI, were computed for that hour. It is calibrated acoustically using an estrous reference source, which is placed over the microphone, measurement of hear from 34 dB (A) to 130 dB (A) can be carried out with this battery operated instrument.

Pollution Control Standard for Noise

Area	*Day dB (A) (L_{eq})*	*Night dB (A) (L_{eq})*
Silence zone	50	40
Residential	55	45
Commercial	65	55
Industrial	75	70

Daytime: 6 am to 9 pm.

Night time: 9 pm to 6 am 18.

Parameters Studies

The parameter universal used in discussion of noise pollution of the environments in L_{eq} the equivalent continuous noise level expressed in dB (A), which is the average rate at which energy is received by the human ear during the period monitored. This parameter can be obtained by a direct objective measurement with an integrating type sound level meter can be computed from a sample recording of the time variation of the noise level during that period. The L_{eq} can be estimated by assuming as Gaussain distribution of noise level using the empirical equation (Tiwari and Kudesia)

$$L_{eq} = ½ (L_{10} + L_{90}) = 1/57 (L_{10} - L_{90})2 \qquad ....(1)$$

where,

L_{10} and L_{90} respectively indicated the level exceeding for 10 per cent and 90 per cent of the time in a record of a noise level in a given interval. Another parameter used if the noise pollution level L_{mp} which can be computed from the time varying noise level using the relation (Patrick, 1977).

$$L_{np} = L_{50} = + (L_{90} + L_{90})/60 (L_{10} - L_{90}) \qquad ....(2)$$

where,

L_{10}, L_{50} and L_{90} respectively indicate the level exceeding for 10 per cent, 50 per cent and 90 per cent of the time in a record of a noise level in a given interval, L_{np} is also expressed in units of dB (A).

While L_{eq} level gives the total energy received by the ear and hence an indicator of the physiological disturbance to the hearing mechanism, L_{np} is also takes into account the variations in the sound signal and hence a better indicator of the pollution in environment for physiological disturbance of human system.

Traffic Noise Index (TNI) is another parameter which indicates the degree or orderliness disorderliness in a traffic flow. This is also expressed in dB (A) and can be computed using the relation (Patrick, 1977).

$$TNI= L_{10} + 4 (L_{10} - L_{90}) - 30 \qquad ....(3)$$

Result and Discussion

The parameter and L_{10}, L_{50}, L_{90}, L_{eq}, L_{np} and TNI as computed using equations 1, 2 and 3 for Sirmour Chowk location (Traffic Zone) are given in (Table 7.5).

Same methods were using other different location such as G.M.H. and S.G. Hospital (Table 7.4), Industrial location (Table 7.1), Residential Location (Table 7.3) and Commercial location (Table 7.2).

These tables are showed the four months average noise level in these sites. The average noise level for the six locations of the computed parameter for Rewa town area are given in Table 7.3. The ambient air quality standard in respect of noise (Central Board for Pollution Control Standards for Noise) are given in Table 7.1.

Table 7.1: Hourly Computed Noise Parameter for Industrial Zone (B.T.L., TAMUS, and Other Industrial Location) Winter Season (October 2001 to January 2002) at Rewa

Time, AM and PM	L_{90}	L_{50}	L_{10}	L_{eg}	L_{np}	*TNI*
8–9	77.25	84.50	99.25	96.74	114.56	135.25
9–10	71.75	82.50	93.50	90.92	112.13	128.75
10–11	75.50	82.75	93.00	89.62	105.35	155.05
11–12	82.50	91.50	101.75	98.62	116.92	129.50
12–1	87.50	91.25	106.50	103.33	116.26	133.50
1–2	86.00	88.25	103.00	99.57	110.06	124.00
2–3	74.50	78.75	92.50	89.18	102.15	116.50
3–4	76.00	81.25	91.75	88.22	101.13	109.00
4–5	72.25	80.00	89.50	86.09	102.21	111.25
5–6	78.00	85.25	96.75	93.54	109.86	123.00
6–7	81.75	86.75	103.00	100.29	115.52	136.75
7–8	87.00	82.25	106.50	103.42	108.08	135.00

Table 7.2: Hourly Computed Noise Parameter for Traffic Zone Winter Season (October 2001 to January 2002) at Rewa

Time, AM and PM	L_{90}	L_{50}	L_{10}	L_{eg}	L_{np}	*TNI*
8–9	67.25	75.00	84.50	81.09	71.21	106.25
9–10	71.75	78.25	87.75	84.24	98.51	105.75
10–11	69.75	81.75	84.75	81.19	100.50	99.75
11–12	74.75	78.75	87.25	83.74	93.85	94.75
12–1	70.25	80.00	86.25	82.74	100.26	104.25
1–2	70.00	78.00	86.75	83.29	99.42	107.00
2–3	68.75	75.25	83.25	79.68	93.25	96.75
3–4	53.50	75.25	88.50	92.49	130.67	163.50
4–5	70.50	75.50	86.50	82.99	95.76	104.50
5–6	73.25	81.50	90.25	86.82	103.31	111.25
6–7	77.75	86:00	94.25	90.77	107.03	113.75
7–8	81.00	88.25	96.75	93.22	108.13	114.00

None of the places in the silence zone recorded noise level below 50 dB (A), the prescribed limits set by the C.P.C.B. (Central Pollution Control Board) the lowest value recorded was 44.25 dB (A) and the highest value recorded was 91 in G. M. Hospital area. In residential zone all selected places recorded noise levels that are higher than the limit set by the C.P.C.B. The lowest value recorded was

47 dB (A) and highest value recorded was 87.75 dB (A) in Gandhi Nagar Colony. L_{10} value were above 65.5 to 82.25 dB (A) in all places. In commercial zone all selected places recorded noise levels that are higher than the limit set by the C.P.C.B. the lowest value 50.25 dB (A) and the highest value recorded was 96.25 (99–Prakash Chowk). In Bus Stand, the lowest value recorded was 54.25 dB (A) and the highest value recorded 106.25 dB (A) and the L_{eq} aloe recorded was 82.29 dB (A) to 97.09 dB (A). In Sirmour Chowk (main noise area) the lowest value are 53.00 dB (A) and the highest recorded value was 113.75 dB (A) (average of 4 months). In industrial zone, the lowest value recorded was 63 dB (A) and the highest value recorded was 117.75 dB (A) (average of 4 months).

Table 7.3: Hourly Computed Noise Parameter for Different Commercial Zone Location Winter Season (October 2001 to January 2002) at Rewa

Time, AM and PM	L_{90}	L_{50}	L_{10}	L_{eg}	L_{np}	*TNI*
8–9	55.50	59.50	70.25	66.69	77.87	84.50
9–10	58.25	62.00	73.00	69.44	80.37	87.25
10–11	61.75	66.00	76.00	72.43	83.63	88.75
11–12	52.25	67.25	77.00	75.37	102.21	121.25
12–1	63.75	67.25	78.00	74.43	75.37	90.75
1–2	64.75	69.75	78.75	75.18	87.01	90.75
2–3	64.00	67.75	77.50	73.94	84.28	88.00
3–4	63.75	69.75	79.00	75.45	88.87	94.75
4–5	65.50	73.25	82.25	78.79	94.67	102.50
5–6	68.50	75.75	83.00	79.43	93.75	96.50
6–7	72.75	78.00	87.25	83.68	96.00	100.75
7–8	74.50	81.00	96.50	93.99	111.26	132.50

Table 7.4: Hourly Computed Noise Parameter for Different Residential Zone Location Winter Season (October 2001 to January 2002) at Rewa

Time, AM and PM	L_{90}	L_{50}	L_{10}	L_{eg}	L_{np}	*TNI*
8–9	53.00	57.00	67.50	63.93	75.00	81.00
9–10	54.25	58.25	67.50	63.95	74.42	77.25
10–11	58.75	64.75	73.25	69.68	82.75	86.25
11–12	60.25	71.50	81.00	78.17	99.42	113.25
12–1	57.25	61.25	71.50	67.93	78.88	84.25
1–2	52.75	57.00	65.75	62.21	72.81	74.74
2–3	51.50	55.00	65.50	61.93	72.26	77.50
3–4	56.25	60.00	70.00	66.44	76.90	81.25
4–5	57.75	66.25	74.25	70.77	87.28	93.25
5–6	61.25	69.25	80.75	77.67	95.08	109.25
6–7	64.50	70.50	82.25	78.90	93.50	105.50
7–8	61.25	67.75	79.50	76.21	91.55	108.25

Table 7.5: Hourly Computed Noise Parameter for Silence Zone (G.M.H., S.G. Hospital) Location Winter Season (October 2001 to January 2002) at Rewa

Time, AM and PM	L_{90}	L_{50}	L_{10}	L_{eq}	L_{np}	TNI
8–9	52.25	56.75	66.00	62.44	73.15	77.25
9–10	58.25	63.00	74.50	71.00	83.65	93.25
10–11	47.75	53.95	66.75	64.16	78.76	93.75
11–12	47.50	55.95	66.75	63.62	81.17	94.50
12–1	49.50	55.75	66.00	62.52	76.78	85.50
1–2	55.50	61.00	71.75	68.25	81.65	90.50
2–3	57.00	63.25	74.00	70.57	85.06	95.00
3–4	56.75	63:00	71.75	68.79	81.75	86.75
4–5	62.25	65.50	76.50	72.93	94.73	89.25
5–6	63.00	69.00	82.50	79.42	94.83	111.00
6–7	61.50	69.25	82.75	80.04	98.02	116.50
7–8	55.40	61.50	74.25	71.04	86.11	100.50

Table 7.6: Mean Value of Ambient Noise Parameter for Various Location at Rewa Town Winter Season (October 2001 to January 2002)

Location	Parameter Measured, dB (A)					
	L_{90}	L_{50}	L_{10}	L_{eq}	L_{np}	TNI
Industrial zone	79.17	84.58	98.08	94.96	109.52	128.13
Traffic zone	70.71	79.46	88.06	85.19	100.16	110.13
Commercial zone	63.77	69.77	79.88	76.57	89.61	98.19
Residential zone	57.40	63.21	73.23	69.82	83.32	90.93
Silence Zone	55.55	61.49	72.79	69.57	84.64	94.48

The L_{10} or the percentile exceeding level is the level of sound which is exceeding 10 per cent of the total time of measurement at Rewa town area was varying between 106.5 to 65.5 dB (A). The maximum L_{10} value was observed at B.T.L. factory between 12.00 noon to 1 pm. The minimum value was recorded at the Civil Line area during (2.00 to 3.00 pm).

L_{50} or 50 percentile exceeding level is the level of sound which is exceeding 50 per cent of the total time of measurement was varying between 92.25 to 53.95 dB (A). The maximum value was observed at industrial area (TAMAS) between 7 to 8 ppm. The minimum value of 53.95 dB (A) was observed at G.M. Hospital area during 10–11 am.

L_{90} or 90 percentile exceeding level is level of sound which is exceeding 90 per cent of the total times of measurement of Rewa town, was varying between 87.5 to 47.5 dB (A). The max value was observed at B.T.L. factory area between 12 to 1 pm. The minimum value of 47.5 dB (A) was observed at G.M. Hospital area between 11 am to 12 noon.

L_{eq} is the equivalent continuous noise level measured for each individual hour. The maximum and minimum noise level of each individual hour was observed varying between 103.55 dB (A) to

61.03 dB (A). The minimum value of 61.03 dB (A) as recorded to civil line are (4 months average) during 1.00 to 2.00 pm and maximum value of 103.55 dBA (A) at B.T.L. factory area between 7 pm to 9 pm.

L_{np} is the noise pollution level, which indicates in this area the L_{np} for each individual hour were varying between 118.22 to 72.32 dB (A). The minimum value was observed at the civil line (Residential) area that is 1.00 pm to 2.00 pm the maximum value was observed at B.T.L. (Industrial area) during 7 to 8 pm.

TNI is the traffic noise index that caused annoyance due to vehicular traffic. The TNI value observed at Rewa town area was varying between 136.75 to 77.5 dB (A) (4 months average). The maximum TNI was observed at B.T.L. (Industrial area) 6 to 7 pm respectively. The minimum value was observed at civil line (residential area) 1.00 to 2.00 pm.

From the above it is concluded that the maximum value of L_{10}, L_{50}, L_{90}, L_{eq}, L_{np} and TNI were observed at B.T.L. (Industrial area) from as (12–1 pm) (7–8 pm), (12–1 pm), (12–1 pm), (7–8 pm), (6–7 pm). The minimum value of L_{90}, L_{50}, L_{eq}, L_{np} and TNI were observed at residential zone (2 to 3 pm) Silence zone (11 am to 12 noon), (10 am to 11 am), residential area (1 to 2 pm).

As per the ambient air quality standards in respect of noise the noise level (L_{eq}) should be in the range of 50 dB (A) for all these areas. But the average noise level (L_{eq}) of five location of Rewa town area (Industrial Silence, Commercial Residential traffic) exceeds the limit prescribed by C.P.C.B. during day time. If there is an increase in the industrial activity, correspondingly, there will be increase in traffic activity, which will increase the ambient noise level. It is concluded that the high noise levels recorded in these places are due to:

1. Blowing of air horns at prohibited areas.
2. Operation of poorly maintained vehicles.
3. Poor road conditions in the town.
4. The encroachments of roadsides which cause slow movement of vehicles with frequent use of horns.

It is suggested that the following measures may by undertaken to control the menace of noise pollution.

1. Strict enforcement of existing law to prohibit air horns inside the town.
2. Proper maintenance of the vehicles.
3. Laying good roads and their maintenance.
4. Strict enforcement of the existing law to remove the encroachments on road sides.
5. Planting of trees and other vegetation inside the town on roadsides and/or around the silence zone will reduce the noise level when they reach the people.

By implementing all the above dedicatory measures of some of the individual measures will decrease the average noise level in the power town area.

References

Bansal, S. 1996. Noise level status of Bhopal city. *Poll. Res.*, (15): 107–108.

Government of India 1989. Notification No. G.R.S. 1063 E dated 26–12–89 published in Gazette No. 643, Dated 26–12–89.

Joshi, Gunwant 1994. Ambient noise levels at Indore city. A reconnaissance survey report regional office M.P. Pollution Control Board, Indore, pp. 52.

Joshi, Gunwant and Sunil Kumar 1992. A report on noise pollution monitoring at Bhopal city BHEL township and Govindpura by Regional office M.P. Pollution control board, Bhopal 21.

Kumar, M. Vijay. Noise Pollution Assessment in Pudukkottal, Tamil Nadu. P.G. and Research Department of Environmental Science, Bishop Heber College, Tiruchirapalli.

Mishra, Bhupesh 2002. Noise pollution and its Behavioural effects on human beings in Rewa and Satna region. *Ph.D. Thesis*. School of Environment Biology Deptt., A.P.S.U., Rewa (M.P.).

Nigam, S.P. 1991. Noise pollution: An expert lecture delivered at U.S. Aid workshop on sitting of Industries organised by Environmental planning and co-ordination organization, Bhopal 21–25 February 1991, pp. 33.

Seth, P.C. 1943, Noise quality status of Indore city report by M.P. Pollution Control Board, Bhopal.

Seth, P.C. 1991. Traffic Noise pollution at three important places of Bhopal city. A report by M.P. Pollution Control Board, Bhopal.

Chapter 8
Age-related Changes in Biochemical Parameters in Long Bones of Indian Toad, *Bufo melanostictus*

*B.N. Andia and H.N. Behera**

P.G. Department of Zoology, Berhampur University, Berhampur – 760 007, Orissa, India

ABSTRACT

The present study reports age-related changes in calcium, phosphorus, Ca/P molar ratio and total collagen contents in long bones of Indian toad, *Bufo melanostictus*. The calcium and phosphorus contents in humerus and femur of toads of both sexes decreased with increase in body weight. The Ca/P molar ratio in humerus reached the approximate value of apatitic ratio (humerus-10: 6.3 to 10 : 6.5 and femur-10: 6.2 to 10 : 6.3) when the male weighed about 38–45 grams and the ratio (humerus-10: 6.2 to 10 : 6.4 and femur-10: 6.2 to 10 : 6.3) when the female weighed about 36–43 grams. The total collagen content of the humerus and femur of both male and female toads decreased with increase in body weight.

Keywords: *Bone mineral, Collagen, Ca/P molar ratio, Ageing bone.*

Introduction

Vertebrate bone is considered as one of the most dynamic tissues and is largely composed of collagen on which calcium and phosphates are deposited during the process of calcification. It was suggested that, alkaline phosphatase, by breaking the phosphate esters into inorganic phosphates;

* *Corresponding Author.*

provide booster mechanism of calcification (Bell *et al.*, 1972). It is also known that extra-cellular fluid supersaturates the calcification crystal, hydroxyapatite, the specific mechanism by which calcification uniquely targets the bones (Neuman and Neuman, 1958). Literature regarding the role of calcium and tissue calcification during ageing in animals is rather inadequate. Calcium accumulation and calcification are believed to be associated with ageing but the evidences in its favour are, in some cases, conflicting (Massie *et al.*, 1989). The biochemical parameters would provide informations regarding calcification, an age dependent process (Freydberg-Lucas and Verzar, 1957; Boros-Farkas, *et al.*, 1967). Precisely, in the present study an attempt has been made to find out the age and sex related changes in biochemical parameters (calcium, inorganic phosphorus and collagen) and of Ca/P molar ratio in the long bones (humerus and femur) of *Bufo melanostictus*.

Materials and Methods

Common toads, *Bufo melanostictus*, (Bufonidae) of both sexes used in the present study collected from Baliapal of Balasore district, Orissa (21°41′15″ N and 87°20′10″ E) were maintained in rectangular glass cages (2 ft × 1 ft × 2 ft) covered with wire-netted wooden lid. They were fed with grasshoppers and chopped goat liver on alternate days and water was provided *ad libitum*.

Following an acclimation period of about 2–3 days in the laboratory, the animals were sacrificed. The long bones (humerus and femur) were taken out and cleaned of adherent materials in distilled water and used for estimation of biochemical parameters.

Calcium and Phosphorus

Long bones (humerus and femur) were cut into pieces and washed in distilled water to remove bone marrow. The bone pieces were dried on filter paper and weighed. The samples were then ashed in porcelain crucibles for 1½ hour. The ashed samples were dissolved in 10 ml of 0.1 N HCl. After filtration, aliquots were taken for calcium analysis by the method of Clark and Collip (1925) and for phosphorus estimation following colorimetric method of Fiske and Subba Row (1925).

The calcium and phosphorus contents were expressed as mg/g bone wet-weight. In each case the molar ratio of calcium to phosphorus was also calculated.

Collagen

Weighed quantities of long bones (humerus and femur) were separately hydrolysed in 3 ml of 6 N HCl in sealed tubes at 110°C for 16 hours. The hydrolysed samples were transferred to 50 ml measuring cylinders and neutralized with 5 N NaOH using methyl red as indicator. The hydroxyproline contents of the neutralised samples were estimated following the colorimetric method of Neuman and Logan (1950) as modified by Leach (1960). The hydroxyproline values were converted to collagen figures after multiplying by a factor of 7.46 (Jackson and Cleary 1967). The results were expressed as mg collagen/g bone wet-weight.

Results

Correlation Between Body Weight and Calcium Content

The total body weight and the calcium content in humerus showed negative correlation both in male ($r = -0.679$, $P < 0.001$) and female ($r = -0.783$, $P < 0.001$) toads (Figures 8.1–8.2). Similarly, the total body weight and the calcium content in femur showed significant negative correlation in both male ($r = -0.792$, $P < 0.001$) and female ($r = -0.785$, $P < 0.001$) toads (Figures 8.3–8.4).

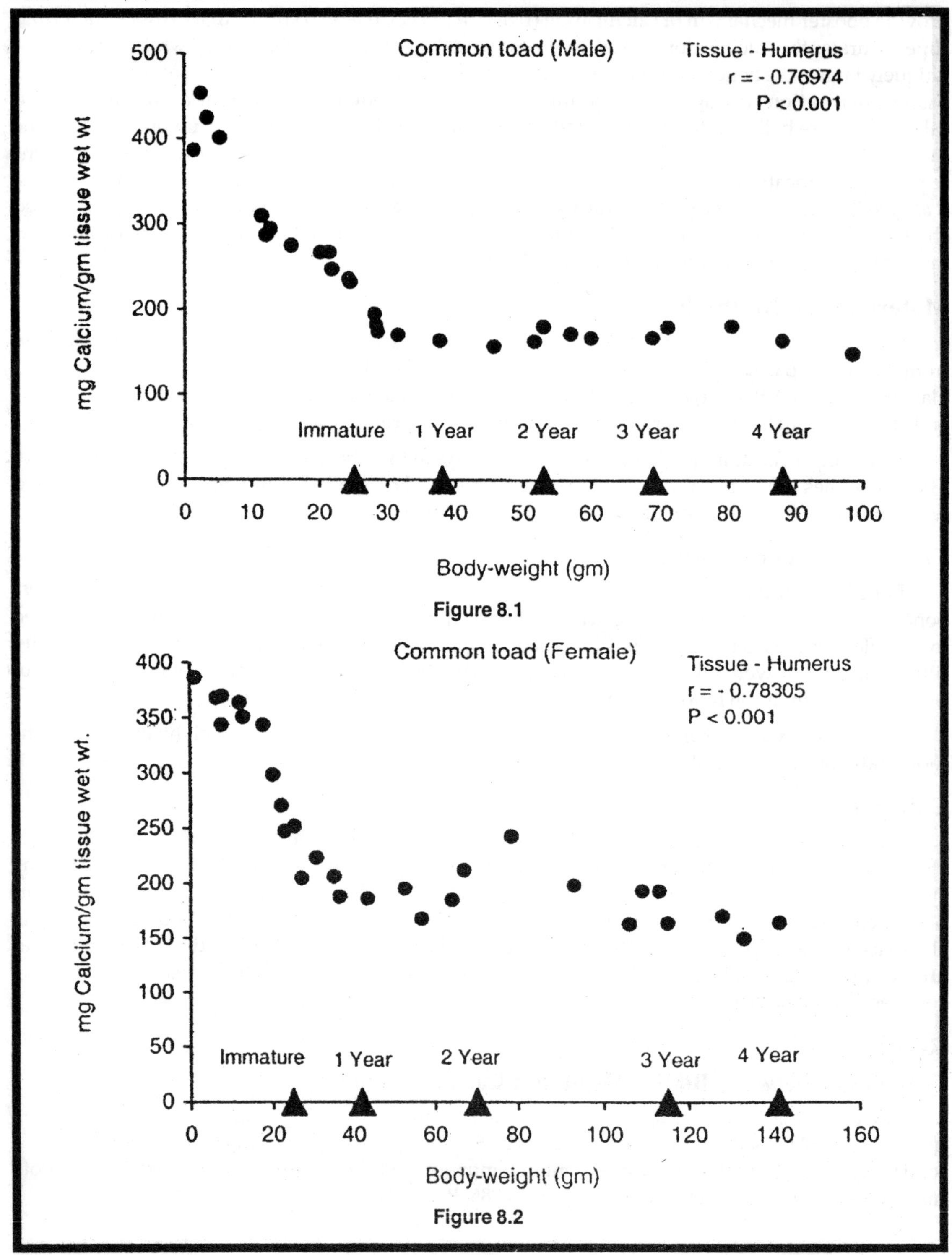

Figure 8.1

Figure 8.2

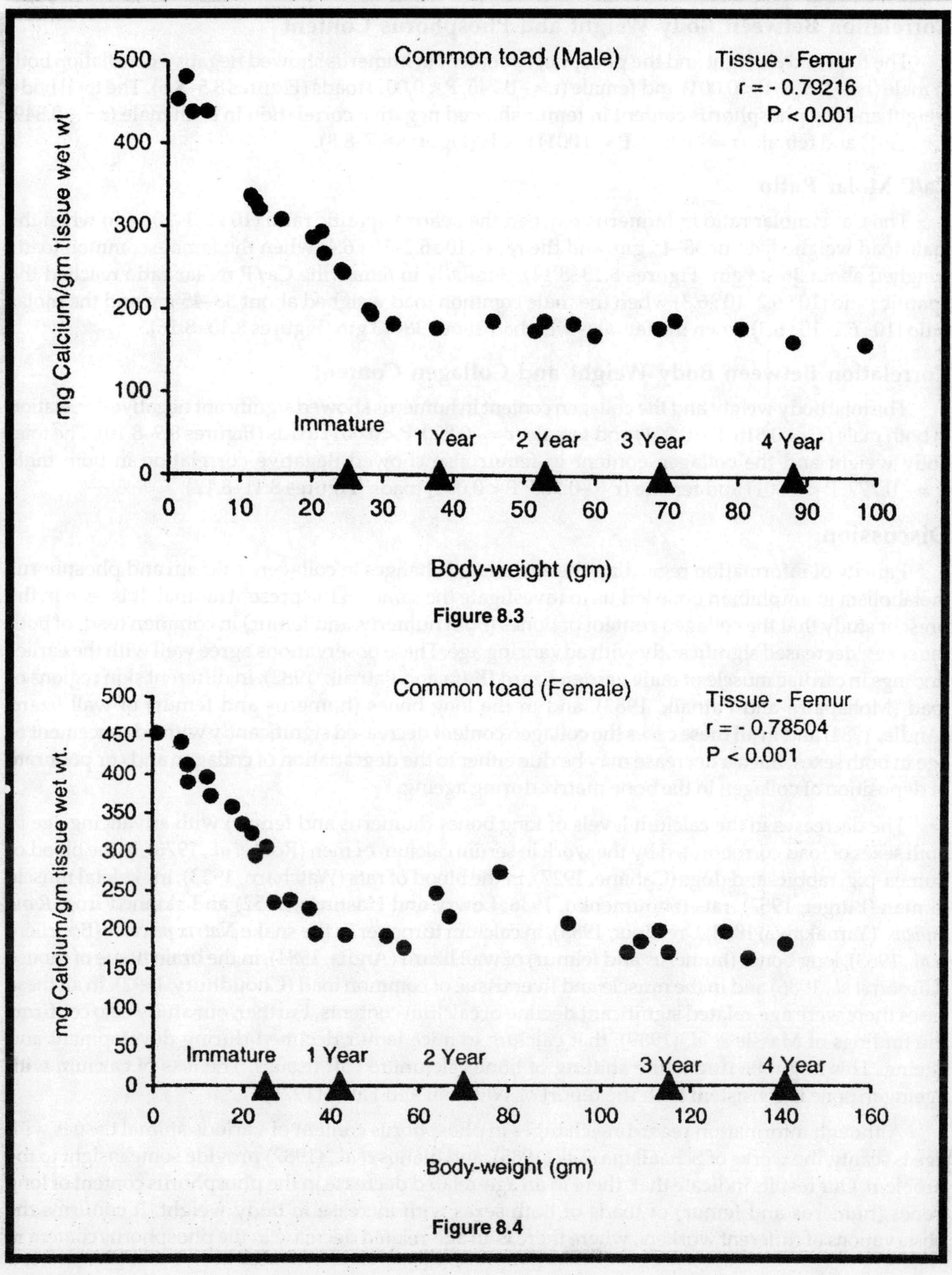

Figure 8.3

Figure 8.4

Correlation Between Body Weight and Phosphorus Content

The total body weight and the phosphorus content in humerus showed negative correlation both in male ($r = -0.857, P < 0.001$) and female ($r = -0.745, P < 0.001$) toads (Figures 8.5–8.6). The total body weight and the phosphorus content in femur showed negative correlation in both male ($r = -0.849$, $P < 0.001$) and female ($r = -0.820, P < 0.001$) toads (Figures 8.7–8.8).

Ca/P Molar Ratio

The Ca/P molar ratio in humerus reached the nearest apatitic ratio (10 : 6.3–10 : 6.5) when the male toad weighed about 38–45 gm. and the ratio (10 : 6.2–10 : 6.4) when the female common toads weighed about 36–43 gm (Figures 8.13–8.14). Similarly in femur the Ca/P molar ratio reached the apatitic ratio (10 : 6.2–10 : 6.3) when the male common toad weighed about 38–45 gm and the molar ratio (10 : 6.2–10 : 6.3) when the females weighed about 36–43 gm (Figures 8.15–8.16).

Correlation Between Body Weight and Collagen Content

The total body weight and the collagen content in humerus showed significant negative correlation in both male ($r = -0.810, P < 0.001$) and female ($r = -0.820, P < 0.001$) toads (Figures 8.9–8.10). The total body weight and the collagen content in femur also showed negative correlation in both male ($r = -0.777, P < 0.001$) and female ($r = -0.785, P < 0.001$) toads (Figures 8.11–8.12).

Discussion

Paucity of information regarding the age related changes in collagen, calcium and phosphorus metabolism in amphibian bone led us to investigate the same on the present animal. It is seen in the present study that the collagen content of bones (both humerus and femur) in common toad, of both the sexes, decreased significantly with advancing age. These observations agree well with the earlier findings in cardiac muscle of male garden lizard (Kara and Patnaik, 1982), in different skin regions of toad (Mohapatra and Patnaik, 1983), and in the long bones (humerus and femur) of wall lizard (Andia, 1984) and in all these cases the collagen content decreased significantly with advancement of age in both sexes. Such a decrease may be due either to the degradation of collagen and/or poor rate of deposition of collagen in the bone matrix during ageing.

The decreases in the calcium levels of long bones (humerus and femur) with advancing age in both sexes of toad corroborated by the work in serum calcium of men (Roof *et al.*, 1976), in the blood of guinea-pig, rabbits and dogs (Cahane, 1927), in the blood of rats (Watcharn, 1933), in skeletal muscle of man (Burger, 1957), rats (Shoumenko, 1936; Lowry and Hastings, 1952) and skinned frog, *Rana pipiens.* (Yarnakawa, 1983; Dresdner, 1983), in calcium turnover in the snake *Natrix piscator* (Bourliére *et al.*, 1963), long bones (humerus and felmur) of wall lizard (Andia, 1984), in the brain tissue of mouse (Gibson *et al.*, 1986) and in the muscle and liver tissue of common toad (Choudhury, 1992). In all these cases there were age-related significant decline of calcium contents. Further, our study also confirms the findings of Massie *et al.* (1989), that calcium in mice femur declined during development and ageing. This might be due to the shifting of bone calcium to soft tissues. The loss of calcium with ageing in bone is consistent with the report of Wilhelmi and Faust (1976).

Although information regarding changes in phosphorus content of various animal tissues with age is scanty, the works of Schaafsma *et al.* (1985), and Blahos *et al.*, (1987) provide some insight to the problem. Our results indicate that, there in an age-related decrease in the phosphorus content of long bones (humerus and femur) of toads of both sexes with increase in body weight. It confirms the observations of different workers, where there is an age-related decrease in the phosphorus content in

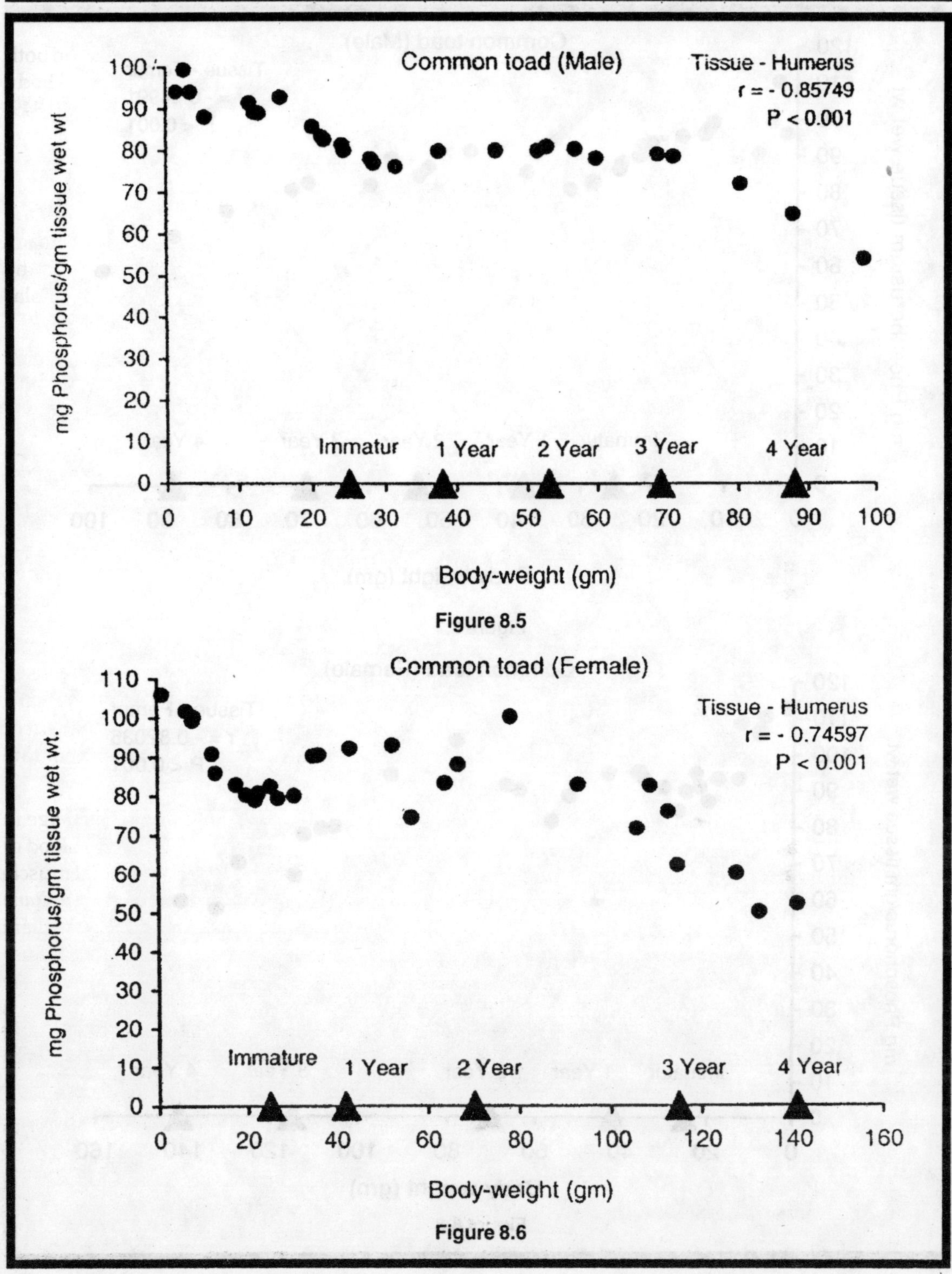

Figure 8.5

Figure 8.6

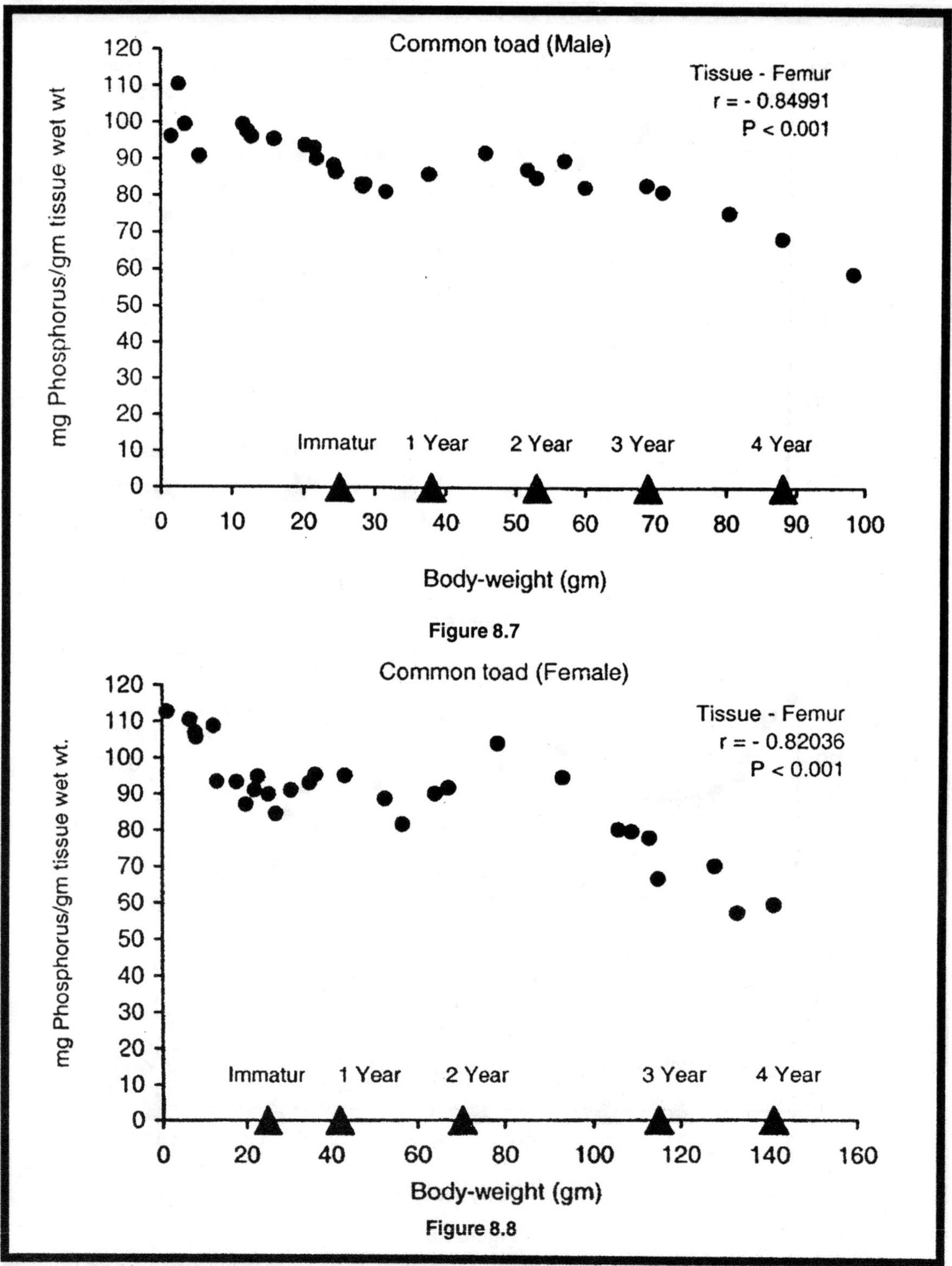

Figure 8.7

Figure 8.8

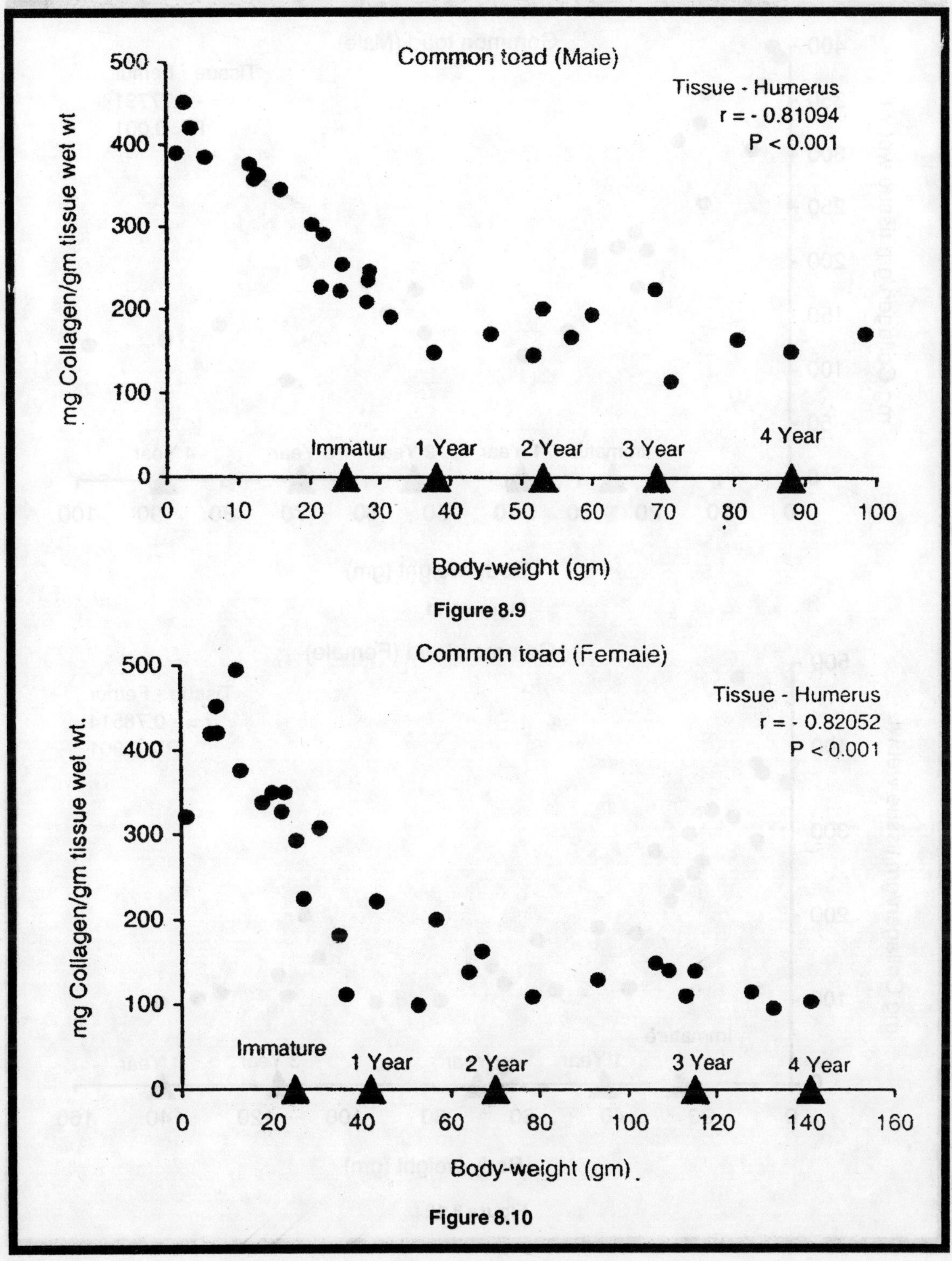

Figure 8.9

Figure 8.10

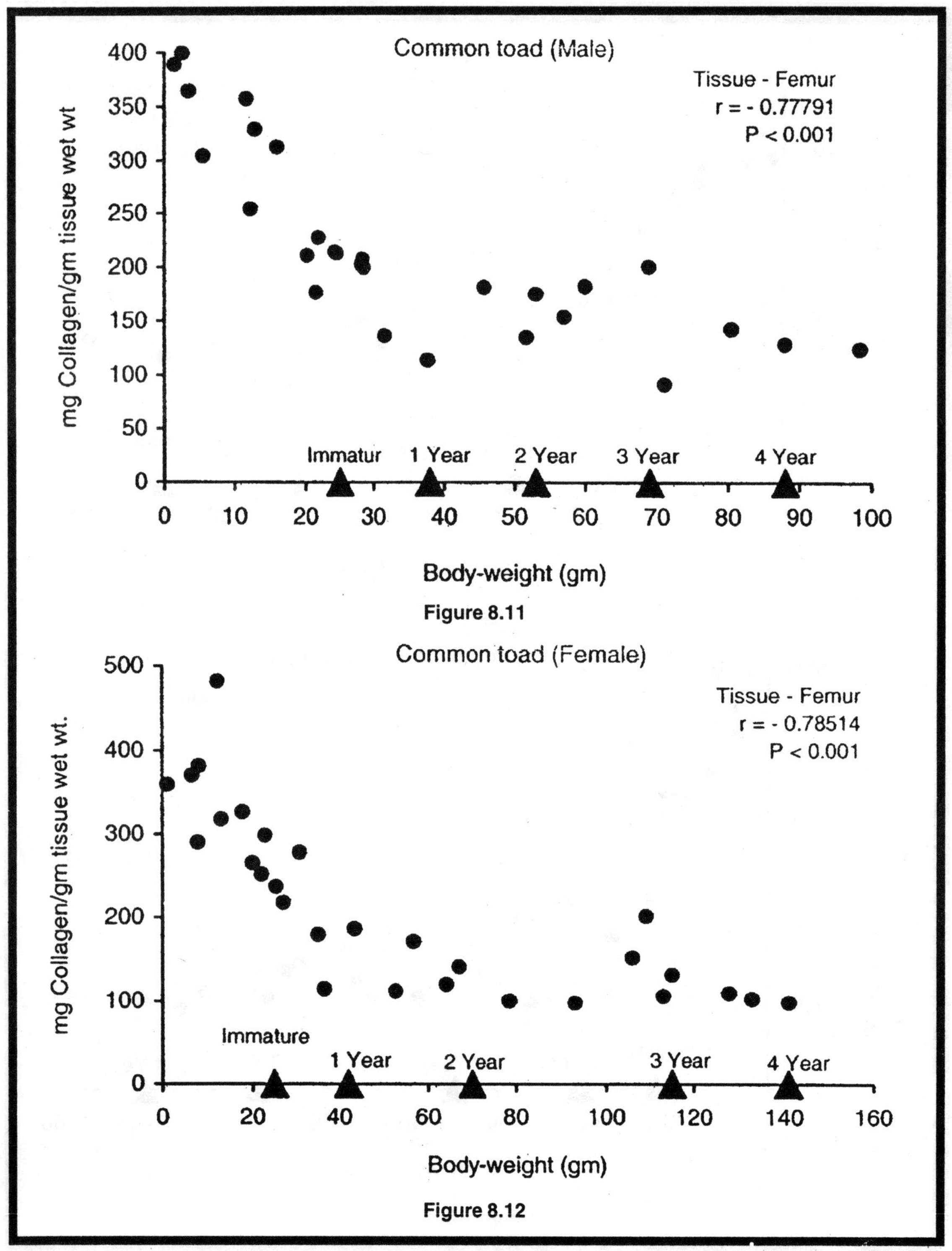

Figure 8.11

Figure 8.12

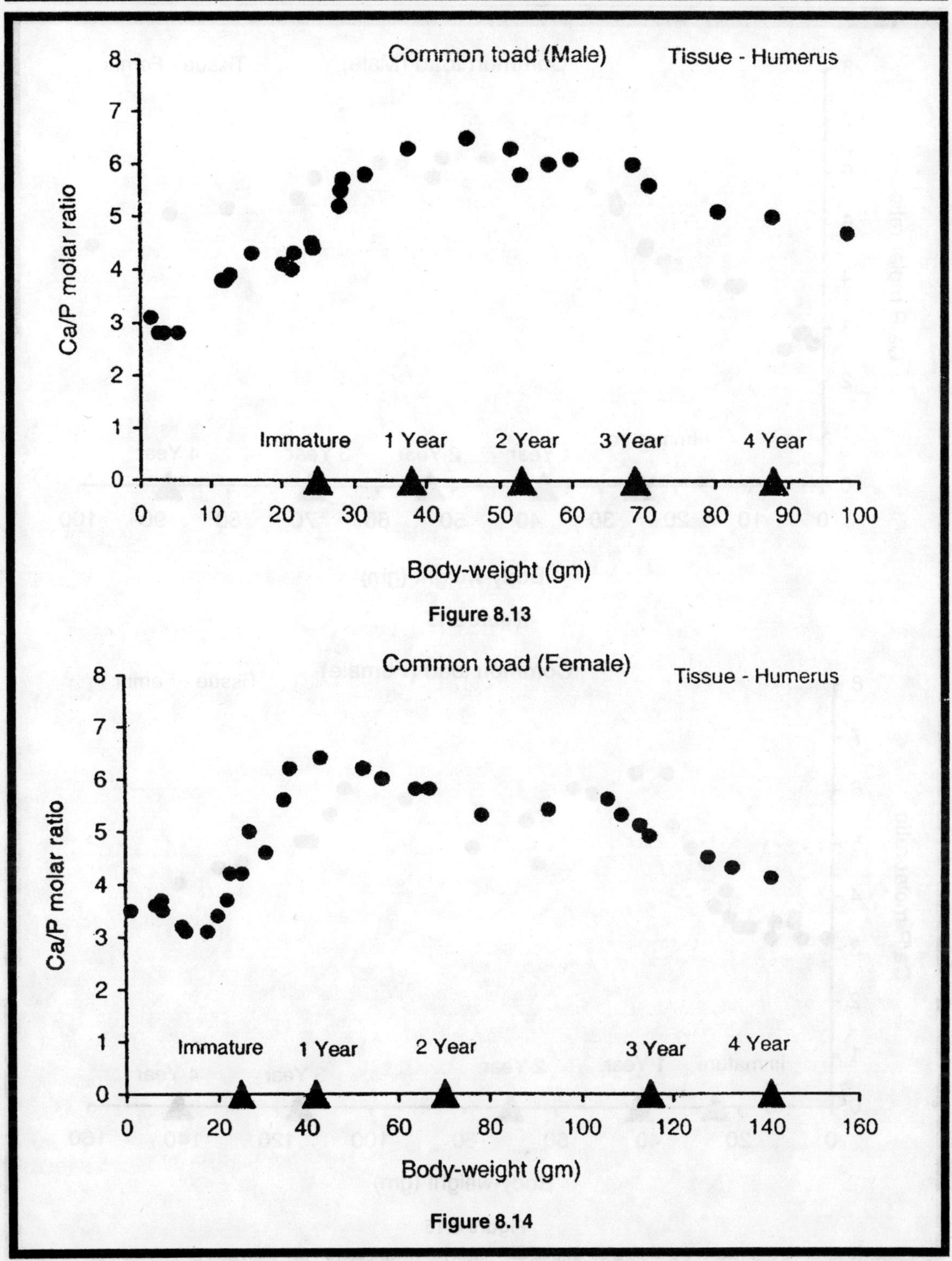

Figure 8.13

Figure 8.14

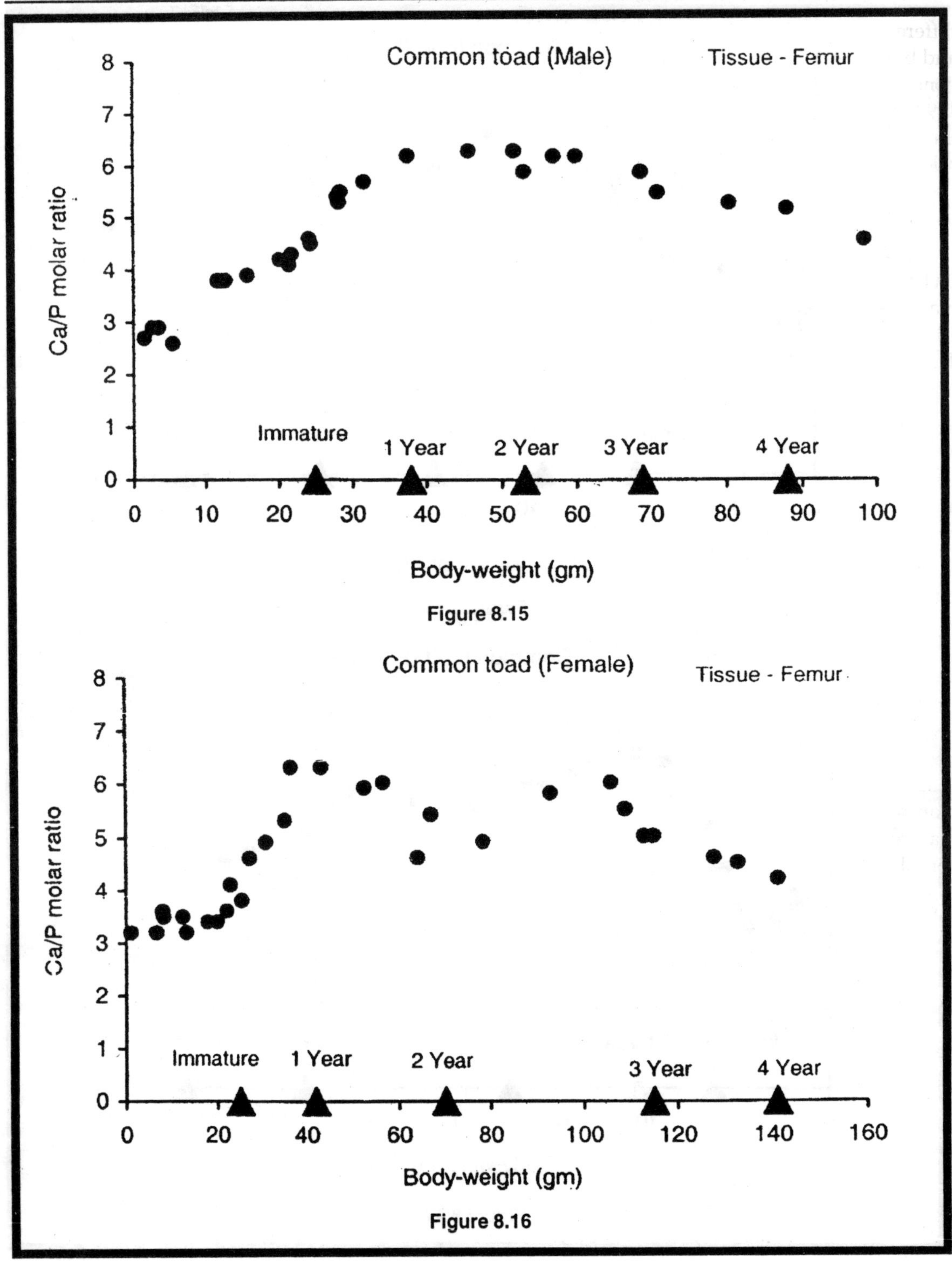

Figure 8.15

Figure 8.16

different tissues in man (Simms and Stolman, 1937), in heart, kidney, liver, intestine, skeletal muscle and brain of rabbits (Koch and Koch, 1913), in the liver and spleen of cattle (Kruger, 1894), in long bones (humerus and femur) (Andia, 1984) and in muscle and liver tissue of common toad (Choudhury, 1992). It is now clear that, during ageing of this species of amphibia, the collagen content is decreased, thus reducing the total template surface. 'This may be one of the reasons of decreased values of calcium and phosphorus contents of bones in aged individuals. More over the change in calcium and phosphorus concentration in various tissues of amphibian with ageing may be due to similar mechanisms.

Calcification in animals is an age dependent process. It is known that calcification of bone occurs on the collagen structure during early stage of development. The deposition of calcium salts occurs mainly in the form of calcium orthophosphates and the molar ratio of calcium to phosphorus in 10 : 6.6. The mechanism of calcification thus seems to be formation of apatite on a collagen template. The 10 : 6.6 apatitic molar ratio of calcium and phosphorus is an indicator of degree calcification of bones.

In the present study, in common toads, *Bufo melanostictus* the Ca/P molar ratio reached the nearest apatitic ratio 10 : 6.3 in humerus and 10 : 6.2 in femur of male at the body weight of 38 grams, that to 10 : 6.2 in humerus and 10 : 6.3 in femur of female at the body weight of 36 grams. At this weight the animal of both sexes attain their sexual maturity and completing 1 year of age in female and 1 + year in male.

The most interesting observation in this study is the gradual shift of Ca/P molar ratio till the common toad weighs 36–43 grams in female and 38–45 grams in male. At this stage Ca/P molar ratio seems to be at peak value and it is to be estimated that, the animals are matured within this range of body weight. This apatitic ratio appears to be maintained till the male common toad weighs 60 grams and female 67 grams and the ratios decreased thereafter. Our observations clearly demonstrate that, the calcification as indicated by Ca/P molar ratio is not intensive, during pre-maturation phase in both sexes. But significant shift in Ca/P molar ratio occurs thereafter due to complete calcification of bones of the species nearest to the maturation weight. As the Ca/P molar ratio is maintained till some range of body weight in both sexes of the animals, these range of animals are treated as the adult matured male or female. Hence for the male common toad of body weight 38–60 grams treated as matured adult male, the female from 36–67 grams treated as matured adult female and both have completed 1–2 years of age.

Acknowledgements

The authors are grateful to Head, P.G. Department of Zoology, Berhampur University (Orissa), for providing with laboratory and library facilities. The State Government of Orissa and the Principal S.R. College; Baliapal are thanked for awarding B.N. Andia, a teacher fellowship for two years.

References

Andia, B.N. 1984. Correlation between skeletochronology and changes in biochemical parameters in wall lizard. *Hemidactylus flaviviridis*, tissues. *M.Phil. Dissertation*, Berhampur University, Orissa.

Bell, G.H., Davidson, J.N. and Emslie Smith, D. 1972. Bone and mineral. In: *Text Book of Physiology and Biochemistry*, 8th Edition. ELBS, Churchill-Livingstone, pp. 208

Blahos, J., Care, A.D. and Sommerville, B.S. 1987. Effect of low calcium and low phosphorus diet on duodenal and ileal absorption of phosphate in chick. *Endocrinol. Enp.*, 21: 59–64.

Boros-Farkás, M., Spichtin, H. and Verzár, F. 1967. Calcification of rat skin induced by dehydrotachosterol. I. Changes in collagen. *Gerontologia*, 13: 129–138.

Bourliére, F., Brocas, J. and Perrin, J. 1963. Age difference in the rate of calcium fixation in a (ring-necked snake) reptile. *Gerontologia*, 7: 153.

Bürger, M. 1957. Altern and Krankheit, 3rd (G. Thieme, Leipzig).

Cahane, M. 1927. Teneur du tissue masculaine et du sang en calcium, magnesium et potassium au point de vue ilikibiologique COR. Soc. Biol., Paris, 96: 1168.

Choudhury, P.K. 1992. Alterations in the calcium and phosphorus contents of some soft tissues of toad, *Bufo melanostictus*, in relation to body size and sex. *M.Phil. Dissertation*, Berhampur University, Orissa, India.

Clark, E.P. and Collip, J.B. 1925. *J. Biol. Chem.*, 63: 461. Quoted in *Hawk's Physiological Chemistry* (Oser, B L.) 14th edition, 1971. Tata McGraw Hill Publication Company, New Delhi, pp. 1133.

Dresdner, K.J. 1983. Cellular calcium uptake and extracellular calcium depletion in frog ventricular muscle. Dissertation Abstracts: April 1984, 44: 3052–B.

Fiske, C.H. and Subba Row, Y. 1925. The colorimetric-determination of phosphorus. *J. Biol. Chem.*, 66: 375.

Freydberg-Lucas, V. and Verzãr, F. 1957. Der calcium softwechsel verschiedener organe bei jungen and alten Tiern Gerontologia, Basel, 195–210.

Gibson, G., Perrino, P. and Dienel, G.A. 1986. *In vitro* brain calcium homeostasis during ageing Mech. Age Dev., 37: 1–12.

Jackson, D.S. and Cleary, E.G. 1967. The determination of collagen and elastin. In: *Methods of Biochemical Analysis*, (Ed.) Glick, D., Vol. 15. Inter Science, New York, pp. 25.

Kara, T.C. and Patnaik, B.K. 1982. Some biochemical changes in the ageing heart of the male garden lizard. *Arch. Gerontol. Geriatr.*, 1: 219.

Koch, W. and Koch, M.L. 1913. Contribution to the chemical differentiation of the central nervous system. *J. Biol. Chem.*, 15: 423.

Kruger, F. 1894. Uberden calcium gehalt der Leberzellen des pindes in seinew verschiederen Entwick lungrstadien, *Z. Biol.*, 31: 392.

Leach, A.A. 1960. Notes on modification of the Neuman and Logan method for the determination of hydroxyproline. *Biochem. J.*, 74: 70.

Lowry, O.H. and Hastings, A.B. 1952. Quantitative histochemical changes in ageing. In: *Problems of Ageing*, (Eds.) Cowdry and Langing. Williams-and Wilkins, Baltimore, pp. 105.

Massie, H.R., Aiello, V.R. and Dewolfe, I.K. 1989. Calcium and calmodulin changes with ageing in C57BL/6J mice. *Gerontology*, 35: 100–105.

Mohapatra, N. and Patnaik, B.K. 1983. Studies on skin collagen of the common indian toad I. Correlation between characteristics of collagen and body weight. *Life Sci. Adv.*, 2: 186.

Neuman, F.N. and Neuman, N.W. 1958. Mechanisms of calcification. In: *The Chemical Dynamics of Bone Mineral.* The University of Chicago Press, Chicago II,USA, p. 169–187.

Neuman, R.E. and Logan, M.A. 1950. The determination collagen and elastin in tissues. *J. Biol. Chem.*, 186: 549.

Roof, B.S., Piel, C.F., Hansen, J. and Fudenberg, H.H. 1976. Serum parathyroid hormone levels and serum calcium levels from birth to senescence. *Mech. Age. Dev.*, 5: 289–304.

Schaafsma, G., Duursma, S.A., Visser, W.J. and Dekker, P.R. 1985. The influence of dietary calcium on kidney calcification and renal function in rats fed high phosphate diets. *Bone* (NY), 6: 155–164.

Shoumenko, L.D. 1936. Variation in the content of some ash element in the blood and organs of white rats at different ages. *Proc. Zool. Biol., Inst. Kharkov Univ. UKR. SSR.*, 3: 119.

Simms, H.S. and Stolman, A.C. 1937. Changes in human tissue electrolytes in senescence. *Science*, 86: 269.

Watcharn, E. 1933. The normal serum calcium and magnesium of the rat. *Biochem. J.*, 27: 1875.

Wilhelimi, G. and Faust, R. 1976. Suitability of the C57 black mouse as an experimental animal for the study of skeletal changes due to ageing with special reference to osteoarthrosis and its responses to tribenoside. *Pharmacology*, 14: 289–296.

Yamakawa, M. 1983. Calcium release from the sarcoplasmic reticulum of skinned frog skeletal muscle fibres. DAI, May'84, 11: 44.

Chapter 9

Distribution of *Salmonella* Antibodies in the Sera of Normal Individuals in Vivekanandha Hospital, Tiruchengode, Tamil Nadu

S. Rajalakshmi, N. Kavitha and S. Uma

Department of Microbiology, Vivekanandha College of Arts and Sciences for Women, Elayampalayam, Tiruchengode – 637 205

ABSTRACT

Enteric fever is epidemic in all parts of India. The present study was carried out to evaluate the antibodies against *Salmonella* antigens in normal hospital patients. The serum samples of 65 cases were collected from Vivekanandha hospital, Elayampalayam, Tiruchengode, Tamil Nadu and 50 were serologically proven as typhoid cases. The qualitative widal tube test was carried out to estimate the antibodies in the serum samples. The cut off titre was 1 : 160 for both O and H antigens. The present study showed antibodies to *Salmonella typhi* more frequently than *Salmonella paratyphi*. In relation to sex, females were frequently affected than males. In relation to age, the antibody titre was lowest in the age group of 41–50 and the titre was highest in the age group of 21–30. There was no history of TAB vaccination as the practice of TAB Vaccination in adult was rare.

Keywords: Salmonella antibodies, Serum agglutination, Widal serum test.

Introduction

Entire fever is epidemic in all parts of India. It is caused by *Salmonella typhi, Salmonella paratyphi* A, B, C. Antigens prepared from these organisms are used in the widal agglutination test to detect the presence of antibodies in patients sera. The widal test was introduced by Widal, Sierd and Ggrunbrawn in 1896. The proportion of typhoid to para-Typhoid A is 10 : 11 and para-Typhoid B is rare. Typhoid fever has estimated to affect 105–200/100,000 population. Isolation by culture method is poor because of the antibiotic therapy given before the diagnosis. Widal test is now widely used in Tamil Nadu, since the endemicity is found to be high.

Classification of *Salmonella*

Salmonella are classified broadly into four subgenera based on the biochemical reactions. Subgenus I is the latest and the most important containing all the species that commonly cause human and animal infections. Other subgenera are of little importance in human disease. Subgenus II is common in reptiles. Subgenus III consists of bacilli formerly designated Arizona and later included in the genus *Salmonella* because of their antigenic similarity.

Serological classification of *Salmonella* is by the Kauffmann-White Scheme. This forms the basis of sero typing of *Salmonella* and depends on the identification of agglutination with structural formulate of O and H antigens of the strains. This scheme gives species status to each serotype.

Antigenic Structure of *Salmonella*

Salmonella possess the following antigens based on which they are classified and identified:

1. Flagellar antigen H
2. Somatic antigen O and
3. A surface antigen Vi.

H Antigens

This antigen present on the flagella is a heat labile protein. It is destroyed by boiling or by treatment, with alcohol, but not by formaldehyde. Since the flagella of *Enterobacter* consists of mainly of a fibrous protein of the myosin-keratin type is referred to as flagellin.

O Antigens

The somatic 'O' antigen is a phospho lipid protein polysaccharide complex which forms integral part of the cell wall. It can be exacted from the bacterial cell by treatment with trichloroacetic acid, as shown by Bovine. When mixed with antisera, O antigens suspension form compact, chalky, granular clumps. O agglutination takes place more slowly and at higher temperature optimum than H agglutination. The O antigen of Salmonella typhi is a polymolecular complex formed of a polysaccharide, a protein and phospholipid, and that the polysaccharide component is responsible for its antigenic specificity.

Vi Antigens

In addition to the H and O antigens some strains, notably the typhoid bacillus contain a further somatic antigen originally described by Felix and Pitt in 1934 as the Vi antigen. The degree of agglutinability of different strains of smooth typhoid bacilli inversely related to the virulence of the strains and also by the presence of a substance which protected the O antigen from the antiserum and was itself antigenic. Because of its association with virulence, it is named as Vi antigen.

Materials and Methods

Among 65 cases of normal hospital patients which includes adults, 12 are males and 38 are females. The serum sample were within 3–10 days of onset of fever.

Precaution

1. Serum may be stored at 4°C for week, when load delays are expected.
2. Kept frozen.
3. Serum should not be inactivated at 50°C.
4. Antigen mixed well before used and must be stored at 4°C.

Techniques

1. Slide agglutination test.
2. Tube agglutination test.

Slide Agglutination Test

Procedure

Glass slide provided in the kit was cleaned and wiped free of water. One drop of undiluted patients serum was added from 1 to 4 circles in the slide and 1 drop of positive control serum in each of the last 2 circles (5 and 6). Antigens–O, H, AH, BH were added in drops to the corresponding circles 1, 2, 3, 4 respectively and O antigen in circle 5 and one of the H antigen in circle 6. The contents of the circle were mixed well with applicator sticks. The slide was rotated for 1 min and observed for agglutination.

In this test, stained *Salmonella* antigens are used to detect specific antibody in the serum from infected patients. The slide test agglutination is used for the primary screening of typhoid infection.

Tube Agglutination Test

Procedure

For each serum sample under test are arranged in 4 rows of 6 tubes each in a rack. Dilution factors–1 : 20, 1 : 40, 1 : 80, 1 : 160, 1 : 320, 1 : 640. About 1.9 ml of the saline was added to the first tube and to the remaining 1 ml of saline was added. 0.1 ml of serum sample was added to the first tube after which it was serially diluted upto 1 : 640. No serum sample should be added to the control. Antigens O, H, AH, BH were added to the corresponding tubes and dilutions. The rack along with the tubes were incubated at 37°C for 16–20 hrs. Agglutination indicates that the patient serum contain antibodies to *Salmonella typhi.*

Results and Discussion

The present study was carried out to evaluate the antibodies against Salmonella antigens at different groups of population. Results established antibodies for *Salmonella typhi* more frequently than *Salmonella paratyphi* and it was found that the antibodies for both H and O were the same. Tube test gave a high degree of perfection than the slide test.

The antibody levels to the *Salmonella antigens* at various titrations present in the normal individual in Vivekanandha Hospital are reported in Table 9.1. The antibody level of the *Salmonella typhi* O and H antigen at variation titration are found to be same and the percentage is calculated as 74.1 per cent. The basal antibody to *Salmonella typhi* H, *Salmonella typhi* O, *Salmonella. paratyphi* A and *Salmonella paratyphi* B are reported in Table 9.2. The basal antibody titre *Salmonella typhi* O and *Salmonella typhi* H were found to be 24 per cent for male and 76 for female respectively for both antigens. A cut-off titre was taken as 1 : 160.

Table 9.1: Titration of *Salmonella* Antibodies in the Normal Individuals of Vivekanandha Hospital (N = 65)

Antigen	Serum Dilution					Total No.	Total %
	1 : 20	1 : 40	1 : 90	1 : 160	1 : 320		
S. typhi H	12	02	01	35	00	50	74.1
S. typhi O	12	02	01	35	00	50	74.1
S. paratyphi A	00	00	00	00	00	00	00
S. paratyphi B	00	00	00	00	00	00	00

Table 9.2: Number of Positive Cases in Normal Individuals of Vivekanandha Hospital (N = 25) (Cut-off = 1 : 160)

Antigen	No. of Positives	Sex			
		Male	Percentage	Female	Percentage
S. typhi O	50	12	24	38	76
S. typhi H	50	12	24	38	76
S. paratyphi A	00	00	00	00	00
S. paratyphi B	00	00	00	00	00
Total	**100**	**24**	**48**	**76**	**152**

To determine the incidence of normal agglutination titre in relation to age. Test showed that 30.76 per cent of the cases belonged to the age group of 0–20, 58.38 per cent to the age group of 21–30, 7.6 per cent at the age group of 31–40 and 6.15 per cent of the age group of 40–50. The widal positivity in the patients will indicate the sensitivity of the test (Table 9.3).

Table 9.3: Incidence of Agglutination Titre in Relation to Age (Cut-off = 1 : 20) (N = 65)

Age in Years	No. of Cases		Total	%	Agglutination in Percentage				
	Male	Female			TH	TO	AH	BH	Total
00–10	00	00	00	00	–	–	–	–	0
11–20	04	16	30.76	30.76	30.76	30.76	–	–	75.2
21–30	09	27	36	55.38	55.38	55.38	–	–	110.76
31–40	01	04	05	7.6	7.6	7.6	–	–	15.2
41–50	02	02	04	6.15	6.15	6.15	–	–	12.3

Widal test results of normal patients in relation to sex showed higher sensitivity for O and H antibodies (Table 9.4). Out of 19 males, 4 were sensitivity to the titre value of 1 : 20, 1 was sensitive to the titre value of 1 : 40. Out of 46 females, 6 were sensitivity to the titre values of 1 : 20, 1 was sensitive to the titre value 1 : 40. The antibody titre was found to be higher in the age group of 21–30 years (55.4 per cent) among which females were frequently affected regarding Tiruchengode.

Table 9.4: Number of Carriers in Normal Individuals in Relation to Sex

Sex	*No. of Sera Examined*	*Serum Dilution of O, H Antibodies*				
		1 : 10	*1 : 20*	*1 : 30*	*1 : 40*	*1 : 50*
Male	19	00	04	00	01	00
Female	46	00	06	00	01	00

There was a demand for fixation of diagnostic titre updating over the basal antibody level with the cut-off titre 1 : 160 or above. The present study showed antibodies for *Salmonella typhi* more frequently than *Salmonella paratyphi* and it was found that the antibodies for both H and O were same. 65 serum samples were collected, out of which 50 were serologically proved to be typhoid fever patients. When screened for the carrier state, 5 out off 12 male and 7 out off 38 female were found to be carriers.

The titre of 1 : 160 for both O and H antigens were high when quantitative tube widal test was carried out to estimate the antibodies.

The antibody titre was found to be highest in the age group of 21–30 Yrs (55.4 per cent) in Tiruchengode among which females frequently affected.

The tube test gave a high perfection than the slide test. The frequency of antibodies against *Salmonella* in normal individuals was reported by various researchers from different parts of the country. This has been compared with the present study. Details are established in the Table 9.5.

Table 9.5: Frequency of Antibodies Against *Salmonella* in Normal Individuals Reported by Various Workers from Different Parts of India Compared with the Present Study

Year	*Workers*	*Cut-off*	*Widal Positive*			
			TO	*TH*	*AH*	*BH*
1999	Present study (Tiruchengode)	1 : 160	50	50	–	–
1998	Nazeema *et al.* (Coimbatore)	1 : 180	5	8	1	1
1996	Shukla *et al.* (Central India)	1 : 80	166	96	–	–
1993	Rasailey *et al.* (Kolkata)	1 : 160	–	–	–	–
1991	Deivanayagam *et al.* (Chennai)	1 : 100	4	4	–	–
1990	Cherian *et al.* (Vellore)	1 : 80	11	2	–	–
1988	Joshi *et al.* (Gujarat)	1 : 80	9	11	–	–
1966	Khan *et al.* (Lucknow)	1 : 20	21	30	–	–
1964	Singh *et al.* (Delhi)	1 : 50	25	42	–	–

Conclusions

Antibodies appear after 1 week of infection. Early treatment with antibodies abrogate antibody production. Anamnestic response may lead to the presence of low level of antibodies in serum. TAB vaccination leads to H antibody rise.

'O' antibody rise indicates recent infection. H antibody is specific for the infectious organisms. The titre of 1 : 100 and above in O and H indicates an active and recent infection. When low titre is observed, repeat the widal test after a week and a four–fold rise indicates active infection.

References

Anantha Narayanan, R. and Jayaram Panicker, C. K.1990. Enterobacteraceal–III *Salmonella*. In: *Textbook of Microbiology*, p. 279–294.

Anonymous 1978. Typhoid and its Br. *Med. J.*, 1: 389–390.

Ashley Miles, A., Graham, S., Wilson-Topley and Wilsons 1966. *Principles of Bacteriology and Immunology*, 5th Edn. p. 248–250, 271, 866–874, 887, 894.

Baker, F.J. and Breach, M.R. 1980. Agglutination and haemolysin tests. *Med. Microbial Techniques*, p. 382–385.

Balbir, Singh. 1964. Normal agglutinins and haemolyphi A in the out patients. *Indian J. Med. Sci.*, 18: 506–512.

Bharesh Patal, Chitnis, O.S. and Shukla. 1997. 100 Years of widal test and its reappraisal in an endemic area. *Indian J. Med. Res.*, 105: 53–57.

Brodie, J. 1977. Antibodies and Aberdeen typhoid out break of 1964: The wide reaction. *J. Hyg. Camb.*, 79: 161–180.

Cherian, T., Sridharam, G., Mohandas, V. and John, T.J. 1990. Prevalence of *Salmonella* typhi O and H antibodies in the serum of infants and preschool children. *Indian Pediatr.*, 27: 293–294.

Collard, P., Sen, R. and Montefiore, D. 1959. The distribution of *Salmonella* agglutinin in sera of health adulats in *Ibadan. J. Hyg. Camb.*, 57: 427–434.

Deivanayagam, N. and Devikumari, D.V. 1991. Prevalance of O and H agglutinins for *S. typhi* in healthy individuals: A cross sectional survey. *Indian Pediatr. J.*, 28: 420–421.

Felix, H. 1968. Antigen-antibodies interaction: Immunochemistry and biosynthesis of antibodies, p. 139–140.

Grumbaum, A.S. 1896. Preliminary note on the use of the agglutinative reaction of human serum for the diagnosis of human serum for the diagnosis of enteric fever. *Lancec*, 2: 800, 80%.

Hughes, M.H. 1955. Enteric fevers and normal Salmonella agglutinins in the gold coast. *J. Hyg. Camb.*, 53: 358–378.

Joshi, P.J., Samuel, A., Dongre, O.S. and Dave, J.P. 1988. Normal agglutinins against enteric group of organisms in healthy subjects. *Indian J. Med. Sci.*, 42: 231–234.

Khan, A.M., Gupta, S.P. and Gupta, N.P. 1966. Antibodies against *Salmonella* in Uttar Pradesh. *Indian J. Med. Sci.*, 45: 697–704.

Kwapinski, J.B. 1965. Antigens. *Methods of Serological Research*, p. 3–8, 33–34, 131–142.

Levine, M.M., Gardos, O., Gilman, R.H., Woodward, W.E., Solisplaza, R. and Waldman, W. 1978. Diagnostic value of the widal test in areas endemic for typhoid fever. *Am. J. Trop. Med. Hyg.*, 27: 795–800.

Pang, T. and Puthucheary, S.D. 1983. Significance and value of the widal test in the diagnosis of typhoid fever in endemic area. *J. Clin. Pathol.*, 36: 471–474.

Parker, M.T. 1990. Enteric Inf: Typhoid and paratyphoid fever. In: *Topley and Wilson's Principles of Bacteriology, Virology, Immunology*, Vol. 3, 8th Edn.

Patel, B.D.S.. Chitinus, Shukla. 1997. *Indian T. of Med. Res.*, 105: 53–57.

Public Health Service Working Party Report. 1961. The detection of the typhoid carrier state. *J. Hyg. Camb.*, 59: 231–247.

Ranjit Sen, and Saxena, S.N. 1969. A critical assessment of the conventional widal test is a diagnosis of typhoid and paratyphoid fever. *Indian J. Med. Res.*, 57: 1–7.

Shipp, C.R. and Rowe, B. (1980). *J. Clin. Pattel.*, 33: 592–597.

Steven and Schroeder S.A. 1968. Interpretation of serological test for typhoid fever. *JAMA*, 206: 839–840.

Verma, J.C. and Gupa, B.R. (1990). *Ind. J. Camp. Microbiology Immunology and Infect. Dis.*, 11: 114–115.

Chapter 10
Antimicrobial Activity of *Premna tomentosa* Willd. to Chosen Pathogenic Bacteria

*J. Matrtin Rathi and S. Gopalakrishnan**

Department of Chemistry, Manonmaniam Sundaranar University, Abishehkepatti, Tirunelveli – 627 012, Tamil Nadu, India

**E-mail: sgkrishrajes@yahoo.co.in*

ABSTRACT

The antibacterial activity of petroleum ether (40°–60°C), benzene, chloroform, methanol and water extracts of the aerial parts of *Premna tomentosa* Willd. (*Verbenaceae*) has been evaluated against six pathogenic strains of gram-positive (*Staphylococcus aureus* and *Streptococcus pyogenes*) and gram-negative bacteria (*Pseudomonas fluorescens, Salmonella typhi, Serratia marcesens* and *Proteus vulgaris*) by using zone of inhibition assay. It is confirmed by Activity Index (AI) test. Activity index was higher (1.48) for benzene extract against the methicillin-resistant *Staphylococcus aureus* followed by methanol extracts (1.44) on *Pseudomonas fluorescens* and *Staphylococcus aureus*. Among all the solvents tested, benzene and methanol have more effects on all the six bacteria than the remaining extracts. Water extract showed antibacterial activity against both gram-positive and gram-negative bacteria tested in this study.

Keywords: Premna tomentosa, Gram-positive bacteria, Gram-negative bacteria, Antibacterial activity.

Introduction

Many efforts have been taken to discover new antimicrobial compounds from various kinds of sources such as soil, microorganism, animals and plants. Drugs of natural origin continue to be important for the treatment of many diseases worldwide. *Premna tomentosa* Willd. (*Verbenaceae*) commonly called as "Krishnapalai" or "Podaganari" is a medicinal plant, extensively used for the treatment of various disorders. Indian Siddha medicine claims a lasting cure for hepatic disorders through oral administration of aqueous extract of *Premna tomentosa* (Shanmugavelu, 1987; Alam, 1993). In traditional medicine, *Premna tomentosa* has also been used to treat stomach disorders and diarrhoea. Furthermore, the leaves possess diuretic properties (Anonymous, 1989). Earlier studies have shown that the methanol extract of *Premna tomentosa* leaves afforded protection against acetaminophen-induced hepatotoxicity in rats (Devi and Devaki, 1998) by its antioxidant property (Devi, 1998).

Antimicrobial activity of *Premna integrifolia* (Kurup and Kurup, 1964); *Premna oligotricha* (Bradshaw *et al.*, 1992; Habtemariam *et al.*, 1992, 1993); *P. schimperi* (Habtemariam *et al.*, 1990) have been determined already. We report here the effect of various solvent extracts of the aerial parts of *P. tomentosa* on the pathogenic strains of two gram-positive bacteria *viz.*, *Staphylococcus aureus* and *Streptococcus pyogenes* and four gram-negative bacteria *viz.*, *Pseudomonas fluorescens*, *Salmonella typhi*, *Serratia marcesens* and *Proteus vulgaris* by using zone of inhibition assay (Jain, 2003). Ampicilin, methicillin and penicillin are used as standard antibiotics.

Materials and Methods

Premna tomentosa aerial parts were collected from Malaiadiputhoor, Tirunelveli District, Tamil Nadu, India during June. The plant was identified by Dr. V. Chelladurai, Research Officer (Botany), Survey of Medicinal and Aromatic Plants Unit–Siddha, CCRAS, Palayamkottai, Tirunelveli District, Tamil Nadu, India and voucher specimen is deposited at the Herbarium of the Department of Chemistry, Manonmaniam Sundaranar University, Tirunelveli District, Tamil Nadu, India (Voucher specimen No. 030). Air-dried aerial parts (400 g) of the plant was extracted with petroleum ether (40°–60° C), benzene, chloroform, methanol water by using a soxhlet apparatus. The last trace of the solvent was removed under reduced pressure distillation and the crude extract was dried in a vacuum desicator and used for the experiments.

Staphylococcus aureus, *Streptococcus pyogenes*, *Pseudomonas fluorescens*, *Salmonella typhi*, *Serratia marcesens* and *Proteus vulgaris* were obtained from All India Institute of Medical Sciences, New Delhi. The antibacterial activity was measured by disc-diffusion method (Vanden Berghe and Vlietinck, 1991). The bacterial cultures were maintained on nutrients agar slants (Himedia, Mumbai) and subcultures were freshly prepared before use. Before streaking, each culture was diluted (1 : 10) with fresh sterile nutrients broth. Plates were prepared by pouring freshly prepared No. 1 medium (Himedia, Mumbai) (20 ml) into 20 mm × 100 mm Petri plates. Inoculum (5 ml) was poured directly over the surface of prepared plates to uniform depth of 4 mm and then allowed to solidify at room temperature.

Test solution (2.5 per cent) of each extract was prepared by dissolving 125 mg of each extract separately in 5 ml of the respective solvents. The extracts were dissolved in the respective solvents and tested for antimicrobial activity using sterile discs (6 mm) (Himedia SD 067). Pure cultures of all bacterial species were inoculated in nutrient agar plate sepamtely. Then the sterile disc-containing plant extract (200 μl) was placed over the seeded agar plates in such a way that there is no overlapping of zone of inhibition. Standard antibiotic discs (ampicillin for *P. vulgaris* and *P. fluorescens*; penicillin for *S. marcescens* and *S. typhi*; and methicillin for *S. aureus* and *S. pyogenes*) were also incorporated as positive control. The plates were kept at room temperature for half an hour for the diffusion of the sample into the agar media then the plates were incubated at 37°C for 24–48 hours. After the incubation period was over, the plates were observed for zone of inhibition (ZI). From the results, activity index was calculated by using the following formula:

$$\text{Activity Index (AI)} = \frac{\text{Inhibition Zone of Samples (IZS)}}{\text{Inhibition Zone of Standard (IZSTD)}}$$

Results and Discussion

The results obtained for the antimicrobial bioassay is presented in Table 10.1. From the results it is very clear that gram-positive bacteria, *S. pyogenes* is resistant to petroleum ether, benzene and chloroform extracts of *P. tomentosa*. But, the methanol and water extracts were inhibiting its growth (9.1 and 10.2 mm for methanol and water extracts, respectively). However, all the extracts (except petroleum ether) of *P. tomentosa* showed bactericidal effects against another gram-positive bacteria *S. aureus*. Among the extracts tested, benzene showed maximum effect (15.3 mm) followed by methanol, chloroform and water (Tables 10.1 and 10.2). Many plant extracts showed strong anti-bactericidal activity against the methicillin resistant *S. aureus* (Tsuchiya, 1996). In contrast, *Stacytarpheta indica* (Linn.) Vahl. (*Verbenaceae*) methanol extract showed no effect on *S. aureus* (Wiart *et al.*, 2004). Among the four gram-negative bacteria tested, in the present investigation *P. fluorescence* showed the highest sensitivity in the methanolic extract followed by water, petroleum ether, benzene and chloroform extracts. For *P. vulgaris* also the highest activity is noticed for the methanolic extract. *P. vulgaris* shows similar activity in petroleum ether, benzene and chloroform extracts.

Table 10.1: Inhibition Zone (IZ) (mm) of Solvent Extracts of *P. tomentosa* and Standard Antibiotics on Pathogenic Bacteria

Solvents/Standards	*Test Bacteria*					
	P. fluorescens	*P. vulgaris*	*S. typhi*	*S. marcesens*	*S. aureus*	*S. pyogenes*
Petroleum ether (40°–60°C)	7.0	11.0	–	7.7	–	–
Benzene	7.0	–	–	12.0	15.3	–
Chloroform	7.0	9.5	–	11.0	14.1	–
Methanol	16.0	13.0	8.0	11.0	14.8	9.1
Water	11.0	11.0	8.0	8.9	9.4	10.2
Ampicillin/Methicillin/Penicillin	20.0	9.0	9.2	22.0	10.3	36.5

Table 10.2: Activity Index (AI) of Solvent Extracts of *P. tomentosa* and Standard Antibiotics on Pathogenic Bacteria

Solvents/Standards	*Test Bacteria*					
	P. fluorescens	*P. vulgaris*	*S. typhi*	*S. marcesens*	*S. aureus*	*S. pyogenes*
Petroleum ether (40°–60°C)	122	1.22	–	0.35	–	–
Benzene	–	–	–	0.55	1.48	–
Chloroform	1.05	1.05	–	0.50	1.37	–
Methanol	1.44	0.87	0.87	0.50	1.44	0.23
Water	1.22	0.87	0.87	0.40	0.91	0.28

Indian plants such as, *Andrographis paniculata* and *Plumbago zeylanica* were found to exhibit maximum inhibition effect on *Pseudomonas aernginosa* and *Staphylococus aureus* at lower concentrations

(Jeevan Ram, 2004). *Vitex trifolia* Linn. (*Verbenaceae*) extracts completely inhibited the growth of both gram-positive and gram–negative bacteria, except for *S. typhi* (Hernández, 1999). In the present study, *S. aureus* shows higher activity in methanolic extract and *S. pyogenes* shows higher activity in aqueous extract. *S. marcesence* growth was inhibited by all the five extracts of *P. tomentosa*. It has been reported that *Premna integrifolia* (Kurup and Kurup, 1964); *P. schimperi* (Habemariam *et al.*, 1990) and *P. oligotricha* (Bradshaw *et al.*, 1992; Habemariam *et al.*, 1992) were active against gram-postive bacteria. An opposite trend was also recorded in *P. oligolricha* (Habemariam *et al.*, 1993).

When we consider the activity index, it is clear from the results that both the methanol and water extracts showed antibacterial activity against all the tested bacteria. The difference in the activity purely depends upon the different secondary metabolites present in the different solvent extracts of *P. tomentosa*. This indicates that the secondary metabolites act as antibacterial compounds, which either inhibit or kill the bacteria by different mechanisms. In gram-negative bacteria, the cell wall is covered with lipopolysaccharide along with proteins and phospholipids (Burn, 1988). It may act as a barrier to access the plant secondary compounds into the peptidoglycan layer of the cell wall. This explains the resistance of the gram-negative strains to the lytic action of most of the extracts exhibiting less antimicrobial effects on these strains. We cannot predict such a kind of mechanism in this study, because *P. tomentosa* shows antibacterial activity against both the gram-positive and gram-negative bacteria.

Antibacterial activity of *P. tomentosa* aerial parts extract can be assumed to be useful for curing infectious diseases. Methanol and water extracts shows the highest activity against almost all the tested microorganisms. Hence it is essential to isolate and identify the active principles present in these extracts as a powerful antiseptic. This supports the traditional use of *P. tomentosa* as an effective antiseptic.

Acknowledgement

One of the authors (JMR) wishes to thank the UGC, New Delhi and the authorities of St. Mary's College, Tuticorin for selecting her under FIP programme.

References

Alam, M., Joy, S., Susan, T. and Usman Ali, S. 1993. Anti-inflammatory activity of *Premna tomentosa* Willd. in albino rats. *Ancient Science*, 13: 185–188.

Anonymous. 1989. *The Wealth of India Raw Materials*. Publications and Information Directorate, Council of Scientific and Industrial Research, New Delhi.

Bradshaw, J.D., Gray, A.I. and Waterman, P.G. 1992. *Planta Medica*, 58(1): 109–110.

Burn, P. 1988. Amphitropic proteins: A new class of membrane proteins. *Trends in Bioche. Sci.*, 13: 79–83.

Devi, K. and Devaki, T. 1998. Protective effect of *Premna tomentosa* on acetaminophen induced hepatitis in rats. *Med. Sci. Res.*, 26: 785–787.

Devi, R., Anandan, T., Devaki, T., Apparananthanm, A. and Balakrishna, K. 1998. Effect of *Premna tomentosa* on rat liver antioxidant defense system in acetaminophen intoxicated rats. *Biome. Res.*, 19: 339–334.

Habtemariam, J.D., Gray, A.I. and Waterman, P.G. 1990. A novel antibacterial diterpene from *Premna schimperi*. *Planta Medica*, 56(2): 187–189.

Habtemariam, S., Gray, A.I. and Waterman, P.G. 1992. A novel antbacterial sesquiterpene from *Premna oligotricha*. *J. Nat. Prod.*, 56: 140–143.

Habtemariam, S, Solomon, Gray, A.I., Alexandar, I., Waterman, P.G. and Peter, G. 1993. A new antibacterial sesquiterpene from *Premna oligotricha*. *J. Nat. Prod.*, 56(1): 140–143.

Hernández, M.M., Heraso, C., Villarreal, M.L., Vargas-Arispuro, I. and Aranda. 1999. Biological activities of crude plant extracts from *Vitex trifolia* L. (*Verbenaceae*). *J. Ethnophar.*, 67(1): 37–44.

Jain, N. and Sharma, M. 2003. Broad spectrum antimycotic drug for the treatment of ringworm infection in human beings. *Current Science*, 85(1): 30–34.

Jeevan Ram, L., Bhakshu, Md. and Venkata Raju, R.R. 2004. *In vitro* antimicrobial activity of certain medicinal plants from Eastern Ghats, India, used for skin diseases. *J. Ethnophar.*, 90(2&3): 353–357.

Kurup, K.K. and Kurup, A.A. 1964. Antibiotic substance from the root bark of *Premna oligotrichta*. *J. Nat. Prod.*, 56(1): 140–143.

Shanmugavelu, M. 1987. *Siddha Cure for Diseases*. Tamil Nadu Siddha Medical Board Publications, Chennai.

Tsuchiya, H., Sato, M., Miyazaki, T., Fujiwara, S., Tanigaki, S., Ohyama, T., Tanaka, T. and Iinuma, M. 1996. Comparative study on the antimibacterial activity of phytochemical flavanones against methicillin-resistant *Staphylococccus aureus*. *J. Ethn. Pharmacology*, 50: 27–34.

Vanden Berghe, D.A. and Vlietinck, A.J. 1991. Screening methods for antibacterial and antiviral agents from higher plants. In: *Methods in Plant Biochemistry: Assay for Bioactivity*, (Eds.) Dey, P.M., Harborne, J.B., Hostettman, K. Academic Press, London, p. 47–69.

Wiart, C., Mogana, S., Khalifah, S., Mahan, M., Ismail, Buckle, M., Narayana, A.K. and Sulaiman, M. 2004. Antimicrobial screening of plants used for traditional medicine in the state of Perak, Peninsular Malaysia. *Fitoterapia*, 75: 68–73.

Chapter 11

Effect of Polyvinyl Pyrrolidone and Dimethyl Sulphoxide on Ethyl Cellulose Transdermal Patches of Verapamil Hydrochloride

*S.K. Sahoo, S. Chatterjee, D. Sahu, S.K. Mishra and B.B. Barik**

University Department of Pharmaceutical Sciences, Utkal University, Vani Vihar, Bhubaneswar – 751 004, Orissa

ABSTRACT

The transdermal patches of verapamil hydrochloride were prepared with ethyl cellulose (EC) and polyvinyl pyrrolidone (PVP) as polymers and dibutyl phthalate as plasticizer. Different concentrations of penetration enhancer, dimethyl sulfoxide (DMSO) were added to the patches and their effects were studied. Films were evaluated by studying various parameters like thickness, weight variation, tensile strength, drug content, skin irritation, flatness, scanning electron microscope (SEM) and *in vitro* skin permeation study through rat skin. The *in vitro* release data were analyzed in computer. Permeation co-efficient and permeation enhancement factors were calculated. SEM study reveals that, drugs are uniformly coated and distributed throughout the film. *In vitro* skin permeation study shows that there is increase in permeation as the concentration of permeation enhancer in the film increases. The combination of DMSO, PVP and EC definitely gave rise to better formulations, which increases the permeation of drug across the biological membrane in a controlled manner.

Keywords: *Verapamil hydrochloride, Transdermal, Polymer, Plasticizer, Penetration enhancer.*

Corresponding Author.

Introduction

Verapamil hydrochloride is a calcium channel blocker, which is effective in the treatment of angina, hypertension and arrhythrnia. But it is substantially metabolized in the liver when administered orally by showing only 20 per cent bioavailability (Anonymous, 1996). In order to avoid such problem, Verapamil hydrochloride has been incorporated into transdermal patches (Kulkarni, *et al.*, 2002) using EC and PVP as polymers and the effect of penetration enhancer, DMSO on transdermal permeation of the drug has also been studied in the present investigation. The patches were prepared on mercury substrate by dissolving EC alone and in combination with PVP in chloroform and dibutyl phthalate was included as plasticizer. DMSO was added to the patches in different concentrations.

Materials

Verapmil hydrochloride was obtained as gift sample from Torrent Pharmaceuticals. Ahmedabad. EC and PVP were obtained from Central Drug House Pvt. Ltd., Mumbai. Chloroform IP and dibutyl phthalate from Qualigens Pvt. Ltd., Mumbai and DMSO from Ranbaxy Laboratories Ltd., SAS Nagar. All other reagents were of analytical grade.

Method

Preparation of Matrix Film (Chowdary and Naidu, 1991)

The films were prepared by dissolving EC alone and in combination with PVP in Chloroform (10 ml). Dibutyl phthalate in concentration of 40 per cent W/W of polymer was incorporated as plasticizer and 10ml of the polymer solution was poured into the mercury substrate in a petridish. Patches with DMSO (at the concentration of 2.5, 5 per cent) were made separately with EC alone and in combination with PVP (Table 11.1). The drug concentration per unit area of the patch was 2.27 mg/cm^2. The patches were dried in controlled, manner over 24 h. By keeping a funnel inverted over the petridish. The dried films were taken off and stored in a desiccator. Aluminum foil was used as backing membrane. The optimum concentration of plasticizer was found to be 40 per cent and in greater concentration film became very soft.

Table 11.1: Data Showing the Formulations of Transdermal Drug Delivery Systems

Formulation Code	*Polymers Used (mg)*	*Plasticizer (Dibutyl Pthalate) (mg)*	*Drug (Verapamil Hydro Chloride) (mg)*	*Enhancer (DMSO) (% v/v)*
F1	EC–500	200	50	**
F2	EC–500	200	50	2.5
F3	EC–500	200	50	5
F4	EC + PVP (500+50)	200	50	**
F5	EC + PVP (500 + 50)	200	50	2.5
F6	EC + PVP (500 + 50)	200	50	5

**: Denotes absence of DMSO.

Preparation of Skin

Young healthy rats aged about 6 to 8 weeks were taken and killed by cervical dislocation. The abdominal skin was carefully separated from the body by keeping the dermis intact. Skin was washed with distilled water to remove the adhering fat. The skin so obtained was examined microscopically for the presence of any damage.

Evaluation of Film

The prepared films were evaluated by studying various parameters like thickness, weight variation, tensile strength, drug content (Manvi, *et al.*, 2003), skin irritation (Kulkarni, *et al.*, 2002; Kapoor, *et al.*, 2002), water vapor transmission (WVT) (Chowdary and Naidu, 1991), flatness, SEM and *in vitro* skin permeation study through rat skin (Manvi, *et al.*, 2003).

Film Characteristics

After preparation of films, various characteristics like thickness, weight variation, tensile strength and drug content were determined (Table 11.2).

Table 11.2: Data Showing the Film Characteristics

Formulation Code	*Thickness (μm) (n = 6)*	*Weight Variation (mg/cm^2) (n = 6)*	*Tensile Strength (gm/cm^2) (n = 3)*	*Drug Content (%)*
F1	2.08 (0.040)	32.83 (0.40)	183.43 (0.58)	88.41
F2	2.17 (0.051)	40.50 (0.54)	135.603 (0.53)	92.44
F3	2.23 (0.081)	49.34 (0.51)	135.53 (0.78)	92.75
F4	2.18 (0.04)	35.66 (0.51)	184.05 (0.97)	93.68
F5	2.21(0.04)	45.33 (0.51)	148.36 (0.54)	93.37
F6	2.26 (0.051)	53.66 (0.51)	155.83 (0.339)	93.68

Skin Irritation Test

The skin irritation test was performed on 10 numbers of healthy albino rats. Drug free film (control) and drug loaded film (test) of various formulations were adhered on unbraided skin of the rat with the help of adhesive tape. The control and test patches were placed on left and right dorsal surface respectively. The patches were removed after 24 h. with help of an alcohol swab and the skin was examined for appearance of any erythemaric and ederna (Table 11.3).

Table 11.3: Data Showing the Results of Skin Irritation Test

Formulation Code	*Control*	*Test*
F1	X	+
F2	X	++
F3	X	++
F4	X	++
F5	X	++
F6	X	++

X: Denotes absence of skin irritation; +: Denotes very mild erythema; ++: Denotes mild erythema.

Flatness Test

Ships of definite length (1.5 cm) was taken on a plane surface and cut lengthwise into several pieces and the constriction of cut piece was determined by comparing with that of the original length. Flatness was determined by measuring the constriction of the strips and 0 per cent constriction implies to 100 per cent flat. So Constriction % = $(L_1 - L_2)/L_2 \times 100$

where,

L_1: Final length of each strip

L_2L Initial length.

All prepared films were found to be 100 per cent flat (Table 11.4).

Table 11.4: Data Showing the Results of Flatness Test

Formulation Code	*Film Length (cm)*	*Amount of Constriction*	*Flatness (%)*
F1	1.5	0	100
F2	1.5	0	100
F3	1.5	0	100
F4	1.5	0	100
F5	1.5	0	100
F6	1.5	0	100

WVT Test

Washed and dried glass vials of equal dimension containing 3 gms. of fused calcium chloride were used as transmission cells for water vapour transmission test. The circular pieces of the film of known thickness were fixed over the beam of glass vials. The vials were kept in a desiccator containing saturated solution of Potassium chloride thereby, maintaining the humidity at 84 per cent, which was measured with a Hygrometer. The vials were taken out from the desiccator and weighed at regular intervals of 12 h upto 3 days. The experiment was carried out in triplicate and the average values were calculated (Table 11.5).

Table 11.5: Data Showing the Results of WVT

Time (hrs.)	*Amount of Water Vapour Gained in Various Formulations (mg/cm²)*					
	F1	*F2*	*F3*	*F4*	*F5*	*F6*
0	0	0	0	0	0	0
12	0.017	0.017	0.014	0.014	0.017	0.024
24	0.033	0.034	0.032	0.03	0.034	0.049
36	0.046	0.048	0.45	0.044	0.049	0.07
48	0.066	0.065	0.059	0.057	0.071	0.091
60	0.085	0.08	0.074	0.07	0.089	0.112
72	0.104	0.096	0.09	0.087	0.101	0.132

SEM Study

SEM study shows that drug particles are uniformly distributed throughout the transdermal patches and well coated with the polymer (Figures 11.2 and 11.3).

In vitro Skin Permeation Study

The contents of donor and receptor compartments were separated by placing freshly prepared rat skin in between them. The skin was mounted in such a way that, stratum corneum side of the skin was constantly remained in an intimate contact with transdermal film in the donor compartment and the receptor compartment contained phosphate buffer solution (pH 7.4) at 37±1°C. The solution was stirred at constant speed of 500 rotations per minute. The *aliquots* of 1 ml were withdrawn at regular time interval and replaced with the fresh medium. The withdrawn samples were analyzed by UV-Spectrophotometer after required dilution and a graph was plotted between cumulative per cent drug release per unit area (mg/cm^2) Vs square root time (minutes) from which permeation co-efficient and permeation enhancement factors were calculated (Table 11.6 and Figure 11.1).

Table 11.6: Data Showing the Permeation Enhancer Factor of Various Formulations

Formulation Code	*Permeation Co-efficient (P)*	*Permeation Enhancer Factor*
F1	0.003765	Control
F2	0.0037	0.9327
F3	0.00509	1.35
F4	0.00397	1.05
F5	0.00665	1.76
F6	0.006935	1.84

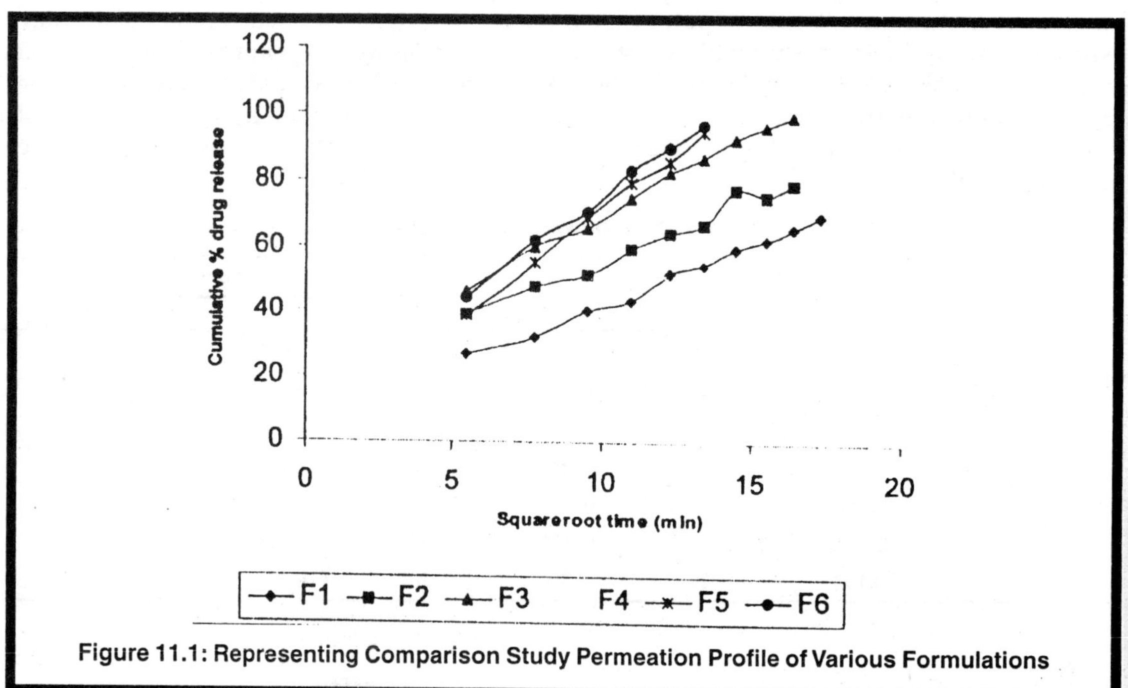

Figure 11.1: Representing Comparison Study Permeation Profile of Various Formulations

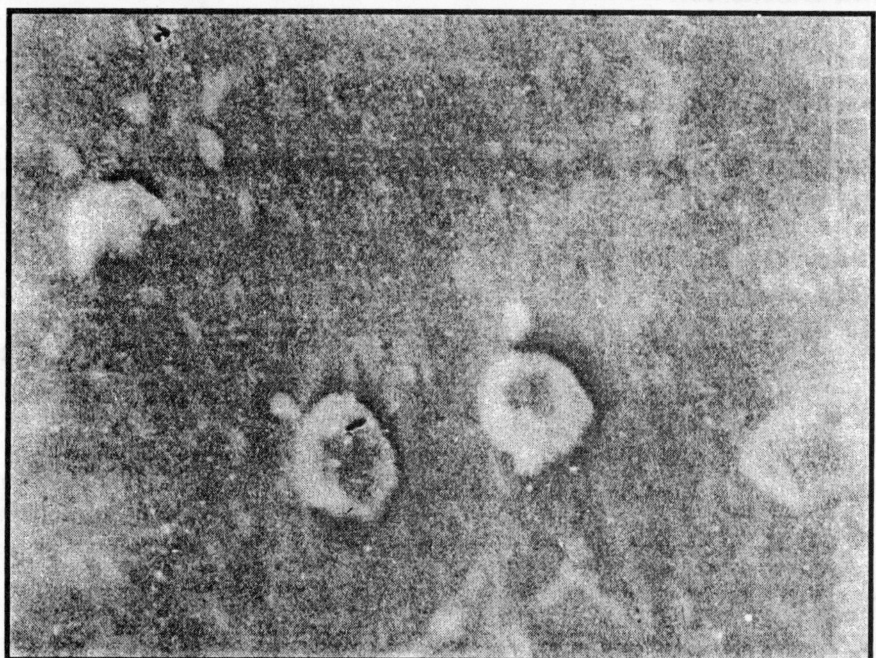

Figure 11.2: Scanning Electron Microscope Study of Prepared Patches (Low magnification)

Figure 11.3: Scanning Electron Microscope Study of Prepared Patches (High magnification)

Results and Discussion

The films prepared by mercury substrate method were found to be more uniform in shape, size and thickness. All prepared films were mechanically tough to permit handling during application over the skin. The absence of skin irritation in the control patches showed that DMSO used in the above formulation had no adverse effect on the skin. Water vapor transmission test indicated that all films were permeable to water vapor. Water vapour transmission followed zero order kinetics. Scanning Electron Microscope study reveals that the drug was uniformly coated and distributed throughout the films. *In vitro* skin permeation study showed that, with inclusion of PVP and DMSO, the permeability of drug increases and the order of permeability enhancement factor was F6 > F5 > F3 > F4 > F2 > F1, which means the inclusion of PVP increases the permeability of EC patches and DMSO increases the permeability of the drug through the intact skin. Therefore we may conclude that the combination of DMSO, PVP and EC definitely gave rise to better formulations. which increases the permeation of drug across the biological membrane in a controlled manner.

Acknowledgements

The authors wish to thank the authorities of Torrent Pharmaceuticals. Ahmedabad for supplying the gift sample of Verapamil hydrochloride and University Department of Pharmaceutical Sciences, Utkal University, Bhubaneswar for providing the necessary facilities to carry out this research work.

References

Anonymous. 1996. *Martindale: The Extra Pharmacopoeia, The Complete Drug Reference,* 31st Edn. Pharmaceutical Press, London, pp. 961.

Chowdary, K.P.R. and Naidu, R.A.S. 1991. Studies on permeability of ethyl cellulose films for transdermal use. *The Eastern Pharmacist*, Sept., p. 119–121.

Kulkarni, R.V., Mutalik, S. and Hiremath, D. 2002. Effect of Plasticizers on the permeability and mechanical properties of Eudragit films for transdermal application. *Indian J. Pharm. Sci.*, 64(1): 28–31.

Kapoor, A., Lewis, S., Venkatesh and Udupa, N. 2002. Sodium alginate patches containing nicotine with different rate controlling membranes. *Indian Drugs*, 39(8): 415–418.

Manvi, F.V., Dandagi, P.M., Gadad, A.P., Mastiholimath, V.S. and Jagatdeesh, T. 2003. Formulation of a Transdermal drug delivery system of Ketotifen Fumarate. *Indian J. Pharm. Sci.*, 65(3): 239–243.

Rama Rao, P. and Diwan, Prakash V. 1997. Influence of casting solvent on the permeability of ethyl cellulose free films for transdermal use. *The Eastern Pharmacist.*, Sept., p. 135–137.

Chapter 12

Effect of Scrotal Heating on the Reproductive Organs of the Laboratory Rat, *Rattus norwegicus*

W. Vidyarani

Department of Life Sciences, Manipur University, Canchipur – 795 003, Manipur

ABSTRACT

The effect of scrotal heating on the testis and epididymis were investigated. Immersion of the scrotum into hot water bath at 52–53°C for 15–20 minutest day for 14 days induced severe atrophic changes in the testis and epididymis. The seminiferous tubules were depopulated; only sertoli cells and spermatogonia were discernible in the tubules; tunica propria appeared collapsed. Leydic cells were atrophied. Epididymides in treated rats also exhibited regressive changes in the tubules; the changes includes flattening of the epithelial cell layer, pycnosis of cell nuclei and an increase in the fibro muscular tissue; lumen was devoid of spermatozoa and secretions.

Keywords: Heat stress, Testis, Epididymis.

Introduction

The deleterious effects of heat on scrotal mammals were reported as early as 1898 by Felizet and Branca in cryptorchid men. Howerver, it was Crew (1922), who for the first time postulated that the antispermatogenic changes noticed during Cryptorchidism were due to the exposure of the testes to the increased temperature of the abdomen. Since then his observations have been confirmed in several mammalian species including rat (Fukui, 1923; Bowler, 1967; Fahim, Fahim Z., Der, Hall and Harman, 1975), mouse (Hardberger, 1950), guinea pig (Young, 1927), rabbit (Oloufa, Bogart and Mekenzie, 1951), dog (Fukui, 1923), ram (Moreira, 2001), monkey (Venkatachalam and Ramanathan, 1962) and man (Fukui, 1923) using various methods of heat application such as warm water, warm air, warm paraffin, sunlight and archlight. The present report deals with the effect of scrotal heating on the reproductive organs of the laboratory rat.

Materials and Methods

Ten adult male rats weighing 200–225 g were employed in this investigation. Animals were maintained on pelleted diet (Hindustan Lever Ltd.) and water *ad libitum*. They were divided into two groups and treated as follows:

Group I: Untreated Controls

Group II: Rats were exposed to heat in scrotal region/day for 14 days and were killed 24 hr after the last treatment.

The rats were given heat treatment under pentobarbitol anaesthesia (Nembutal, 25 mg/kg body weight). Scrotal regions were immersed in a water bath at 52–53°C for 15–20 minutes daily for 14 days. Animals were killed under ether anaesthesia and testes and epididymides were removed and fixed in Bouins' fluid for 24 hr. Tissues were embedded in paraffin, sectioned at 6μ and stained with Harris haematoxylin and eosin.

Observations

Histology of the Testis

Testis of untreated controls exhibited full spermatogenic activity (Figure 12.1). By contrast, testes of rats exposed to heat treatment showed severe regression changes in the seminiferous tubules with suppression of spermatogenesis. The seminiferous tubules were totally depopulated and were lined by a ring of sertoli cells and speramatogonia; the lumen contains dehisced cells. Tunica propria appeared collapsed. Leydig cells were atrophied (Figure 12.2).

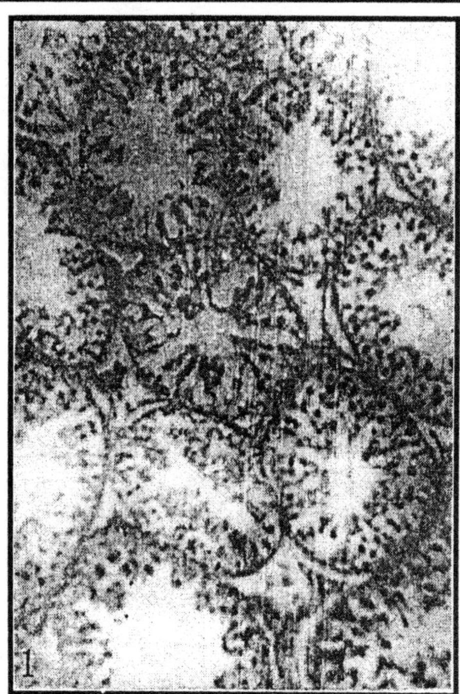

Figure 12.1: Testis of an Untreated Control (Group I)

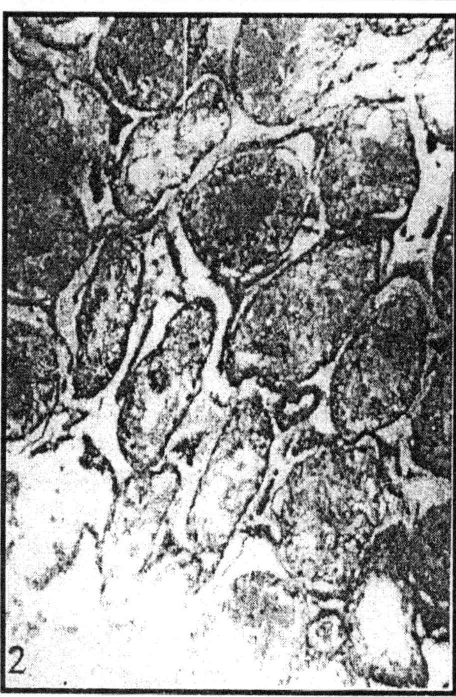

Figure 12.2: Testis of a Rat Killed 14 Days After Daily Exposure to Heat i Scrotal Region (Group II)

Histology of Epididymis

The epididymis of untreated controls presented normal histological features (Figure 12.3). In rats killed 14 days after exposure to the heat treatment, the organ exhibited severe degenerative changes. There was flattening of epithelial cell layer, pycnosis of cell nuclei and an increase in the fibro muscular storma. The lumen of the tubule was devoid of secretion and spermatozoa (Figure 12.4).

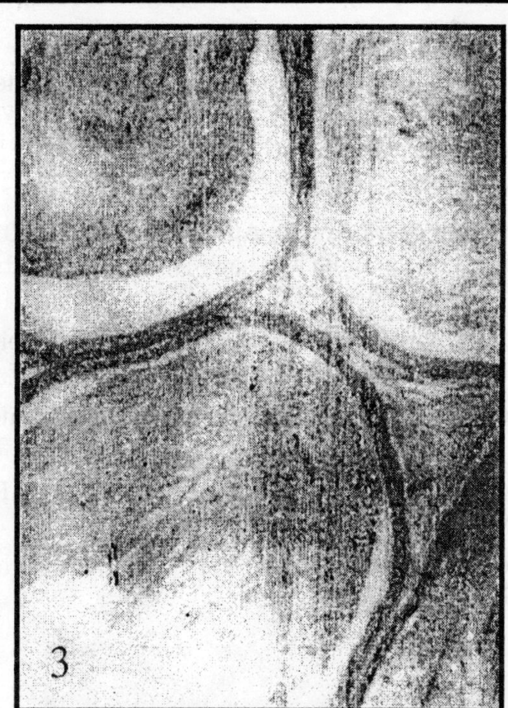

Figure 12.3: Cauda Epididymidis of a Control (Group I)

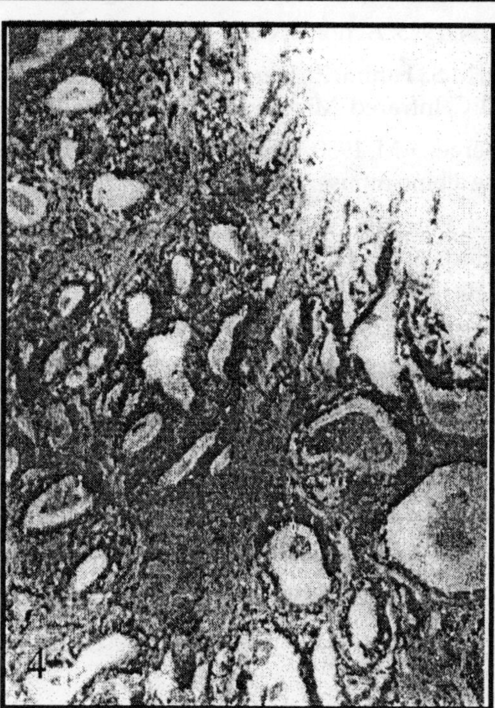

Figure 12.4: Cauda Epididymidis of a Rat Killed 14 Days After Daily Exposure to the Beat in Scrotal Region (Group II)

Discussion

The present study indicates that exposure of the scrotum to heat induces marked degenerative changes in the testis with inhibition of spermatogenesis. This is in agreement with the 'finding in the same or other species (Fukui, 1923; Young, 1927; Venkatachalam and Ramanathan, 1962; Bowler, 1967; Fahim, Fahim Z., Der, Hall and Harman, 1975). The present results also indicate that heat treatment caused severe atrophic changes in the epididymis. It is well known that spermatogenesis in mammals is a heat sensitive phenomenon and that increasing the temperature of the testis above an optimal level causes meiotic abnormalities and/or destruction of the cell stages involved (Van Demark and Free, 1970). Thus it is probable that suppression of spermatogenesis as noticed in the present study is caused due to the artificially increased temperature of the testis.

References

Bowler, K. 1967. The effects of repeated temperature applications to the testis on fertility in male rats. *J. Reprod. Fertility*, 14: 171.

Crew, F.A.E. 1922. A suggestion as to the cause of the aspermatic condition of the imperfectly descended testis. *J. Anat.*, 56: 98.

Felizet, G., and Branca, A. 1898. Histologie der testicule ectopique. *J. Anat. Physiol.*, Paris, 38: 329.

Fukui, N. 1923. Action of body temperature on the testicle. *Japan. Med. World*, 3: 160.

Fahim, M.S., Fahim, Z., Der R., Hall D.G. and Harman J. 1975. Heat in male contraception (Hot water 60°C, Infrared, Microwave and Ultrasound). *Contraception*, 11(5): 549–562.

Hardberger, F.M. 1950. Effects of constant high temperature (32.2°C) on the testes and spermatozoa of the albino mouse. *Proc. Louisiana Acad. Sci.*, 13: 35.

Moreira, Emerson Pinto, Moura, Arlindo de Alencar Araripe and Araujo, Airton Alencar de. 2001. Effects of scrotal insulation of testis size and semen criteria in santa ines hairy sheep raised in the state of Ceara, Northeast of Brazil. *Rev. Bras. Zootec.*, 30(6): 1704–1711.

Oloufa, M.M., Bogart, R., and Mekenzie, F.F. 1951. Effect of environmental temperature and the thyroid gland in the male rabbit. *Oregon, State Coil. Agr. Expl. Sta., Tech. Bull.*, 20.

Venkatachalam, P.S. and Ramanathan, K.S. 1962. Effects of moderate heat on the testes or fats and monkeys. *J. Reprod. Fertility*, 4: 51.

Van Demark, N.L. and Free, M.J. 1970. Temperature effects In: *The Testis*, (Eds.) A.D. Johnson, W.R. Gomes and N.L. Van De mark. Academic Press, New York, Vol. 3, pp. 233.

Young, W.C. 1927. The influence of high temperature on the guineapig testes. I. Histological changes and effects on reproduction. *J. Exptl. Zool.*, 49: 459.

Chapter 13
Magnesium in Scalp Hair and Fingernails in Relation to Different Parameters

Rita Mehra and Meenu Juneja

Department of Pure and Applied Chemistry, Maharshi Dayanand Saraswati University, Ajmer – 305 009, Rajasthan, India

ABSTRACT

Magnesium concentration in hair and nails of male subjects, working in different sections of roadways and carriage workshop living in Ajmer of Rajasthan was determined by Atomic Absorption Spectrophotometer. Results are discussed in relation to age, food habit, smoking, drinking habits, use of hair cosmetics, hair color and health effects. Correlation of Mg was observed with non-vegetarian type food, black hair and hypertensive subjects. The other factors such as cosmetics use, liquor intake and smoking habit were non-significant.

Keywords: Personal characters, Magnesium, Hair color, Atomic absorption spectrophotometer.

Introduction

Magnesium (Mg), the eighth most abundant element in the earth's crust, has the ability to replace calcium (Ca). In human body it occurs in muscles, liver, kidney and in extra cellular fluids, besides the skeleton. It is important for stimulation of muscles and nerves including heart. It acts as a stabilizer of plasma membranes, intracellular membranes and nucleic acids and serves as a catalyst in physiological activities (Belitz and Grosch, 1999). Magnesium is absorbed by active transport by blood to bones and other tissues in the body. About 50 per cent of the ingested Mg is excreted through kidney (Vidya, 1996). Magnesium deficiency causes many diseases such as hypomagnesemia, chronic kidney disease, acidosis, diabetic coma, weakness, dizziness, distension of the abdomen, conclusive seizures, uncontrolled muscle twitching, heart failure and death. Magnesium may play a role in several cardiovascular disorders including cardiacarrhythmias, ischemic heart disease, cardiomypethy and sudden death. It is reported that fatal effect of hypermagnesemia is due to reduction in calcium levels. (Hagerty, 1995; Goetz, 1990; Considine, 1989). The best food sources of magnesium are cereals, legumes, nuts, meats, milk and dairy products (Stanfield, 1992).

Hair and nails are biological materials which, due to their growth and property to retain metals, reflect both the environmental and the biomedical history of an individual (Georgescu *et al.*, 1997). During the last three decades, much interest has been shown in metal concentrations in hair and nails to diagnose nutritional deficiencies or toxicities of metals (Jamall and Jaffer, 1987; Wilhelm *et al.*, 1991; Nowak and Chmielmicka, 2000). The idea of hair and nail analysis is very inviting because they are easily sampled, analyzed and are protein tissues with very low metabolic activity (Valkovic, 1988).

There is no information on determination of metal levels in body and in relation to personal and environmental parameters in the state of Rajasthan, with very few studies reported in India. This has been the driving force behind taking up this study. In our earlier works we have given the concentrations of lead, cadmium, copper, iron, nickel, zinc, manganese and calcium in hair (Mehra and Juneja, 2002 a,b,c). In this article, we report the magnesium concentration in hair and nails and their variation with food, smoking, drinking habits, use of hair cosmetics and hair color of subjects of different age groups.

Materials and Methods

Hair samples of male subjects of Ajmer and surrounding places were randomly collected from the posterior vertex of the head, as close to the scalp skin as possible, using a pair of clean stainless steel scissors. The samples, each weighing one gram, were thoroughly mixed to ensure homogeneity. They were then stored in closed plastic bages until the day of analysis. A questionnaire was also filled for obtaining the personal and medical history of the subjects as per the recommendations of World Health Organisation. The information required to be filled in the performa *viz.*, sex, age, hair color, personal habits (smoking, drinking and type of food habit), place of residence, occupation, possible metal exposure and use of hair and nail cosmetics etc. for categorization and investigating a correlation of metal concentrations, if any.

For collection of nail samples, volunteers were asked to wash their hands thoroughly with water and soap, followed by drying with a clean towel or tissue paper. Nails were cut from fingers with non-contaminated scissors. All nail samples were also sealed in plastic bags prior to analysis.

The hair samples were cut into pieces of about 1 cm prior to washing. Samples prewashed with nonionic detergent were soaked in deionized water for 10 minutes. This was followed by soaking in acetone to remove external contamination and rinsed alternatively with deionised water and again with acetone three times. Then the samples were dried at 110° C for 1 hour and stored in dessicator (Chatt and Katz, 1988).

For washing of nail samples, the samples first scrapped and cleaned off dust particles with nonionic detergent, were washed following a standardized washing procedure (Gammelgaard *et al.*, 1991). This was followed by soaking in acetone to remove external contamination, rinsing five times with deionized water and drying in an oven at 110° C for 30 minutes and stored in dessicator.

Wet Acid Digestion and Preparation of Water Clear Solution

Samples were soaked in 10 ml of 6 : 1 mixture of concentrated nitric acid and perchloric acid and then heated at 160–180°C until a clear solution is obtained. Digested samples were transferred to 100 ml volumetric flask and diluted with 0.1 N nitric acid.

Analysis

The concentration of Mg was determined by Atomic Absorption Spectrophotometry (Perkin-Elmer AAS model 250) with graphite furnace and air acetylene flame. The main instrumental parameters set for the estimation of magnesium by atomic absorption spectrophotometer were as follows: wavelength: 285.2 nm, bandwidth: 0.5 nm and lamp current: 4 mA.

Results and Discussion

Literature reveals that less emphasis has been laid on the determination of the magnesium concentration and its probable relationship with factors associated with personal characteristics. In this article, we have assayed the magnesium concentrations in human hair and nails and observed its variation to investigate the possible relationship between age, food habit, smoking habit, drinking habit, hair, color and use of hair cosmetics.

The magnesium levels as a function of different age groups shown in Table 13.1 exhibit significant hair magnesium levels in non-vegetarians of 41–50 years age group and nails magnesium levels in all age groups. Besides being an essential nutrient important in building and maintaining bones alongwith calcium, the relatively high amount of magnesium in meat, fish etc. can be ascribed as a possible reason (Stanfield, 1992). Student t test (parametric test) at $P < 0.05$ reveal significant levels of Mg in hair and nails of smoking subjects of higher age groups which requires further investigation on similar subjects. Data reveals that no trend could be obtained in subjects taking liquor which is therefore not of primary significance. A large amount of magnesium present in the body in view of its requirement leads to significantly high levels of the magnesium in the controls studied in relation to hair applied with cosmetics (Belitz and Grosch, 1999).

Hair color was observed to be a factor of primary significance in the study with relation to magnesium concentrations. The results of the parametric test applied with respect to black hair color reveal that dark colored hair contained greater concentration of magnesium as also reported by Watanabe *et al.* (1992) and Schroeder and Nason (1969).

In our work we have also observed significant magnesium concentrations in blood pressure patients (MgH and MgN being 297.42±106.74 and 570.86±299.09 in hypotensive subject; MgH and MgN being 482.61±306.76 and 786.21±109.23 in hypertensive subjects). Thus more randomised clinical trails are required to be carefully performed to reach to a conclusion between magnesium levels in the body and the hypertension. An association between magnesium deficiency and ischemic heart disease has also been demonstrated in few epidemiological studies. Patients with ischemic heart disease have a higher incidence of magnesium deficiency as compared to controls (mean magnesium concentration in µg/g ± S.D. are 223.14±86.06 in hair and 432.80±129.99 in nails). The insignificant magnesium concentrations with respect to ischemic heart disease (mean 166.91 µg/g±36.08 in hair and 259.26 µg/g±86.59 in nails) obtained in our study are also in agreement to the reported studies (Nath, 2000). It can thus be suggested that magnesium deficient individuals be administered magnesium supplement routinely but after well designed trails.

Acknowledgements

The first author gratefully acknowledge the University Grants Commission for financial assistance. We also thank Dr. P.K. Seth, Director ITRC, Lucknow, Dr. K. Lal, Director NPL, New Delhi and Dr. H.N. Saiyed, Director and Dr. D.J. Parikh, Deputy Director NIOH, Ahmedabad for their invaluable cooperation in metal analysis of the samples.

Table 13.1: Variation of Mean Magnesium Concentration (μg/g±SD) in Hair (MgH) and Nails (MgN) in Different Age Groups

Subjects	Hair				Nail			
	21–30	31–40	41–50	51–60	21–30	31–40	41–50	51–60
Vegetarian	277.85±140.27 (n=45)	347.23±265.43 (n=45)	188.31±41.18 (n=45)	236.15±80.54 (n=45)	578.19±149.29 (n=45)	670.09±229.18 (n=45)	399.37±113.58 (n=45)	439.19±156.18 (n=45)
Non Vegetarian	302.25±51.73 (n=45)	301.55±128.24 (n=45)	222.07*±47.81 (n=45)	258.57±192.28 (n=45)	800.03*±430.62 (n=45)	816.78±432.16 (n=45)	452.26*±120.01 (n=45)	570.84*±319.99 (n=45)
Smokers	249.26±89.89 (n=45)	413.61*±231.14 (n=45)	360.84*±195.35 (n=45)	255.53±175.71 (n=45)	679.95±484.7 (n=45)	872.94*±351.29 (n=45)	788.4*±304.88 (n=45)	361.94±306.13 (n=45)
Non Smokers	258.91±133.68 (n=45)	217.55±41.96 (n=45)	171.97±30.51 (n=45)	235.97±88.23 (n=45)	803.37±338.95 (n=45)	699.36±308.27 (n=45)	666.66±231.41 (n=45)	570.06±76.81 (n=45)
Liquor Users	–	326.28±156.89 (n=45)	384.53*±272.09 (n=45)	183.09±19.53 (n=45)	–	817.17±339.14 (n=45)	771.16*±337.91 (n=45)	306.59±287.54 (n=45)
Liquor Non users	313.76±190.9 (n=45)	380.09±214.01 (n=45)	257.72±117.69 (n=45)	274.68±158.52 (n=45)	616.76±256.29 (n=45)	732.1±291.09 (n=45)	596.18±248.65 (n=45)	646.48±256.4 (n=45)
Dye Users	–	339.08±146.35 (n=45)	184.01±44.04 (n=45)	–	–	–	–	–
Henna Users	175.95±127.6 (n=45)	188.7±46.4 (n=45)	194.35±50.79 (n=45)	–	–	–	–	–
Controls (Non dye, Henna Users)	395.37±212.26 (n=45)	335.46±272.01 (n=45)	202.33± 24.01 (n=45)	229.19±75.3 (n=45)	–	–	–	–
Black Hair	360*±221.84 (n=45)	405.43*±227.54 (n=45)	226.05*±69.29 (n=45)	179.17±15.41 (n=45)	–	–	–	–
Brown Hair	254.11±130.53 (n=45)	295.78±86.42 (n=45)	201.37±72.39 (n=45)	269.86±160.51 (n=45)	–	–	–	–
Mixed Hair	–	402.14±213.82 (n=45)	244.67±87.12 (n=45)	–	–	–	–	–
White Hair	–	–	155.8±106.03 (n=45)	–	–	–	–	–

n: Number of samples; *: Values significant at $P < 0.05$.

References

Belitz, H.D. and Grosch, W. 1999. *Food Chemistry*, 2nd Edn. Springer Verlag Publishers, Germany, pp. 397.

Chatt, A. and Katz, S.A. 1988. The biological basis for trace element in hair. In: *Hair Analysis: Applications in Biomedical and Environmental Sciences*. VCH Publishers, New York, p. 28–42.

Considine, D.W. 1989. *Van Nostrand's Scientific Encyclopedia*, 7th Edn., Vol. 2. Van Nostrand Reinhold Publishers, New York, p. 1758–1760.

Gammelgaard, B., Peters, K. and Menne, T. 1991. Reference values for the nickel concentration in human fingernails. *J. Trace Elem. Electrol. Health Dis.*, 5: 121–123.

Georgescu, R., Pantelica, A., Salagean, M., Cracium, D., Constantinescu, O. and Frangopol, P.T. 1997. Instrumental neutron activation analysis of the hair of metallurgical workers. *J. Radioanal. Nucl. Chem.*, 224(1–2): 147–150.

Goetz, P.W. 1990. *The New Encyclopedia Britannica*, Vol. 7. Encyclopedia Britannica Inc., Chicago, pp. 678.

Hegarty, V. 1995. *Nutrition, Food and the Environment*. Eagan Press, USA, p. 265–266.

Jamall, I.S. and Jaffer, R.A. 1987. Elevated iron levels in hair from steel mill workers in Karachi, Pakistan. *Bull. Environ. Contam. Toxicol.*, 39: 608–614.

Mehra, R. and Juneja, M. 2004. Biological monitoring of lead and cadmium in human hair and nail and their correlations with biopsy materials, age and exposure. *Indian J. Biochem. Biophy.*, 41: 63–66.

Mehra, R. and Juneja, M. 2003. Occurrence of calcium in human hair and nails and some parametric influences, *Chem. Environ. Res.*, 12(1&2): 165–172.

Mehra, R. and Juneja, M. 2003. Adverse health effects in workers exposed to trace/toxic metals at workplace. *Indian J. Biochem. Biophy.*, 40: 131–135.

Nath, R. 2000. *Health and Disease: Role of Micronutrients and Trace Elements*. A.P.H. Publishing Corporation. New Delhi, p. 377–402.

Nowak, B. and Chmielmicka, J. 2000. Relationship of lead and cadmium to essential element in hair, teeth and nails of environmentally exposed people. *Ecotoxicol. Environ. Saf.*, 46: 265–274.

Schroeder, H.A. and Nason, A.P. 1969. Trace metals in human hair. *J. Invest. Dermatol.*, 53: 71–78.

Stanfield, P.S. 1992. *Nutrition and Diet Therapy: Self-Instructional Modules*, 2nd Edn. Jones and Bartlett Publishers, London, pp. 92.

Valkovic, V. 1988. *Human Hair, Vol. 2: Trace Element Levels*. CRC Press, Florida, pp. 1.

Vidya, C. 1996. *A Text Book of Nutrition*. Discovery Publishing House, New Delhi, p. 124–126.

Watanabe, H., Yashiba, Y., Kono, K., Watanabe, M., Inoue, S., Tanioka, Y., Dote, T., Orita, Y., Umebayashi, K. and Nagaie, H. 1992. Fluorine and other trace elements in hair by X-ray fluorescence analysis. *Fluoride*, 25(2): 55–65.

Wilhelm, M., Hafner, D., Lombeck, I. and Ohnesorge, F.K. 1991. Monitoring of cadmium, copper, lead and zinc status in young children using toenails: comparison with scalp hair. *Sci. Total Environ.*, 103: 199–207.

Chapter 14

Heavy Metal Pollution in River Tunga, Bhadra and Tungabhadra at Kudali Near Shimoga, Karnataka

*G. Suresha**, *K. Ramadas** *and E.T. Puttaiah***

**Department of Botany, Technical Resource Center, Sri Dharmasthala Manjunatheswara College, Ujire – 574 240, South Canara Dist., Karnataka, India*
***Department of P.G. Studies and Research in Environmental Science, Kuvempu University*

ABSTRACT

Heavy metal pollution is of serious concern in these days. In India the disposal of untreated wastes from, tanneries electroplating units, mine processing and smelters industries have become a major threat to the environmental quality. These wastes containing heavy metals such as chromium, cadmium, iron, aluminium, manganese, nickel cobalt, copper, molybdenum, zinc, lead etc. are discharged directly to the rivers. The rivers a major source of irrigation and domestic water supply are used as repositories for the disposal of industrial effluents along with sewage and other waste material.

Keywords: *Heavy metal pollution, River Tunga, Bhadra and Tungabhadra.*

Introduction

The heavy metals are non-degradable and create an environmental stress and often flow through the different trophic levels causing a deleterious effect on the living organism, Rao (1979).

A good number of industries are located in the vicinity of Shimoga and Bhadravathi, the two fast growing cities of Karnataka, which discharge their effluents into river Tunga and Bhadra respectively. These two rivers confluence at 'Kudali' village and form river Tungabhadra, which is 14 kilometers away from Shimoga and 16 kilometers from Bhadravathi, 'Kudali' refers to the confluence point. Hence the village is named as 'Kudali'.

The industries namely 'The Mysore paper mills', Vishweswaraya iron and steel industries', Electroplating and smelters are situated in Bhadravathi and electroplating, leather tanning, paper-recycling industries situated at Shimoga.

The effluents from these industries are discharged into river Bhadra and Tunga respectively. Further the domestic wastewater also adds to this.

A perusal of available literature has revealed that there is no much scientific work with respect to heavy metal contamination of Tungabhadra river. However a preliminary survey on heavy metal pollution of river Bhadra has been carried out by Parameshwara Naik *et al.* (2000).

Thus with this background it is proposed to study the concentration of heavy metals in river Tungabhadra at the confluence point of river tunga and Bhadra.

Materials and Methods

The present work was carried out for a period of six months that is from November 2002 to March 2003. Three sampling stations were identified and they include S_1 (Upstream of Tunga river 1 km away before confluence with river Bhadra), S_2 (Upstream of Bhadra river 1 km away before confluence with river Tunga) and S_3 (Down stream of river Tungabhadra 2 kms. away from confluence point) (Figure 14.1). From all these three sampling stations water samples were collected once in 15 days and brought to the laboratory for the analysis of physicochemical characteristics. Some parameters *viz.* Air temperature, water temperature, pH, dissolved oxygen and sampling for B.O.D. were carried on spot of collection.

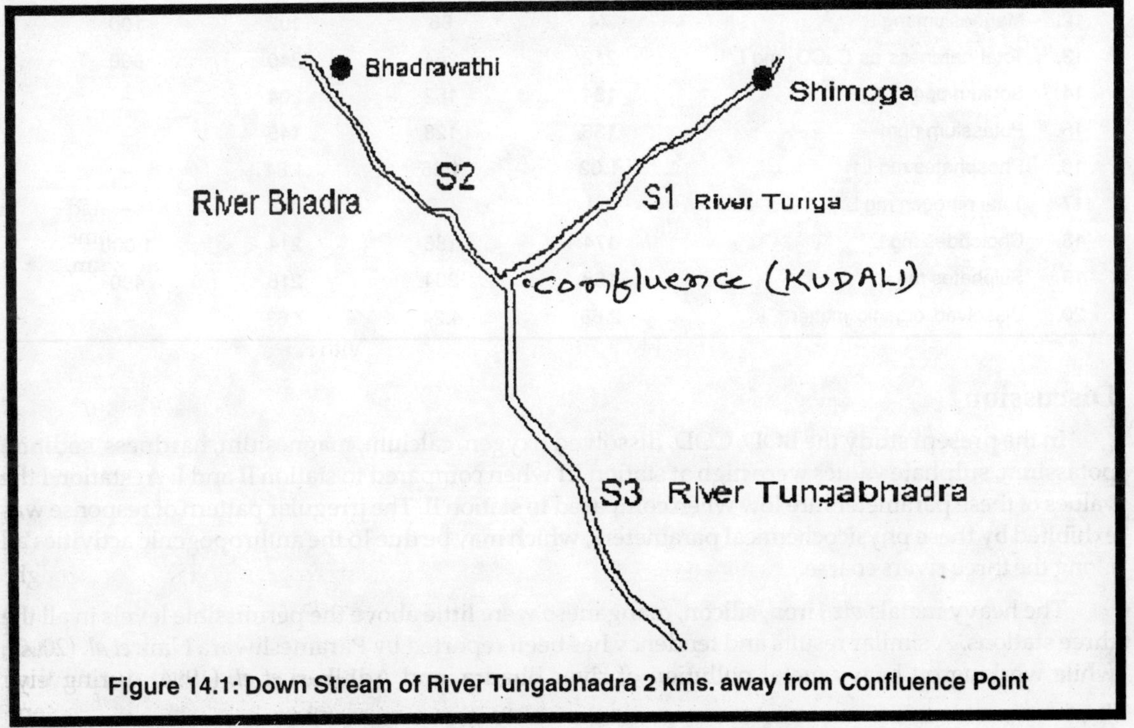

Figure 14.1: Down Stream of River Tungabhadra 2 kms. away from Confluence Point

Standard methods of APHA (1992), Saxena (1990) were followed for the analysis of physicochemical characteristics. Heavy metal composition was determined using atomic absorption spectrophotometer (Chemito).

Results

The values of physicochemical parameters and heavy metal composition of three sampling,stations with ISI tolerance limits are presented in Tables 14.1 and 14.2 respectively.

Table 14.1: Physicochemical Characteristics of Water Samples from Tunga, Bhadra and Tungabhadra Rivers

Sl.No.	Parameters	S_1	S_2	S_3	Tolerance Limits
1.	Air temperature °C	26	25.8	26.4	–
2.	Water temperature °C	24.2	24.6	25.2	–
3.	Colour	Colourless	Light Brown	Colourless	–
4.	Turbidity NTU	5.0	12.0	9.2	25
5.	Electrical conductivity µ mhos cm^{-1}	254.6	248.4	156.2	–
6.	pH	7.8	8.1	8.0	6.5–8.5
7.	Total solids mg L^{-1}	1284	1316	1524	–
8.	Dissolved oxygen mg L^{-1}	6.4	6.0	6.6	–
9.	Biochemical oxygen demand 5d 20°C	16.2	14.0	18.4	30.0
10.	Chemical oxygen demand mg L^{-1}	186.2	176.8	195.8	–
11.	Calcium as $CaCO_3$ mg L^{-1}	186	214	281	200
12.	Magnesium mg L^{-1}	74	86	102	100
13.	Total hardness as $CaCO_3$ mg L^{-1}	212	264	240	600
14.	Sodium ppm	184	162	204	–
15.	Potassium ppm	136	128	145	–
16.	Phosphates mg L^{-1}	1.02	1.66	1.84	–
17.	Total nitrogen mg L^{-1}	1.1	0.9	1.2	–
18.	Cholorides mg L^{-1}	174	186	214	1,000
19.	Sulphates mg L^{-1}	168	204	218	400
20.	Dissolved organic matter	2.68	4.24	7.63	–

Discussion

In the present study the BOD, COD, dissolved oxygen, calcium, magnesium, hardness, sodium, potassium, sulphate values were high at station III when compared to station II and I. At station I the values of these parameters are low when compared to station II. The irregular pattern of response was exhibited by these physicochemical parameters, which may be due to the anthropogenic activities all along the three rivers coarse.

The heavy metals *viz.*, iron, silicon, manganese were little above the permissible levels in all the three stations. A similar results and tendency has been reported by Parameshwara Naik *et al.* (2000), while working on heavy metal pollution of river Bhadra, and Adhikari *et al.* (1994) during their studies on possibilities of using Calcutta sewage effluents and sludge for irrigation, heavy metal contents in soil and water.

Table 14.2: Heavy Metal Composition in River Tunga, Bhadra and Tungabhadra.

Sl.No.	Parameters	S_1	S_2	S_3	Tolerance Limits
1.	Iron (Fe)	0.34	0.16	0.42	0.10
2.	Nickel (Ni)	0.01	0.014	0.028	0.50
3.	Manganese (Mn)	2.072	2.086	2.095	2.00
4.	Cobalt (Co)	0.006	0.009	0.011	0.20
5.	Lead (Pb)	0.008	0.007	0.009	0.5
6.	Molybdenum (Mo)	0.004	0.085	0.019	0.005
7.	Aluminium (Al)	0.005	0.004	0.007	1.0
8.	Copper (Cu)	0.002	0.002	0.001	0.2
9.	Silicon (As SiO_2)	0.74	1.14	0.86	–
10.	Chromium (Cr)	0.0052	0.0036	0.0048	0.05
11.	Zinc (Zn)	0.0088	0.088	0.0066	5.0

All the values are expressed in terms of ppm.

Apart from this the river Bhadra receives effluents from Vishweshwaraya Iron and Steel Industries Ltd., Mysore paper mills, Mysore Sugar mills, tanneries, smelters, electroplating units at Bhadravathi. River Tunga also receives wastes from smelters, garages, electroplating and domestic sewage. This might be the possibility of increase in the concentration of some heavy metals. In nut shell the, river Bhadra and Tungabhadra are polluted and river Tunga is comparatively less polluted.

References

APHA, AWWA and WPCF 1992. *Standard Methods for the Examination of Water and Wastewater.* APHA, Inc. New York.

Adhikari, S., Mitra, A., Gupta, S.K. and Banerjee, S.K. 1994. Possibilities of using Calcutta sewage effluents and sludge for irrigation and manorial purposes, Part I: Physicochemical characteristics and impact of heavy metal contents on soil, and water. *Proc. Ind. Nat. Acad.*, 60(6): 541–552.

ISI (Indian Standard Institution) 1985. *Standards for Discharge of Industrial Effluents*, IS 2490.

Panda, A.K., Mihtra, S., Muralidhar, J. and Sahoo, B.N. 1996. Distribution of heavy metals in a reservoir ecosystem of an iron ore mining area. *J. Ind. Pollution Control*, 12(2): 145–147.

Parameswara Naik, T. Suresha, G. and E.T. Puttaiah 2000. A Preliminary survey on Heavy metal pollution in river Bhadra near Bhadravathi town. *J. Eco. Env. and Cons.*, 6(4): 489–490.

Rao. 1979. *India's Water Wealth.* Orient Longman Ltd., New Delhi.

Sharma, B.K. and Kaur, H. 1995. *Environmental Chemistry.* Goel Publication House, Meerut.

Chapter 15

Chemical Impact on the Histological Studies on the Thyroid in the Freshwater Fish *Channa orientalis* (Sch.)

*S.V. Deshmukh** and *K.M. Kulkarni***

**Department of Biology, Rural Institute (Agriculture), Amravati – 444 603*
***Environmental Physiology Laboratory, P.G. Department of Zoology, Government Vidarbha Mahavidyalaya, Amravati – 444 604*

ABSTRACT

Thiourea treated thyroid gland showed sign of hyperactivity (increase in height of follicular eptithelium). Hypofunction (non vacuolated colloid) and drastic atrophic changes (fibrosis and cyst formation).

Keywords: *Thyroid gland, Thiourea, Hyperactivity, Hypofunction.*

Introduction

Thiourea is an antithyroid drug, known to inhibit the synthesis of thyroxine. Due to its effects, animals can accumulate iodide in normal manner, but the iodination of thyroxine in the gland is blocked and thyroxine can not be synthesized. Thiourea made it easy to establish the functional relationship of thyroid gland with the pituitary gland and goands. This goitrogen compound provides a mean of chemical thyroidectomy useful in fishes. Since the surgical removal of thyroid gland is not possible because of the diffused structure of the gland, the method of chemical thyroidectomy by administrating thiourea has been employed in this study.

In fishes, the thyroid follicles lie scattered along the blood vessels under pharynx. In chondrichthyean fishes, some teleosts and lungfishes, the follicles of thyroid are arranged into distinct glandular mass. In some teleosts thyroid follicles may disperse from pharyngeal region to brain, kidney, eyes, spleen etc. In lampreys and bony fishes, the thyroid follicles tend to be dispersed along the ventral aorta.

The hormone that derived from thyroid gland is referred as thyroid hormone. The function of thyroid gland is to elaborate, store, and discharge secretions. Its basic function is to concentrate iodine and synthesize the thyroid hormone. The thyroid has greater capacity to store its secretions. Thyroid gland is composed of cyst or oval like follicles of variable sizes. Each follicle is lined by single layer of cuboidal to low columnar epithelium. Cavity a follicle contains storage product of secretory epithelium called colloid. It is the only gland which stores its product outside the cells. When the gland is inactive there is a tendency for colloid to accumulate and for the epithelium to become low cuboidal to squamous or flattened. When the gland is overactive the colloid storage are depleted and the epithelium becomes columnar. The thyroid depleted and the epithelium becomes columnar. The thyroid follicles have a remarkable ability to trape inorganic iodine which can be stored and incorporated into hormones which are stored in the follicle cavity. But iodide concentrating activity of the thyroid gland is regulated by different control system.

TSH with preincubation of thyroid cell with iodide depressed, iodide concentration. On the other hand pretreatment with iodide depressed iodide pump activity as well as diminished the stimulatory effect of TSH added thereafter (Sherwin and Tong, 1974). According to Sinha and Singh (1990) extrathyroidal conversion of T4 to T3 increases during spawning phase as compared to prespawning phase in fresh water catfish *Clarias bastrachus* during its annual reproductive cycle. Melandar and Sundler (1972) suggested that in rat intrathyroid mast cells participate in the process of thyroid hormone secretion through effect of substrate released from these cells upon stimulation by TSH. The process of synthesis of thyroid hormones are same in fishes as in higher vertebrates. The thyroid hormones are thyroxine (tetraiodothyronine) and triiodothyronine. The follicular cells synthesize a protein thyroglobulin which is present in colloid. Thyrogobulin is an important as subsequent synthesis of thyroid hormone. Valenta *et al.* (1982) suggested that intracellular thyroglobulin iodination stimulates rat thyroids. In response to endocrine stimulation, follicular cells engulf the colloid by phagocytosis. The colloid within the endocytotic vesicle is enzymatically degraded to yield thyroid hormones which are released from follicular cells into the extracellular space where they enter the abundant capillaries.

The first demonstration of thyroxine synthesis by the teleostean thyroid gland was in platyfish *Xiphophorus maculatus* (Berg and Gorbman, 1953) and the rate of turnover of 131 I–labelled thyroxine and triiodothyronine in adult frogs and toads is temperature dependent (Dowling *et al.*, 1964). The thyroid function is regulated by thyrotrophic hormones, which are released by teleostean pituitary. The increase in quantity and activity of follicle is only due to continuous stimulation of thyroid gland by TSH. This result in colloidal content and cells became columnar in shape and increased in size. Where as in absense of TSH synthesis of thyroid hormone is minimum as a result the follicular cells become, flattened and lumen remained enlarged and full of colloid.

Many effect of thyroid hormone have been described in teleosts however its exact physiological role is still not fully known. Thyroxine had also been claimed to increase common excretion by goldfish (Hoar, 1958). Teatments with thyroxine leads to silvering of the skin in many teleosts (Pickford and Atz, 1957). According to various workers thyroxine stimulate the growth in many teleosts. Thyroxine also stimulates the formation of bone in various teleosts and sturgeon (Gorbman. 1969). It plays an important role in regulation of annual, gonadal and body weight cycle of birds (Hohn, 1961; Thapliyal,

1969, 1978, 1980, 1981; Assenmacher, 1973; Assenmacher and Hallageans. 1980; Assenmacher and Jallageas, 1980; Thapliyal *et al.*, 1982, 1983; Lal and Thapliyal, 1984). Lal and Thapliyal (1984) suggested that in migratory Redheaded bunting *Emberiza bruniceps*, certain levels of thyroid hormones are required for full growth and coupling of annual gonadal and body weight cycles. Thyroid hormones also plays an important role in photorefractory state of the migratory redheaded Bunting *Emberiza brunicep*.

From the above evidences it can be concluded that in fishes thyroid plays an important role during osmoregulation, accelerating metabolic rate, during formation of bone, silvering of skin, during full growth. However its exact role in a fish *Channa orientalis* is still not known. Certain drugs inhibit thyroid function by anatagonizing formation of thyroid hormones. But De Escober and Del Rey; (1962) stated that thiourea do not depress the peripheral deiodination of thyroid hormones in rat. According to Das and Isichei (1988) hypothyroidism cause structural changes in the thyroid cell membrane which would interfere with the trapping of iodine and shedding of thyroxine in the hypothyroid patients in human beings. The thyroid follicles have remarkable ability to trap inorganic iodine, which can be stored and incorporated into hormones, which stores into follicular cavity. But the antithyrid drug thiourea causes inhibition of iodine incorporation into tyrosine. But the action of thiourea on histology of thyroid gland are not still recorded. Therefore an attempt has been made to study the effect of antithyroid drug thiourea on the morphology and histology of the thyroid gland.

An attempt therefore has been made to clarify this point further by subjecting the fish *Channa orientalis* (Present study) to thyroidectomy. But here the surgical thyroidectomy is not possible because of the diffused structure of the gland. So chemical thyroidectomy had been tried with the help of antithyroid drug thiourea.

Materials and Methods

The species which has been selected for the present work is of economic value and readily available throughout the year in market, in the rivers and water bodies, nearby Amravati region. Locally the fish is known as "DOK" and is a common edible fish in the region. The fishes (*Channa orientalis*) were acclimatized to laboratory conditions for two weeks according to APHA/AWWA/WPCF (1975) standard methods.

In the present experiments the animals were treated with high and low concentration of thiourea solution.

Two hundred and twenty four young, matured acclimatized animals were selected having similar sizes and weights. The animals were segregated into three groups.

Group I (Total 112 fishes)

As control group which is further subdivided into group IA and group IB having 56 animals in each group. Fishes received injections of fish saline (0.7 per cent).

Group II

As high concentration group for thiourea, in which fishes received intramuscular injections of 3 per cent thiourea (dose–0.3 mg/gm of body weight) at alternate day (total 56 fishes). Thiourea was prepared in 0.7 per cent NaCl. Solution.

Group III

As low concentration group for thiourea in which fishes received intramuscular injections of 2 per cent thiourea (dose–0.2 mg/gm of body weight) at alternate day (total 56 fishes).

Group IA used an control for group II and group IB is used as control for group III.

For studying histological changes in glands, fishes from groups–I, II, III were sacrified by decapitation at the interval of 24, 48, 72, 96, 168, 360 and 720 hours. At sacrifice lower jaws with thyroid were decalcified in 6 per cent solution of formic acid in 5 per cent formaldehyde for about a week. The tissues were dehydrated by passing through various grades of alcohol, cleared in xylene and a embedded in peraffin wax (56–58 c). Blocks were prepared and sectioned at 4– 6 μ and stained with following techniques.

Lower jaw with thyroid were stained with Ehrlich haematoxylin-eosin technique.

Assessment of Thyroid Activity

Assessment of thyroid activity was measured by microscopic examination of transverse sections (5–6 μ) of lower jaws in both control and experimental groups. This was supplemented by measurements in the following parameters.

1. Average number of follicles in a section (5 to 6 section/ fish).
2. Average follicular cell height was calculated by measuring 5 to 6 cells per follicles at random in each of the 4 to 5 follicles per section. Thus measurements were carried out in 5 sections per animal. This average epithelial cells height were calculated of animal's thyroid gland.
3. Average size of the follicles were estimated by measuring diameter of 5 to 6 follicles per section and 5 sections per animal. And thus the mean diameter of follicles was found out of the animal.
4. Average number of cells per follicles were counted random follicle per section. This procedure was followed for 5 follicles per section and 5–6 slides per animal tissue, yielded the average number of cells per follicle.

The data obtained after various measurements were tested for statistical significance. The experimental values were statistically evaluated by student's "t" formula.

Results and Observations

Thiourea Treated Fish Thyroid

Histology of Thyroid Exposed to High (0.3 mg/gm of body weight) Dose Exposure (Table 15.1; Figure 15.1)

Thyroid exposed to thiourea exhibited marked histological changes. The follicles were increased in number from 28.412 (Control) to 144.252 at 360–720 hours exposure (Table 16.1). Follicles 80. became irregular in shape. The diameter of follicles observed was 59.662 μ in control group, but decreased to 32.814 μ, 29.655 μ, 20.502 μ, 19.488 μ at the interval of 24–48 h, 72–96 h, 168–360 h, 360–720 h exposure respectively. The height of follicles increased from 3.970 μ in control group to 4.931 μ (24–48 h), 5.104 μ (72–96h) 5.933 μ (168–360 h) and 6.155 μ (360–720 h). But the number of cells per follicles showed a significant reduction. Percent changed values are noted in Table 15.1 of the cells per follicles were 36.201, –44.822, –64.942, –68.133 after 24–48, 72–96, 168–360 and 360–720 hours of exposure respectively. In some follicles cells lining it multiplies and projected into the follicular cavity. Nuclei became pycnotic. In some follicles loss of colloidal material was observed. Vacuolation in the cytoplasm of secretory cells was also noticed. The follicles were found closely packed and widely distributed in the lower jaw region.

Table 15.1: Effect of Thiourea Treatment on the Thyroid Gland (Histological Parameters) of the Fish *Channa orientalis* (High Dose –0.3 mg/gm of Body Weight)

Experimental Period (h)	*Follicles Per Section*	*Cell Height (μ)*	*Follicular Diameter*	*Cells per Follicles*
Control	28.412 ± 2.706	3.970 ± 0.252	59.662 ± 3.166	34.814 ± 3.496
24–48	59.022*** ± 5.702	4.931* ± 0.301	32.814** ± 2.011	22.210** ± 1.922
	(107.742)	(24.182)	(–45.922)	(–36.201)
72–96	82.531 NS ± 5.530	5.104*** ± 0.270	29.655 NS ± 1.708	19.221*** ± 1.923
	(109.592)	(28.462)	(–50.302)	(–44.822)
168–360	119.302*** ± 5.622	5.933* ± 0.330	20.502* ± 1.877	12.210* ± 1.464
	(320.077)	(–49.377)	(–65.633)	(–64.942)
360–720	144.252 NS ± 5.774	6.155*** ± 0.387	19.488 NS ± 1.513	11.099 NS ± 1.416
	(407.920)	(54.911)	(–67.344)	(–68.133)

P values: *: < 0.1, **: < 0.01; ***: < 0.001.

NS: Non Significant, ( ): Parenthesis figures are percentage change.

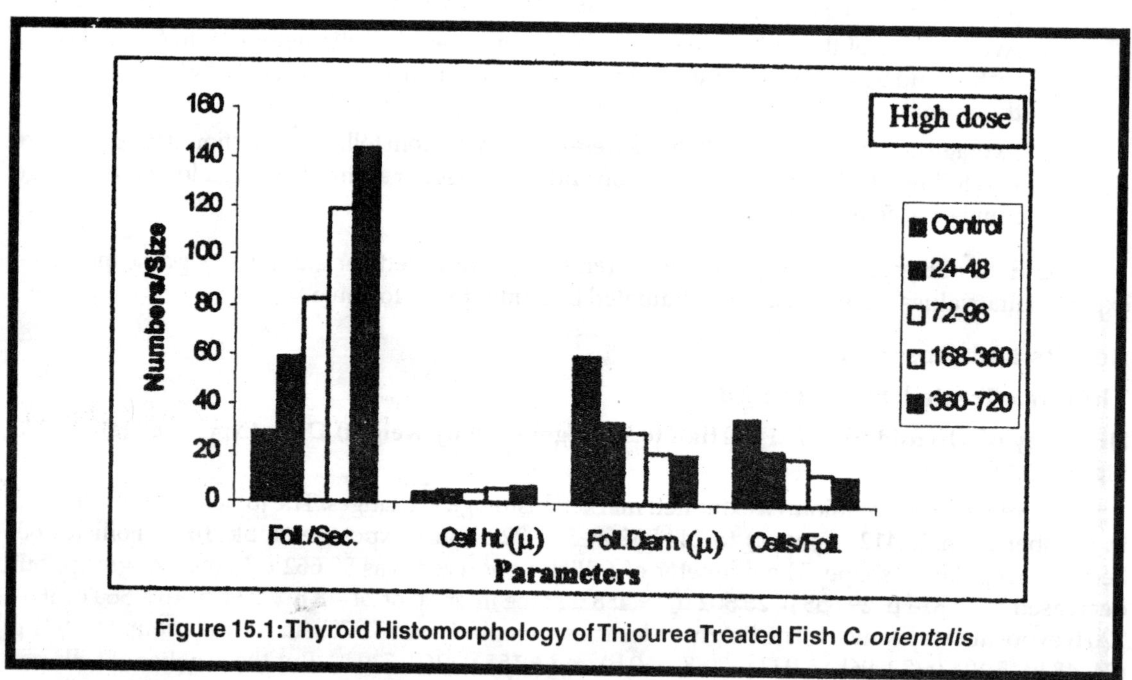

Figure 15.1: Thyroid Histomorphology of Thiourea Treated Fish *C. orientalis*

Histology of Thyroid Exposed to Low (0.2 mg/gm of body weight) Dose Exposure (Table 15.2; Figure 15.2)

Some irregular shaped follicles were observed. Follicle showed increase gradually in number from 28.403 in control to 54.940 (24–48 h); 74.850 (72–96h); 81.460 (168–360 h); 103.954 (360–720). The follicular diameter was reduced from 59.613 μ (control) to 21.402 μ upto 360–720 hours of exposure. The secretory cells increased in height from 4.003 μ in control group to 4.233 μ (24–48 h);

4.872 μ (72.96 h); 5.22 μ (168–360) h; 5.955 μ (360–720 h) in the experimental group. The cells per follicle decreased significantly from 34.809 in control group to 19.387 upto 360 to 720 hours of exposure (Table 15.2). Follicular cells showed hypertrophy. Colloid losses from some of the follicles. In some follicular cells vaculoes were observed in the cytoplasm with pycnotic nuclei. In some follicles follicular epithelial cells invaded into follicular cavity. Follicles were distributed into the lower jaw. Loss of colloidal material from follicular cavity was observed. Vascularity of the thyroid gland in general increases. Blood cells were also observed in the follicles. Thus thyroid gland shows hypertrophy, hyperplasia and some what hypermia, after thiourea treatment.

Table 15.2: Effect of Thiourea Treatment on the Thyroid Gland (Histological Parameters) of the Fish *C. orientalis* (low dose –0.2 mg/gm of body weight)

Experimental Period (h)	*Follicles Per Section*	*Cell Height (μ)*	*Follicular Diameter*	*Cells per Follicles*
Control	28.403 ± 2.724	4.003 ± 0.264	59.631 ± 3.152	34.809 ± 3.485
24–48	54.940 NS ± 4.142	4.233** ± 0.236	46.711 NS ± 2.244	25.812* ± 2.382
	(93.430)	(5.495)	(–21.666)	(–25.846)
72–96	74.850*** ± 4.632	4.872** ± 0.221	36.582 NS ± 2.233	24.402** ± 2.071
	(163.528)	(21.708)	(–38.652)	(–29.897)
168–360	87.460** ± 5.188	5.224** ± 0.284	26.701** ± 2.202	21.354** ± 2.144
	(207.925)	(–30.502)	(–55.207)	(–38.653)
360–720	103.954 NS ± 6.200	5.955 ± 0.286	21.402** ± 2.192	19.387*** ± 2.102
	(265.996)	(48.763)	(–64.109)	(–44.304)

P values: *: < 0.1, **: < 0.01; ***: < 0.001.

NS: Non Significant, ( ): Parenthesis figures are percentage change.

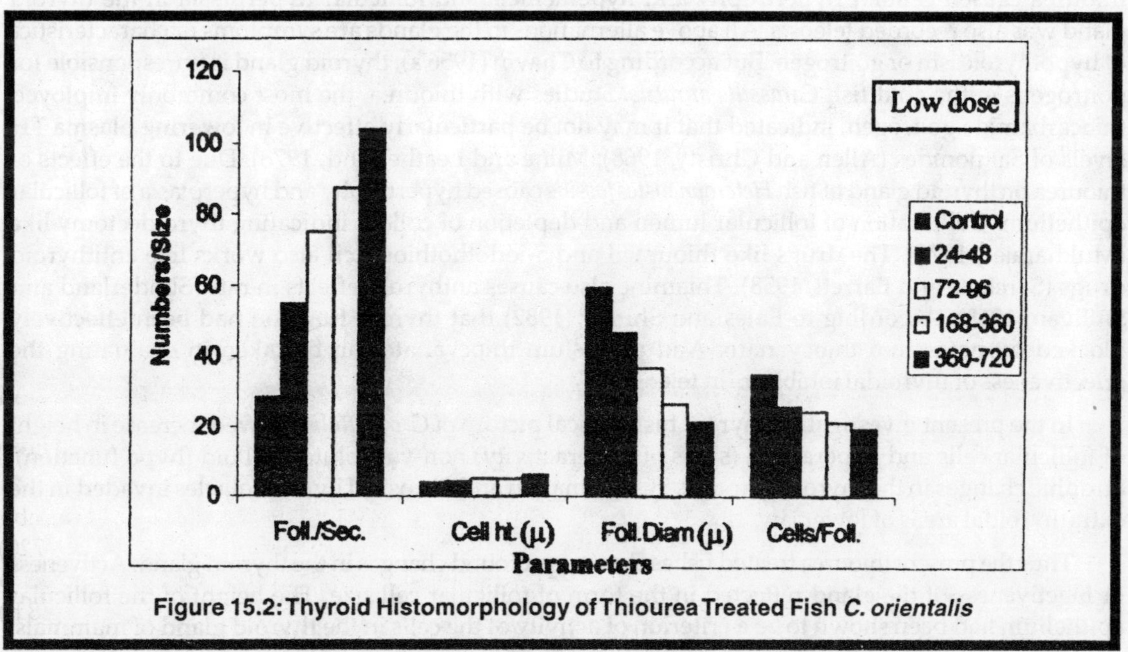

Figure 15.2: Thyroid Histomorphology of Thiourea Treated Fish *C. orientalis*

Discussion

In pisces including teleosts, thyroid, gland is situated in the lower jaw. In most teleost thyroid gland is not in an encapsulated form but it had diffused structure, including *Channa orientalis*. It is not always compact and encapsulated as the follicles are also found in the kidney, spleen, even in the skin (Baker, 1959); Srivastava and Sathyanesan, 1969). So its surgical removal is not possible. The main function of thyroid gland is to produce thyroid hormone by accumulating iodine and its union with tyrosine. Secretion of follicular cells lining the follicles which are secretory in function. But Malander and Sundler (1972) suggested that, in rat, intrathyroidal mast cells participate in the process of thyroid hormone secretion.

Thyroid hormones have different functions and are varied from species to species, such as in human beings it accelerates the basal metabolic rate, in amphibians, it plays important pole during reproduction, fall of growth in birds. According to Darrs and Kuhn (1982), the hypophysis and the thyroid gland of *Ambystome mexicanum* may release optimal amount of hormones necessary for metamorphosis following proper, stimulation but the TRH can not functions, as releasing hormone in this respect. But exact role in pieces is still not clear especially in teleosts. To study the exact, function of thyroid hormone in teleosts, the present investigation had been made.

Thyroid gland regulates the organs, organ systems and metabolic processes any disturbance in the secretion of thyroid gland influences the functioning of organ, organ system as well as metabolic processes.

The thiourea had been used as an antithyroid drug (Pitt. Rivers, 1950) to achieve "Chemical thyroidectomy" and its effects on the thyroid gland of teleosts have been studied by s several workers (Olivereau, 1954; Prasada Rao, 1969). Many workers done their work upon the effect of antithyroid drug thiourea upon thryroid gland of teleosts is reviewed by Pickford and Atz. (1967); Kinnear (1960); Belsare (1965); Rao (1969); Oehadrai (1971). According to all above investigations antithyroid drug *i.e.* thiourea caused cellular hypertrophy and hyperaemeia and follicular hyperplasia in the thyroid gland was also recorded teleosts. All above alternations in the glands are symptoms or characteristics of hypothyroidism or goitrogen. But according to Chavin (1956 a), thyroid gland is unresponsible for goitrogens in the gold fish *Carassills auratus*. Studies with thiourea, the most commonly imployed thiocarbamide goitrogen, indicated that it may not be particularly effective in lowering plasma TH levels of Salmonides (Allen and Christy, 1968); Milne and Leatherland, 1978). Due to the effects of thiourea on thyroid gland of fish *Heteropneustes fossilis* caused hypertrophy and hyperplasia of follicular epithelium. Obliteration of follicular lumen and depletion of colloid indicating thyroidectomy like (Mukharajee, 1975). The drugs like thiouracil and 5-iodothothiouracil also works like antithyroid drugs (Sarclone and Barrett; 1958). Thiamine also causes anthyroid effects in rats (Slinderland and Sullivan; 1968). According to Eales and Shirley (1982) that thyroid function had been effectively blocked by potassium thiocyanate. And potassium thiocyanate can be taken in evaluating the effectiveness of thyroidal inhibitors in teleosts.

In the present investigation thyroid histological picture of *C. orientalis* showed increase in height of follicular cells and hyperaemia (signs of hyperactivity) non-vacuoluted colloid (hypo function), atrophic changes in the thyroid gland *i.e.* cyst formation and fibrosis. Thyroid follicles invaded in the extrathyroidal areas of lower jaw.

Thus the present thiourea treated fishes showed profound changes in the thyroid gland. Activeness or inactiveness of the gland reflected in the form of follicular cell size. The height of the follicular epithelium had been shown to be a criterion of activity of the cells in the thyroid gland of mammals, reptiles, amphibians and fishes (Uotila, 1939; Maity, *et al.*, 1973; Uhlenhuth, *et. al.*, 1945; Stolk, 1951). In Indian fruit bat *Rousettus* after antithyroid drug propylthiouracil treatment causes the increase in height of follicular cells. Lumen of the follicles contained less colloid and numerous vacuoles at the peripheral region of colloid which indicated hyperactive stage of thyroid (Shrikhande, 1983). Similar histological changes in the thyroid were observed following an antithyroid drug propylthiouracil administration (Gorbman and Bern, 1974; and Turner and Bagnara, 1976). The active thyroid gland showed tall columnar epithelium, while in the inactive thyroid gland with reduced function showed squamous type of epithelium of follicular cells (Gorbman and Bern, 1962). The follicular epithelium of *Heteropneustes fossilis* is squamous but after treatment with thiourea the height of the epithelium increased (Mukherjee; 1975).

Other than the structure of follicular cells, the condition of colloid o follicles is also most important. Appearance of vacuoles in the colloid has been shown to be a sign of the thyroid activity and a high rate of absorption of the colloid from According to Pickford and Atz (1957) the presence of numerous vacuoles in the colloid expresses high activity and their reduction or absence indicates proportional or total inactivation of the gland. Thiourea treated *Heteropneustes fossilis* showed numerous vacuoles in the colloid of the thyroid gland which due to quick absorption of thyroid hormone from the colloid of the follicles. Many follicles became acolloida, their nuclear size reduced and thus gland become atrophied in the fish *Heteropneustes fossilis* (Mukherjee, 1975). Atrophy of thyroid follicles after treatment with thiourea was also reported in *Phoxinus phoxinus* (Barrington and Matty, 1955). Thus the size and number of vacuoles in the colloid has been found increase in correlation with that of the follicular epithelial height (Mukherjee, 1975). Thiourea treated fish showed non-vacuolated colloid suggests reduced thyroid function.

References

Allen, D.M. and M. Christy (1978). Thiourea does not block visual pigment responses to prolactin in trout. *Vision Res.*, 18: 859–860.

APHA, AWWA and WPCF (1975). *Methods for Examination of Water and Wastewater*, 14th ed. Am. Publ. Hlth. Asso. Washington.

Assenmacher, I. (1973). In: *Avian Biology*, Vol. 3, (Eds.) D.S. Farner and J.K. King. Academic Press, New York, pp. 183.

Assenmacher, I. and M. Hallageas (1980). In: *Environmental Endocrinology*, (Eds.) Assenmacher and D.S. Farner. Springer Verlag, Berlin, pp. 52.

Assenmacher, I. and M. Jallageas (1980). Adaptive aspects of endocrine regulation in birds. In: *Hormones, Adaption and Evolution*, (Eds.) S. Ishii, T. Hirano, and M. Wada. Springer Verlag, Berlin/New York, p. 93–102.

Barrington, E.J.W. and A.J. Matty (1955). The identification of thyrotrophin secreting cells in the pituitary gland of the minnow (*Phoxinus phoxinus*). *Quart. J. Micr. Sci.*, 96: 193–201.

Chavin, W. (1956a). Thyroid distribution and function in the goldfish, *Carassius auratus* L., as determined by the uptake of tracer doses of radioiodine. *Anat. Rec.*, 124: 272.

Darras, V.M. and E.R. Kuhn (1982). Effects of TRH, Bovine TSH and Pituitary Extracts on thyroidal T4 Release in *Ambystoma mexicanum*. *Gen. Comp. Endocrinol.*, 51: 286–291.

Das, S.C. and U.P. Isichei (1988). Hypothyroidism and its Effects on thyroid issue and plasma lipids. *Indian J. Exp. Biol.*, 26: 92–94.

De Escobar, G.M. and F. Escobar Del Rey (1962). Influence of Thiourea, potassium perchlorate and thiocyanate and graded doses of prophylthiouracil on thyroid hormones metablism in thyroidectomized Rats, Isotoically, Equillibrated with varying Doses of Exogenous Hormone. *Endocrinology*, 71: 906–913.

Dowling, J.T., D. Razevska and D.J. Goodner (1964). Metabolism of thyroid hormone in Frogs and Toads. *Endocrinology*, 75: 157.

Eales, J.G. and S. Shirley (1983). Influence of potassium thiocyanate on thyroid function of Rainbow trout *Salmo qairdneri. Gen. Comp. Endocrinol.*, 51: 39–43.

Gorbman, A. (1969). Thyroid Function and its control in Fishes. In: *Fish Physiology*, Vol. II: The Endocrine System, (Eds.) W.S. Hoar and D.J. Randall. Academic Press, New York and London, p. 241–274.

Gorbman, A. and H.A. Bern (1974). *Textbook of Comparative Endocrinology*. Wiley Eastern Private Ltd., pp. 114.

Gorbaman, A. and H.A. Bern (1962). *A Test Book of Comparative Endocrinology*. John Wiley and Sons Inc., New York, pp. 468.

Hoar, W.S. (1958). Effect of Synthetic thyroxine and gonadal steroids in the metabolism of goldfish. *Canad. J. Zool.*, 36: 113–121.

Hohn, E.O. (1961). Endocrine glands, thymous and pineal body. In: *Biology and Comparative physiology of Birds*, Vol. 2, (Ed.) A.J. Marshall. Academic Press, New York, pp. 87.

Lal, P. and J.P. Thapliyal (1984). Photorefractoryness in the migratory red headed bunting. *Proc. 1st Intern. Symp. Environ. and Horm*, (Eds.) Ischii, B.K. Follett and A. Chandola. Springer Verlag, Berlin (In press).

Melandar, A. and F. Sundler (1972). Significance of thyroid mast cells in thyroid Hormone secretion. *Endocrinology*, 90: 802–807.

Milne, R.S. and J.F. Leatharland (1978). Effect of ovine TSH Thiourea, ovine prolactin and bovine growth hormone on plasma thyroxine and triiodothyronine levels in rainbow trout (*Salmo gairdneri*). *J. Comp. Physiol.*, 124: 105–110.

Mukheriee, A. (1975). Effects of thiourea treatment on thyroid and ovary of the catfish *Heteropneustes fossilis* (Bloch.). *Indian. J. Exp. Biol.*, 13: 327–332.

Pickford, G.E. and J.W. Atz. (1957). *The Physiology of the Pituitary Gland of Fishes*. Zoological Society, New York, pp. 613.

Sherwin, J.R. and W. Tong (1974). The actions of iodide and TSH on thyroid cells showing duel control system for the iodide pump. *Endocrinology*, 94: 1465.

Shrikhande, D.Y. (1983). An experimental analysis of certain pituitary–adrenal gonadal relationship in Indian fruit bat *Rousettus lesshenaulti. Ph.D. Thesis*. Nagpur University, Nagpur.

Sarclone, E.J. and H.W. Barret (1958). The antithyroid activity of 5-iodo-2-thiouracil. *Endocrinology*, 63: 143–150.

Sinha, N. and T.P. Singh (1990). Extrathyroidal conversion of T4 to T3 in a freshwater catfish, *Clarias batrachus* during prespawning and spawning phases of its annual reproduction cycle. *India J. Exp. Biol.*, 28: 680–682.

Slingerland, D.W. and J.J. Sullivan (1968). An antithyroid Effect thiamine. *Endocrinology*, 82: 895–897.

Thapliyal, J.P. (1969). Thyroid in avian reproduction. *Gen. Comp. Endocrinol, Suppl.*, 2: 111–122.

Thapliyal, J.P. (1978). Reproduction in Indian birds. *Pavo*, 16: 151–161.

Thapliyal, J.P. (1980). Thyroid in reptiles and birds. In: *Hormones, Adoption and Evolution* (Eds.) S. Ishii, T. Hirone and M. Wada. Japan Sci. Soc. Press Tokyo, Springer Varlag, Berlin, pp. 241–250.

Thapliyal, J.P. (1981). Presidential address, endocrinology of avian reproduction. In: *Proceedings, 68th Session of Indian Sci. Congr., Sec. Zoo. Entomol. and Fish.*, p. 1–11.

Thapliyal, J.P., A.K. Pati, V.K. Singh and P. Lal (1982). *Gen. Comp. Endocrinol.*, 46: 325.

Turner, C.D. and Bagnara, J.T. (1976). *General Endocrinology*, 6th Edition. W.B. Sanders Company, Philadelphia.

Velenta, S.J., W.C. Florsheirn and B.S. Sharma (1982). Acute Effects of iodine on the stimulated rat thyroid. *Endocrinology*, 111: 1721–1727.

Chapter 16
Thrombocytopenic Effect of Buprenorphine in Mice

Dhriti Banerjee and Nirmal Kumar Sarkar

*Postgraduate Department of Zoology, Presidency College, Kolkata – 700 073
West Bengal, India*

ABSTRACT

Chronic administration of the synthetic opioid analgesic, buprenorphine was found to considerably decrease the count of platelets of blood in laboratory mice following 8 weeks but not earlier durations of treatment, which indicated a slow and moderate thrombocytopenic effect of the drug, somewhat comparable to a similar kind of effect of some other opioids, reported in case of human subjects.

Keywords: *Buprenorphine, Platelets, Thrombocytopenic action.*

Introduction

Thrombocytopenia is a haematological disorder characterised by a marked decrease of the number of platelets in blood, which may lead to a delay in the clotting of blood and a concomitant increase of the bleeding time following injuries (Firkin *et al.*, 1989). A few earlier reports indicate that opioid drugs like morphine and diacetylmorphine may occasionally produce a thrombocytopenic effect in human subjects. A case-report described a sudden development of thrombocytopenia in a 23 years old woman who had been treated with morphine for a week after an adrenalectomy at the Hermann Hospital, Houseton, USA; the disorder was cured after two weeks of withdrawal of morphine treatment (Cimo *et al.*, 1982). Five young, male intravenous abusers of diacetylmorphine, unacquainted with one another but presenting with a similar pattern of bleeding disorder, were admitted to the Boston City Hospital, Massachusetts, USA, during a span of four months; all of them were found to be thrombocytopenic upon examination of their blood samples (Adams *et al.*, 1978). However, whether the synthetic opioid, buprenorphine, which was originally prepared to be used as an analgesic drug but was later known to be abused by some of the drug addicts of a few countries (Singh *et al.*, 1992; Robinson *et al.*, 1993; Nigam *et al.*, 1994), has any thrombocytopenic effect or not remains unexplored. An earlier report described that a mild anaemia developed in some human volunteers following daily injections of buprenorphine for a period of five weeks (Jasinski *et al.*, 1978). The present authors had earlier observed a marked leucopenia in mice treated with buprenorphine for a period of two months (Banedee and Sarkar, 1997). However, the count of platelets of blood was not determined in either of the aforesaid studies. The present study aims at exploration of the issue in laboratory mice following prolonged buprenorphine treatment. The study has also been extended to the post-treated animals to ascertain if the drug-induced changes could revert to normal over time, upon withdrawal of the drug.

Materials and Methods

Healthy and adult (8 weeks old), male, albino mice of the Swiss strain were taken for the study. Only male animals were taken in analogy with the fact that human drug abusers are generally males and seldom females. The mice were daily given an intraperitoneal injection of buprenorphine hydrochloride (Unichem Laboratories Ltd., Mumbai; batch no. G 52436) at a dose of 300 µg/Kg body weight for eight consecutive weeks. The dose was followed after that used by some other workers in their behavioural study in buprenorphine-treated rats (Macenski *et al.*, 1994). The control mice were given simple saline (0.9 gm per cent) injections. In both experimental and control animals, injections were given during 10.00 to 11.00 a.m. and the volume of the injected fluid was limited to 7.5 ml/Kg body weight. All animals were maintained without any injection after those eight weeks.

The experimental mice were anaesthetised under ether vapour and blood samples were collected by cardiac puncture following 2, 4, 6 and 8 weeks of buprenorphine treatment and also after 2 and 4 weeks of drug withdrawal. Blood platelet count was determined using a haemocytometer following the sodium citrate method (Dasgupta, 1977). Besides, blood smears were drawn, air-dried and stained with 10 per cent Giemsa's stain buffered to pH 7.0 (Dacie and Lewis, 1984) for microscopic examination of platelet-clumps. Whenever the treated or the post-treated mice were used for study, the corresponding control animals were also sacrificed side by side.

Results

No significant difference was noted after 2, 4 and 6 weeks of buprenorphine administration between the counts of blood platelets of the treated and the corresponding control groups of mice. However, the count was found to be considerably decreased below the corresponding control value after 8 weeks of treatment. The counts noted in the post-treated mice after either 2 or 4 weeks of withdrawal of the drug did not differ significantly from that recorded in the corresponding control groups. Platelet counts recorded in various treated, post-treated and control groups of mice are shown in Table 16.1.

A microscopic examination of blood smears revealed that small clumps of platelets were scattered amidst other blood cells in all treated, post-treated and control groups of mice. However, some additional and distinctive features were observed only in that particular group of mice which received buprenorphine treatment for 8 weeks. In this group, the smears occasionally showed a rosette-like adherence of red blood cells around a neutrophil and also, the so-called phenomenon of 'platelet satellitism' (Williams *et al.*, 1986) or adherence ofplatelets to a neutrophil. Both red blood cells and platelets were occasionally found to adhere to the same neutrophil (Figure 16.1).

Table 16.1: Blood Platelet Counts in Buprenorphine-treated, Post-treated and Corresponding Control Groups of Mice (five mice per group) (all values represent mean ± S.D.)

Group of Mice Under Study	*Platelet Count (million/c.mm of blood)*
Treated (2 weeks)	1.23 ± 0.08
Corresponding controls	1.16 ± 0.07
Treated (4 weeks)	1.31 ± 0.09
Corresponding controls	1.34 ± 0.11
Treated (6 weeks)	1.17 ± 0.06
Corresponding controls	1.24 ± 0.08
Treated (8 weeks)	0.85 ± 0.05
Corresponding controls	1.27 ± 0.09
Post-treated (2 weeks after withdrawal)	1.21 ± 0.06
Corresponding controls	1.26 ± 0.06
Post-treated (4 weeks after withdrawal)	1.30 ± 0.09
Corresponding controls	1.22 ± 0.06

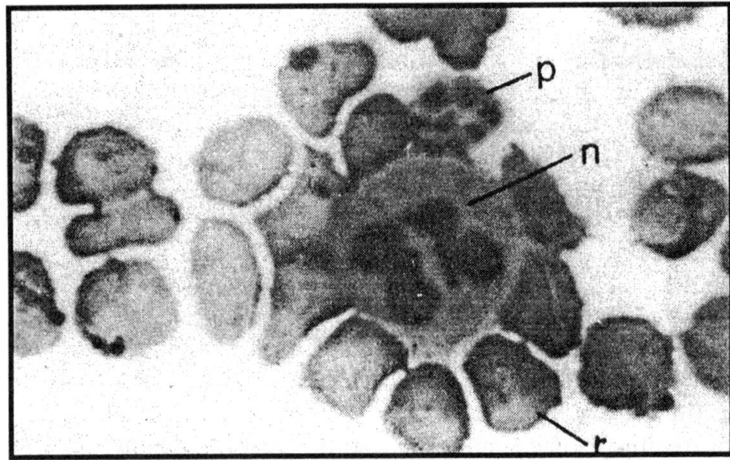

Figure 16.1: Platelet Satellitism (p) and Adherence of Red Blood Cells(r) to a Nuetrophil(n) in a Blood Smear of a Mouse Treated for 8 Weeks with Buprenorphine

Discussion

The present study reveals that buprenorphine may exert a thrombocytopenic effect in mice only at an advanced stage (8 weeks) but not at earlier stages (2, 4 or 6 weeks) of treatment. However, the drug effect was not only slow in onset but also moderate in nature, as evidenced from the observed platelet counts. The finding of platelet satellitism provides an additional support to our inference on a thrombocytopenic action of buprenorphine, because platelet satellitism has occasionally been observed in human patients suffering from thrombocytopenia (Dacie and Lewis, 1984; Williams *et al.*, 1986).

In essence, the present study indicates that buprenorphine shares a thrombocytopenic action with such other opioids as morphine and diacetylmorphine (Adams *et al.*, 1978; Cimo *et al.*, 1982). The opioids might have triggered the formation of antiplatelet autoantibodies, which in turn caused destruction of a considerable number of platelets in the circulating blood (*idem*). Finally, it may be suggested that the platelet counts of blood of human abusers of buprenorphine should be periodically monitored and wherever necessary, appropriate clinical measures are to be followed.

References

Adams, W.H., Rufo, R.A., Talarico, L., Silverman, S.L. and Brauer, M.J. 1978. Thrombocytopenia and intravenous heroin abuse. *Ann. Int. Med.* 89: 207–211.

Banerjee, D. and Sarkar, N.K. 1997. Haematological changes in buprenorphine-treated mice. *Folia Biol.*, 45: 157–162.

Cimo, P.L., Harnmond, J.J. and Moake, J.L. 1982. Morphine-induced immune thrombocytopenia. *Arch. Int. Med.*, 142: 832–834.

Dacie, J.V. and Lewis, S.M. 1984. *Practical Haematology*, 6th Edn. Churchill Livingstone, Edinburgh, p. 22–83.

Dasgupta, C.R. 1977. *Haematological Technique*, 4th Edn. Publishing Syndicate, Kolkata, p. 40–43.

Firkin, F., Chesterman, C., Penington, D. and Rush, B. 1989. *deGruchy's Clinical Haematology in Medical Practice*, 5th Edn. Oxford University Press, Oxford, p. 360–405.

Jasinski, D.R., Pevnick, J.S. and Griffith, J.D. 1978. Human pharmacology and abuse potential of the analgesic buprenorphine. *Arch. Gen. Psych.*, 35: 501–516.

Macenski, M.J., Schaal, D.W., Cleary, J. and Thompson, T. 1994. Changes in food-maintained progressive-ratio responding of rats following chronic buprenorphine or methadone administration, *Pharmacol. Biochem. Behav.*, 47: 379–383.

Nigam, A.K., Srivastava, R.P., Saxena, S., Chavan, B.S. and Sundaram, K.R. 1994. Naloxone-induced withdrawal in patients with buprenorphine dependence. *Addiction*, 89(3): 317–320.

Robinson, G.M., Dukes, P.D., Robinson, B.J., Cooke, R.R. and Mahoney, G.N. 1993. The misuse of buprenorphine and a buprenorphine-naloxone combination in Wellington, New Zealand. *Drug Alc. Depend*, 33: 81–86.

Singh, R.A., Mattoo, S.K., Malhotra, A. and Varma, V.K. 1992. Cases of buprenorphine abuse in India. *Acta Psych. Scand.*, 86: 46–48.

Williams, W.J., Beutler, E., Erslev, A.J. and Lichtman, M.A. 1986. *Hematology*, 3rd Edn. McGraw-Hill, New York, p. 9–24.

Chapter 17

Environmental Induced Changes in the Biomodal Gas Exchange and Haematology of Facultative Air-breathing Fish, *Mystus punctatus* (Jerdon)

R. Devika, G.M. Natarajan, N. Parthi, P. Esther Joice, P. Palanisamy N. Bhuvaneshwari, G. Sasikala and S. Binu Kumari***

Post Graduate and Research Department of Zoology,
Government Arts College (Autonomous), Coimbatore – 641 018,
**Dublin-9, Ireland, **Chikanna Govt. Arts College, Tirupur – 641 602*

ABSTRACT

Mystus punctatus extracted 69.23 per cent of O_2 from water and 30.77 per cent from air. Air-exposure, starvation, temperature and hypoxic exposure induced marked changes in the bimodal respiration. Air-exposure also altered the haematological parameters.

Keywords: Mystus punctatus, Temperature, Air-exposure, Haematology.

Introduction

Bimodal respiration, the capacity to exchange respiratory gases in both air and water, occurs in many invertebrates and lower vertebrates. Classifications based on the usage of bimodal breathing are

---

* *Corresponding Author.*

an efficient way of circumscribing the ranges of physical and ecological factors like affecting respiration. These usually begin by distinguishing between amphibious and aquatic air-breathers. Amphibious air-breathers range from species such as climbing perch (*Anabas*) that are spontaneously active on land for periods ranging from second to hours. Aquatic air-breathers do so while remaining in water. Among the aquatic air-breathing fishes are species that do this facultatively, that is, when aquatic will not sustain aquatic respiration.

Mystus punctatus breathers air occasionally and is a facultative air-breather. At time, however, the fish will remain voluntarily submerged for extended periods, though the duration of submergence and the accompanying physiological adjustment are not known (Ghosh, 1984). Therefore the present investigation was undertaken to study the environment induced changes in the bimodal gas exchange and blood characteristics of *Mystus punctatus*.

Materials and Methods

Healthy live specimens of *Mystus punctatus* of various body sizes collected from the rivers and streams of Plaghat district of Kerala were held in cement cisterns with running water (28–30°C). They were fed on alternate days for 15 days and were fasted for at least 12 hr before experiments. The fish were acclimatized to the experimental conditions in the respirometer atleast overnight. Job's (1953) continuous flow apparatus was used to study the aquatic O_2 uptake through gills. O_2 uptake for unit time through aquatic route was estimated with the help of the equation: $VO_2 = wf(O_2 in - O_2 out)$, where $VO_2 = O_2$ uptake (Mg O_2/h–1) and wf = water flow (ml/min). For studying the aerial respiration, respirometers were designed involving the principles of manometric techniques. O_2 uptake from air and water was also measured simultaneously when the fish was in water with access to air using respiratory chambers designed by Natarajan (1972). The effect of air-exposure on the bimodal respiration was studied by siphening the fresh water from tank without handling the fish. After air-exposure, fish were returned to individual respirometers. Fish starved for 96 hr were studied for bimodal respiration. The effect of increase of temperature of the medium on the bimodal respiration. The effect of increase of temperature of the medium on the bimodal O_2 uptake was observed by increasing slowly the temperature of the thermostated apparatus. The effect 96 hr hypoxic acclimation (2.5 MgO_2l^{-1}) on the bimodal O_2 consumption was also studied.

Blood from control and air-exposed fish were collected and RBC, Hb, PCV, MCV, MCH, MCHC and O_2 capacities were determined following the methods of Blaxhall and Daisley (1973). Blood glucose was estimated by Folin-Wu method (Klontz and Smith, 1968). The level of significance was studied by applying student 't' test (Daniel, 1987).

Results

The air-bladder in *Mystus punctatus* is of physostomous type and has a well modified and complicated structure. When confined to an aquarium *Mystus punctatus* occasionally visits the surface to fill the gill chamber with air. The frequency of this process varied a great deal, but the average duration of one air-breathing cycle, the time lapsed from intake of air to complete expiration, was from 20–30 min, or in rare cases one hour. Shorter periods of 30–45 min were also observed.

The fish is a facultative air-breather. Fish with an average weight of 14.33 g (Table 17.1) extracted 69.23 per cent from water and 30.77 per cent from air. One-hour air exposure resulted in 38.84 per cent aerial consumption and 70.15 per cent aquatic O_2 uptake. 96 hr starvation increased the aquatic O_2 (70.15 per cent) and decreased the role of air-bladder (29.85 per cent). Increase of water temperature (35°C) increased the role of air-bladder (34.18 per cent) and decreased the gill respiration (65.83 per cent). 96 hr hypoxic exposure increased the aquatic O_2 uptake (70.19 per cent) and decreased the air-bladder role (29.81 per cent).

Table 17.1: Environmental Induced Changes in the Bimodal Respiration of *M. Punctatus*

Fish Weight (g)	*Treatment*	*Aquatic*	*Bimodai Respiration ($mlO_2\ h^{-1}$)*		*Aquatic*	*Aerial*	*Aquatic/ Aerial (%)*
14.33±0.33	Control (28°C)	1.04±0.01	0.46±0.09	1.5±0.01	69.23	30.77	2.37
90.3±0.29	1 hr air exposure	0.58±0.01	0.37±0.06	0.94±0.05	61.16	38.44	1.57
10.06±0.27	96 hr starvation	0.85±0.01	0.38±0.04	1.23±0.02	70.15	29.85	2.39
9.78±0.31	35°C	1.49±0.01	0.78±0.02	2.26±0.03	65.83	34.18	1.94
9.85±0.30	96 hr hypoxia	1.34±0.02	0.57±0.08	1.91±0.02	70.19	29.81	2.36

Table 17.2: One Hour Air-exposure on the Haematology of *M. Punctatus*

Blood Parameters	*Normal (Male + Female)*	*Air-exposed*	*% Changes*	*P*
RBC (X 10^6/cmm)	2.80±0.21	1.79±0.36	– 36.07	< 0.01
Hb (g/100ml)	6.30±0.19	5.24±0.30	– 16.83	< 0.01
PCV (dl)	29.40±0.88	23.50±0.94	– 20.07	< 0.05
MCV (fl)	105.00±2.10	131.28±4.70	+ 25.03	< 0.05
MCHC (%)	21.43±3.45	22.30±3.10	+ 4.06	< 0.001
IMCH (pg)	22.50±3.45	29.27±5.02	+ 30.09	< 0.05
O_2 capacity (Vol+)	7.88±0.06	6.55±0.05	– 16.83	< 0.01
Blood glucose (mg/100ml)	67.18±6.00	59.20±5.14	– 11.88	0<0.01

Values are $\bar{X}$ ± S.E. of 6 observations.

One hour air exposure decreased the RBC (–36.07 per cent), Hb (–16.83 per cent), PCV (–20.07 per cent), O_2 capacity (–16.83 per cent), and blood glucose (–11.88 per cent). However MCV (+25.03 per cent) and MCH (+30.09 per cent) increased significantly.

Discussion

Mystus punctatus is a less efficient air-breather. Air breathing frequency typically increases with temperature (Arguilar *et al.*, 2000); increased temperature means increased metabolism and secondary effects of temperature on tissues and organ systems (*e.g.*, pH changes, reduced HbO_2 affinity, increased heart and ventilatory activity and a lowered O_2 extraction efficiency). Facultative air-breathers elevates their respiratory rates in hypoxia and temperature (Vijayalakshmi, 1996). Facultative air-breathers doubtlessly increases the survival probability of fish which are forced to endure periodic exposure to water unsuitable for aquatic respiration (Takasusuki, Fernandes and Severi, 1998). During tropical dry season, for example, fish that typically occur in normoxic flowing streams may become trapped for weeks or months in stagnant ponds that are hypoxic and hypercarpic (Natarajan, 1972). Moreover, the inhabitants of some tropical ponds may have to endure hypoxia every night irrespective of season.

Mystus punctatus breathers air frequently but is not obligatory air-breathers. Starvation, temperature, hypoxic exposure and induced air exposure stress induces bimodal gas exchange machinery to cope with the changed environmental conditions. The role of air-bladder in enduring the stress is significant.

Similar changes were also reported for *Periophthalmodon schlosseri* (Ishimatsu *et al.*, 1999; Aguilar *et al.*, 2000).

Results of the present study also demonstrate that air-exposure stress, has a pronounced effect on certain blood parameters. These changes were proportional to the duration of air-exposure. The percent increase or decrease in certain blood parameters can be compared with the facultative air-breathing fish, *Notopterus notopterus* (Vijayalakshmi, 1996). The increased mean cell volumes during air-exposure "stress" could have been due to the swelling of red blood cells, or to the release of large red blood cells into the general circulation. Although the effects of "stress" could have been attributed to hypoxia and (or) CO_2.

It is also possible that lactic acid may be involved. This is possible as the blood glucose concentration fell after 1 hr air-exposure. In Amphipnous which is an obligate air-breather 20 per cent fall in glucose level was notice 5½ hr air exposure (Singh *et al.*, 1976). Few papers have been concerned with the effects of air-exposure or similar stress on the blood parameters of the fish. Generally, obligate air-breathers are consistent in maintaining a steady balance in the blood values in prolonged air-exposure stress than facultative air-breathers (Ishimatsu *et al.*, 1999). A long-term study involving many more blood parameters may expose the adaptive strategies of this species in combating air-exposed stress.

Acknowledgements

The authors are grateful to Government of Tamil Nadu and Bharathiar University for all the helps.

References

Agruilar, N.M., Ishimatsu, A., Ogawa, K. and Huat, K.K. 2000. Aerial ventilatory responses of the mudskipper, *Periophthalmodon schlosseri*, to altered aerial and aquatic respiratory gas concentration. *Comp. Biochem. Physiol.* 127: 285–292.

Blaxhall, P.C. and Daisley, K.W. 1973. Routine haematological methods for use with fish blood. *J. Fish. Biol.*, 5: 771–782.

Daniel, W.W. 1987. *Biostatistics: A Foundation for Analysis in the Health Science.* John Wiley and Sons, New York.

Klontz, G.W. and Smith, L.S. 1968. In: *Methods of Animal Experimentation,* (Ed.) Gay, W.I. Academic Press, New York.

Ishimatsu, A., Agruilar, N.M., Ogawa, K., Hishida, Y., Takeda, T., Oikawa, S., Kanda, T. and Huat, K.K. 1999. Arterial blood gas levels and cardiovascular function during varying environmental conditions in a mudskipper, *Periophthalmodon schlosseri. J. Exp. Biol.*, 202: 1753–1762.

Natarajan, G.M. 1972. Studies on the respiration of *Anabas scandens* (Cuvier). *M.Sc. Thesis*, I.C.A.R., New Delhi.

Singh, B.R. Thakur, R.N. and Yadav, A.N. Changes in the blood parameters of an air-breathing fish during different respiratory conditions. *Folia. Haematol.*, 2: 216–225.

Takasusuki, J., Fernandes, M.N. and Severi, W. 1998. The occurrence of aerial respiration in *Rhinelepis strigosa* during progressive hypoxia. *J. Fish. Biol.*, 52: 369–379.

Vijayalakshmi, P. 1996. Studies on some aspects of bimodal gas exchange, haematology and nitrogen excretion in the South Indian feather-back, *Notopterus notopterus. Ph.D. Thesis*, Bharathiar University.

Chapter 18

Investigation of Drinking Water Quality of Basavanhole Tank with Reference to Physico-chemical Characteristics

J. Narayana, R. Purushothama, B.R. Kiran, K.P. Ravindra Kumar and E.T. Puttaiah

Department of Environmental Science, Kuvempu University, Shankaraghatta – 577 451, Karnataka

ABSTRACT

The article deals with the analysis of physico-chemical parameters of a Basavanahole tank at Sagar taluk, during February 2004 to October 2004. The present study was to assess the drinking water quality by examining various physico-chemical characteristics of this tank. The water temperature of this water body ranged from 23–32°C. pH was alkaline in nature in an the months. Electrical conductivity was between 72–124.16 µmhos/cm. While, turbidity was deviated from 9.3–20.3 NTU. Total dissolved solids ranged from 46.08–79.3 mg/L; free carbon dioxide was between 4.4–13.2 mg/L. Dissolved oxygen and chloride content of water samples varied from 2.36 to 15.4 mg/L and 8.0 to 18.43 mg/L respectively. Cations such as calcium and magnesium were between 3.36–10.94 and 1.80–4.68 mg/L. However, total hardness fluctuated between 20.0 to 44.0 mg/L. Total alkalinity was ranged from 30–60 mg/L, total acidity was between 2.5–10.0 mg/L. Nevertheless, Sodium and potassium ions varied from 1.24–3.8 and 1.9–3.1 mg/L respectively. Other water quality parameters *viz.*, BOD, Phosphate, Nitrite, Nitrate, sulfate were within the permissible limit. Hence, water found suitable for drinking and irrigation purpose, oligotrophic in nature.

Keywords: *Water quality, Basavanahole tank.*

Introduction

Water is one of the most essential item needed by living beings for their survival and growth, maintains an ecological balance between various groups of living organisms and their environment (Santhosh Kumar, 1984) detailed study on the status of pollution of any water body is of much importance because it ultimately helps in the proper management of the water. A physico-chemical approach to monitor water pollution is most common and literature is available on these aspects. Considerable hydrobiological investigations are carried out on man made impoundment's in India (Surendra Kumar and Sharma, 1991; Bahura, 1998). The tank plays an important role in maintaining the water table and is also used by the sagar town people for drinking and fishing. Therefore, this work has been undertaken to determine the physico-chemical status of the waterbody.

Materials and Methods

Basavanahole tank is located in the Sagar town of Shimoga district at 14°15′50′ N latitude and 75°15′51″ East longitude. It covers an area of 14.90 hectare with full water level and 3 to 4 meters depth at the middle of the tank.

The surface water samples were collected at monthly intervals from February 2004 to October 2004. Water temperature was recorded by mercury thermometer. The pH was recorded using pH pen, electrical conductivity by conductivity meter. Chemical analysis for dissolved oxygen, biological oxygen demand, Chloride, sulfate, calcium and magnesium were estimated using titrimetric method. Standard procedure was followed for the analysis of physico-chemical characteristics (APHA, 1998).

Results and Discussion

The results of the water quality analysis are presented in Table 18.1. The water temperature ranged between 23°C and 32°C during the study period. The variation in temperature was varies between summer and rainy seasons. Similarly seasonal variation of temperature also reported by (Jain *et al.*, 1999) The hydrogen ion concentration of water in all months are alkaline (7.1–8.3). Maximum pH of 8.3 was observed in the month of April and minimum of 7.1 in the month of October. The fluctuation in pH have been found to follow changes in sunshine (Lind, 1938) and higher pH values were considered to be indicative of higher rate of productivity. The present study revealed a similar relationship in the tank. pH and dissolved oxygen varied together in the waterbody confirming earlier observations of Bharathi and Kore (1975) and Armugam and Furtado (1980).

Turbidity values ranged from 9.3 to 20.3 NTU maximum turbidity was observed during rainy months due to surface runoff. Minimum values of turbidity observed in March, it may be due to higher temperature, which increases the decomposition of organic matter. Electrical conductivity values fluctuated from 72 to 124.16 µmhos/cm. Dissolved oxygen showed maximum level 12.4 in the month of October, while the minimum DO level 2.36 mg/L was recorded at this tank in July. Solubility of oxygen decreases with increases in temperature (Sabata and Nayar, 1995). Similarly increase in DO is obviously related to decrease in temperature. However present water body showed similar correlation of temperature with DO.

Higher free CO_2 was recorded during August (13.2 mg/L) and lower content in the months of May and June *i.e.*, 4.4 mg/L, higher concentration of free CO_2 is due to, respiration of organisms and absence of photosynthesis. Phenolphthalein alkalinity was absent in all the months. The maximum value of total alkalinity (60 mg/L) was recorded in the month of April and lower concentration of 30 mg/L in August. Philipose (1959) suggested that a water body with alkalinity value higher than 100 mg/L is nutritionally rich. By this standard tank water is oligotrophic. The lowest value of 20 mg/L was observed in the month of May, Kannan (1991) bas classified water on the basis of hardness values in the following manner; 0–60 mg/L soft, 60–120 moderately hard, 120–160 mg/L hard above and 180 is very hard. On the basis of hardness values, this tank water can be included in a soft category.

Table 18.1: Physico-chemical Parameters of Basavanahole Tank (February 2004–October 2004)

Parameter	*Feb*	*Mar*	*Apr*	*May*	*Jun*	*July*	*Aug*	*Sep*	*Oct*	*Range*
Air Temperature °C	34	34	34	31	29	25	30.5	24	28.5	2.5–34
Water Temperature °C	29	31	32	29	23	24	28	23	27.5	23–32
pH	7.5	7.8	8.3	7.7	8.0	7.2	7.3	7.4	6.5	7.1–8.3
Turbidity (NTU)	17.2	9.3	20.1	16.3	14.0	20.3	17.1	11.3	17.9	9.3–20.3
E. Conductivity (μmhos/cm)	104	124.16	119	115	97.0	84.0	72.0	72.0	78	72–124.16
Total Dissolved Solids	66.56	79.36	76.16	73.6	62.02	53.76	46.08	46.08	49.92	46.08–79.3
Dissolved Oxygen	9.6	8.10	2.43	6.48	5.67	2.36	2.43	7.6	15.40	2.36–15.4
Free Carbon Dioxide	8.8	8.8	8.8	4.4	4.4	8.8	13.2	8.8	13.2	4.4–13.2
Chloride	12.76	15.60	14.18	15.60	18.43	14.18	8.0	11.34	11.34	8.0–18.43
Calcium	9.25	10.94	7.57	3.36	5.89	6.73	4.2	5.05	5.05	3.36–10.94
Magnesium	4.12	4.07	3.49	2.83	2.10	4.68	3.78	1.80	2.78	1.80–4.68
Total Hardness	40.0	44.0	32.0	20.0	23.33	36.0	26.0	20.0	24	20.0–44.0
BOD	1.61	1.61	3.40	1.61	1.65	3.46	4.0	2.6	2.13	1.61–4.0
Carbonate	00	00	00	00	00	00	00	00	00	00
Bicarbonate	50.0	40.0	70.0	56.0	60.0	40.0	30.0	40.0	10	30–60
Total Alkalinity	40.0	40.0	50.0	46.0	60.0	40.0	30.0	40.0	10	30–60
Total Acidity	5.0	5.0	2.5	8.33	5.0	7.5	5.0	10.0	5.0	2.5–10.0
Phosphate	0.02	0.01	0.021	0.03	0.02	0.1	0.13	0.1	BDL	0.01–0.13
Nitrite	0.01	BDL	0.02	BDL	BDL	0.01	0.04	0.01	BDL	0.01–0.04
Nitrate	0.11	0.12	0.12	0.19	0.21	0.17	0.23	0.15	0.10	0.11–0.23
Sulfate	16.89	20.73	16.52	18.92	15.20	13.82	7.29	4.2	18.40	7.29–20.73
Sodium	3.1	2.2	3.8	3.0	3.3	3.2	3.1	2.8	1.24	1.24–3.8
Potassium	2.3	1.9	3.1	2.9	2.8	2.6	2.8	2.8	2.8	1.9–3.1

Note: All the parameters are in mg/L, except pH, temperature °C, turbidity (NTU) and electrical conductivity (μmhos/cm). BDL: Below detectable limit.

The amount of total dissolved solids found in experimental sampling is 46.08 mg/L to 79.36 mg/L. TDS in water is mainly due to the presence of bicarbonates, sulfates, chlorides of calcium, magnesium and sodium in the present study the tank contain lower concentration of TDS.

Calcium is very important element influencing flora of ecosystem, which plays potential role in metabolism and growth. The range of it varied from 3.36 to 10.94 mg/L. the maximum calcium concentration 10.94 mg/L was recorded in the month of March and minimum of 3.36 mg/L in May. According to Ohle (1934), values above 25 mg/L indicate calcium rich in water. As per this definition the tank water is poor in calcium ions. The concentration of magnesium varied between

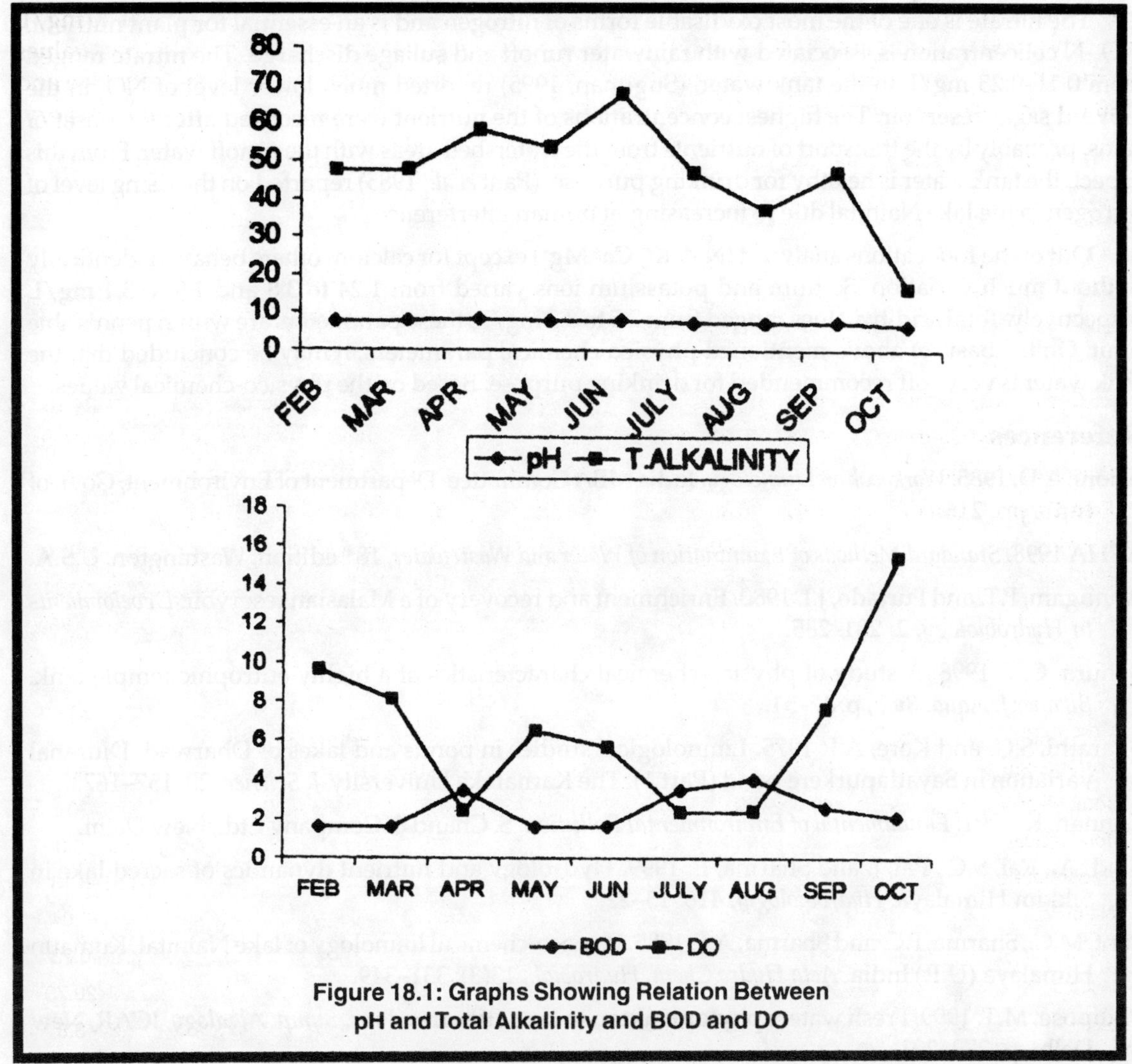

Figure 18.1: Graphs Showing Relation Between pH and Total Alkalinity and BOD and DO

1.80 and 4.68 mg/L. normally these ions are not problematic but at higher concentration increase total hardness of water.

The higher amount of chloride 18.43 mg/L was recorded at this tank in the month of June. Minimum concentration of 8.0 mg/L was noticed in August. (Thresh, *et al.*, 1944) also pointed out that high concentration of chloride is due to large quantity of organic matter in it, which was further supported by (Adoni, 1985) however such a condition is not observed in the present water body.

BOD values in water samples are ranged between 1.61–4.0 mg/L. these values are within permissible limit. Phosphate concentration was low in the water. The maximum of 0.13 mg/l was recorded in the month of August, low values of phosphate was recorded. Phosphate is considered amongst the primary limiting nutrients in ponds and lakes (Schindler, 1971) low values of phosphate have been reported from various Himalayan lakes and reservoirs (Raina and Peter, 1999)

The nitrate is one of the most oxidisable forms of nitrogen and is an essential for plant nutrient. NO_3-N concentration is associated with rainwater runoff and sullage discharge. The nitrate ranged from 0.11–0.23 mg/L in the tank water. (Sugunan, 1995) reported much lower level of NO_3 in the Govind sagar reservoir. The highest concentrations of the nutrient were recorded after the onset of rains, probably by the transport of nutrients from the watershed areas with the runoff water. From this aspect, the tank water is healthy for drinking purpose. (Pant *et al.*, 1985) reported on the rising level of nitrogen in the lake Nainital due to increasing of human interference.

Out of the four cations analyzed (Na^+, K^+, Ca^+ Mg^+) except for calcium others behaved identically without much variation. Sodium and potassium ions varied from 1.24 to 3.8 and 1.9 to 3.1 mg/L respectively. Total acidity values ranged from 2.5 to 8.3 mg/L. these parameters are within pennissible limit. On the basis of above-mentioned physico-chemical parameters, it may be concluded that the tank water is very soft recommended for drinking purpose. Based on the physico-chemical values.

References

Adoni, A.D. 1985. *Workbook on Limnology*. India MBA Committee. Department of Environment, Govt. of India pp. 216.

APHA 1998. *Standard Methods of Examination of Water and Wastewater*, 18th edition, Washington. U.S.A.

Annugam, P.T. and Furtado, J.I. 1980. Enrichment and recovery of a Malasian reservoir. *Developments in Hydrobiology*, 2: 281–285.

Bahura, C.K. 1998. A study of physico-chemical characteristics of a highly eutrophic temple tank. *Bikaner J. Aqua. Biol.*, p. 47–51.

Bharathi, S.G. and Kore, A.P. 1975. Limnological studies in ponds and lakes of Dharwad. Diuranal variation in Sayadapurkere pond (Part 1). The Karnataka University. *J. Science*, 20: 157–167.

Kannan, K. 1991. *Fundamental of Environmental Pollution*. S. Chand & Company Ltd., New Delhi.

Lind, A., Rai, S.C., Pal, J. and Sharma, E. 1999. Hydrology and nutrient dynamics of sacred lake in Sikkim Himalaya. *Hydrobiologia*, 416: 13–22.

Pant, M.C., Sharma, P.C. and Sharma, A.P. 1985. Physico-chemical limnology of lake Nainital, Kumaun Himalaya (U.P.) India. *Acta Hydro Chem. Hydrobiol.*, 13(3): 331–349.

Philipose, M.T. 1999. Fresh waters phytoplankton in land fisheries. *Proc. Symp. Algalogy*, ICAR, New Delhi, p. 272–291.

Raina, H.S. and Peter, T. 1999. Cold water fish and fisheries in the Indian Himalayas lakes and reservoirs. In: *Fish and Fisheries at Higher Altitudes-Asia*, (Ed.) T. Peter. FAO Fish. Tech. Paper No. 385, Rome, p. 64–88.

Sabata, B.C. and Nayar, M.P. 1995. *River Pollution in India: A Case Study of Ganga River*. APH Publishing Corp., New Delhi, pp. 223.

Santhosh Kumar, G. 1984. *Water Supply Engineering*. Khanna Publ., New Delhi.

Schindler, D.L. 1971. Carbon, nitrogen, phosphate and eutrophication in fresh water lakes. *J. Phycol.* 17: 321–329.

Surendrakumar and Sharma, L.L. 1991. Comparative physico-chemical limnology of lakes Pichola and Gatehsagar Udaipur (Rajasthan). *Poll. Res.*, 10(3): 173–178.

Thresh, J.C., Sacking, E.V. and Beale, J.F. 1944. *The Examination of Water Supplies*. Ed. Taylor, E.W.

Sugunan, V.V. 1995. *Reservoir Fisheries of India*. FAO. Fish. Tech. Paper No. 345, FAO, Rome.

Chapter 19

Histological Alterations in Tadpoles of *Bufo Melanostictus*, Exposed to a Sublethal Concentration of Chromium

D. Anusuya, D.J. Prakash and I. Christy

Postgraduate Department of Zoology, Voorhees College, Vellore – 632 001, India

ABSTRACT

Effect of sublethal concentration of heavy metal chromium (0.2 ppm) on tadpoles of Bufo melanostictus was studied. Results reveal extensive histological abnormalities in the internal gills, intestine, liver and eye of tadpoles exposed to chromium, in contrast to that of control tadpoles, showing the severity of chromium toxicity.

Keywords: *Chromium, Tadpole, Histological alterations, Necrosis, Sublethal, Toxicity.*

Introduction

Heavy metals, are more hazardous pollutants of ecosystems with deleterious effects on aquatic biosystems. Rapid industrialization, urbanization and other developmental activities have led to deterioration of the environment (Salunk *et al.*, 1982). Such heavy metal toxicants have been shown to enter and accumulate in aquatic fauna and flora causing heavy modification (Berman and Lal, 1994). Earlier reports reveal that they may disturb cellular functions, altering vital physiological and biochemical mechanisms of animals (Larsen *et al.*, 1976; Shastry and Rao 1981; Muthukrishnan *et al.*, 1986; and Radhakrishniah *et al.*, 1991) because of their inherent ability to form complexes with ligands containing sulphur, nitrogen and oxygen as electron donors (Valee and Ulmer, 1972). The deleterious effects caused by the toxicity of chromium, have been reported and reviewed (Mertz, 1969; Binanchi and Lewis 1986; Anna M. Fan and Harding Barlow, 1987). Extensive information is available on effects of chromium of Ichthyofauna (Venugopal and Reddy, 1961; Khaviraj, 1984 and Srivatsava and Mauria, 1991). There is a paucity of information on such effects in amphibian population excepting few reports (Mckee and Wolfe 1963; Greenhouse 1976; Schowing and Boverio 1975). Further it is also essential to evaluate toxic effects of chromium or any heavy metal on amphibians, since frogs and toads form very important members of food chain in aquatic as well as terrestrial ecosystem (Leuven *et al.*, 1986). It has also been established that amphibian embryos and larvae could be used as sensitive assay system to study and evaluate toxic and teratogenic effects of aquatic pollutants (Birge *et al.*, 1975 and Greenhouse, 1976). Therefore, the present investigation aims to evaluate the histopathological abnormalities caused in some selected tissues of tadpoles of Bufo melanostictus due to chromium toxicity.

Material and Methods

Tadpoles of Bufo melanosictus, ranging from 1.2 to 1.5 cms were collected from near by water bodies and acclimated in bore well water which had the characteristics such as temperature 28° C, pH 8.2, dissolved oxygen 5.2 ± 0.4, for two days before commencing the experiments.

Suitable stock solution of technical grade Potassium dichromate (28.28 gms/litre) was prepared from which required number serial dilutions were made and static bio-assays were conducted as per the procedure (APHA 1974). LC_{50} for 96 hr value was determined following the procedure of Reish and Cshida (1987) and $1/10^{th}$ of LC_{50} value (0.2 ppm) was taken as sublethal concentration as suggested by Kirubakaran (1998) and the same was used in the present study. The experimental animals were kept in chromium containing water (0.2 ppm) while the control tadpoles were kept in bore well water. During the experimental period, the medium was changed once a day and the tadpoles were fed with freshly boiled spinach leaves. The total period of experimentation lasted for 20 days.

At the end of experimental period, tadpoles were sacrificed and fixed in 5 per cent formaldehyde and processed for histological studies. Serial sections of 7 µm thick were taken and stained with haematoxylin and eosin. The sections were identified and photographed.

Results and Discussion

Histological approach is the most valuable tool for assembling the action of toxicant at tissue level providing data concerning tissue damage (Sprague 1971). Visible histopathological abnormalities caused due to toxicity of heavy, metals in animals have been reported earlier (Srivatsava *et al.*, 1982; Ramesh Kumar *et al.*, 1988 and Srivatsava and Maurya 1991). The present study also reveals the extensive damage in the internal gill architecture of treated tadpoles compared to gills of controls.

As observed in fish by earlier workers, there is a clear loss of receptor epithelial layer, showing erosion of epithelial cells with necrosis of tissues in the chromium treated tadpoles in contrast to intact gill structure of control tadpoles. Intestine is the vital organ for digestion and any food contaminated with a toxicant is known to cause damage to the intestinal epithelium (Amita Moitra and Roshan Lal, 1989).

In the present study, chromium exposed tadpoles showed visible alterations in intestinal epithelium. In the control, intestinal epithelium was intact with more granulated and deeply stained cells where as chromium treated tadpoles exhibited thickened, broken and lightly stained mucosal layer with dispersed cells.

Liver is an organ to metabolize and eliminate any toxicant or drug and any histopathological changes observed in liver due to impact of toxicant can be used as a reliable index to evaluate structural and functional damage of target tissues (Shastry and Sharma, 1978) and the extent of its damage is related to the mode of action of toxicant, its accumulation and persistence (Ravi Kumar 1994). In the present study, the liver of control tadpoles showed in intact histoarchitecture with deeply stained nuclei and granulated cytoplasm. In contrast, liver of treated tadpoles revealed enlarged cells, lightly stained cytoplasm, extensive vacuolization and shifting of nuclei to the periphery of some cells. These findings agree with the earlier reports (Bakthavatchalu, 1984 and Narain and Singh, 1991) which showed characteristic abnormalities like liver chord disarrangement, nuclear pycnosis and degeneration of cytoplasm in fishes exposed to various pesticides.

The eyes are in direct contact with the external medium and therefore its exposure to toxicant medium in the treated tadpoles, has resulted in erosion of sclerotic coat, chord layer, and a collapsed lens vesicle in contrast to compact nature of eye structure with intact lens, retinal layer, choroid layer and sclerotic coat of control tadpoles.

Thus the present observations clearly indicate that even a sublethal concentration of heavy metal chromium could result in severe histological impairments of various tissues in tadpoles exposed for a considerable length of time.

References

Anna, M. Fan and I. Harding Barlow 1987. In: *Advances in Modern Environmental Toxicology.* Eds.) M.A. Mehlmon Princeton. Scientific Publishing Co. Inc., p. 87–125.

APHA 1974. *Standard Methods for the Examination of Water and Wastewater.* APHA, AWWA and WPCF, Washington, p. 1134.

Bakthavatchalu, R. 1984. Histomorphology of liver, kidney and intestine of the fish. Anabus testudion exposed to Furon. *Environ. Ecol.*, 2: 243–8.

Berman, S.C. and M.M. Lal 1994. Accumulation of heavy metals Zn, Cu, Cd, and Pb) in soil and cultivated vegetables and weeds grown in industrially polluted fields. *J. Environ Biol.*, 15(2): 107–115.

Bianchi and A.G. Lewis 1986. Mechanism of chromium genotoxicity. *Toxicol. Env. Chem.*, 9: 1–25.

Brige, W.J., Westerman, A.G. and J.A. Black 1975. Sensitivity of vertebrate embroys to heavy metals as a criterion for water quality. Phase III. Use of fish and amphibian embroys as bioindicator organisms for evaluation of water quality. Res. Rep. University of Kentuky, Water Resource Inst., U.S.A. No. 91.

Greenhouse 1976B. Evaluation of the teratogenic effects of hydrozine, methylhydrozine and dimethyl hydrozine on embryos of Xenopus laevis, the South African toad. *Teratology*, 13: 167–178.

Khaviraj 1984. The effect of copper on maximum respiration rate and growth rate of puch perca fluviatilis. L., *Water Research.*, 18: 30–144.

Kirubagaran, R. 1989. Endocrine and neuroendocrine physiology of the cat fish clarious batrachus in relation to mercury treatments. *Ph.D. Thesis*, Centre of Advance Study in Zoology, Banaras Hindu University.

Larson, A. Bengteson and B.E. Stevenberg 1976. Some hematological and biochemical effects of CCI in fish. *Sem. Ser. Soc. Exp. Biol.*, 2:35–46.

Leuvan, R.S.E.W., Denttar, C., Christian, M.M.C. and W.H.C. Heijligers 1986. Effect of water acidification on the distribution pattern and reproductive success of amphibians. *Experentia*, 42: 493–503.

McKee, J.E. and H.W. Wolf 1963. Water quality criteria publication of the California state water resources control board, U.S.A. Pub. No. 3: A revised reprint, 1976.

Mertz, W. 1969. Chromium occurrence and function biological systems. *Physiol. Res.*, 49: 163–234.

Moitra, Amita and Roshan Lal 1989. Effect of sublethal doses of Malathion and BHC on the intestine of fish, *Puntius Sarana. Env. & Ecol.*, 7 2): 412–414.

Muthukrishnan, J., Viswarajan, S. and E. Subbulakshmi 1986. Effect of sublethal concentration of Mercuric Chloride on transformation of food by the fish *Cyprinus Carpio. Environ. Ecol.*, 4: 526–532.

Narain, A.S. and B.B. Singh 1991. Histopathalogical Lessions in Heteropreustes fossils subjected to acute thiodon toxicant. *Acta Hydrochim. Hydrobiologi*, 19: 235–243.

Radhakrishnaiah, K., Suresh, A., Urmilaa, B. and B. Sivaramakrishnan 1991. Effects of mercury and liquid metbolic profiles in the Organs of *Cyprinus Carpio* Lin. *J. Mendal*, 8: 125–135.

Rameshkumar, P.B., Vijayalakshmi, S. and C. Rajamanikam 1988. Toxicity effect of Zinc Sulphate on gill in the freshwater fish *mystus vitatts* Block. Asbst No. 67, National Symposium on Ecotoxicology, Annamalai Nagar.

Ravikumar, P. 1986. Impact of Phosphomidon of histological and biochemical changes in the FW Fish *Oreochromis Mossambicus* Peters. *M.Phil. Thesis*, Annamalai University.

Reish, L. Donald and Philip Cshida 1987. Manual of methods in aquatic environment research, 247, Part–10: short term static bio assays, FAO, Fisheries Technical Paper, Rome.

Salunk, J., Balogh, V.K.O. and E. Berta 1982. Heavy metals in animals of the lake Belaton. *Water Res.*, 16: 1147–1152.

Schowing and Boverio 1979. Influence teratogene du bichlorure demercure surle development. Embryonairede, amphibian Xenophous laevis, Acta, Embryol expt., 1: 39–52.

Shastry, K.V. and S.K. Sharma 1978. Eldrin toxicity on liver of *Channa Punctatus* Block. *Ind. J. Exp. Biol.*, 16: 372–373.

Sprague, J.B. 1971. Measurement of pollutant toxicity to fish. III, Sublethal effects and sate concentration. *Water Res. J.*, p. 245–260.

Srivastava, V.M.S. and R.S. Maurya 1991. Effect of Chromium stress on gill and intestine of *Mystus Vittatus* Block: Scanning E.M. Study. *J. Ecobiol.*, 3(1): 69–71.

Srivastava, V.M.S., Tripathi, R.S. and A.K. Saxena 1982a. Chromium included histopatholgic changes in some tissues of puntius Sophore Hamilton. *J. Biol. Res.*, 2: 67–68.

Vallee, B.L. and D.D. Ulmer 1972. Biochemical effects of Mercury, Cadmium and lead. *Ann. Rev. Biochem.*, 41: 91.

Venugopal, N.B. and S.L. Reddy 1992. Nephrotoxic and hepatotoxic effects of trivalent and hexavalent Chromium in a teleost fish, *Anabas Scandans*, Enzymological and Biochemical changes. *Ecotoxicol. Env. Safety*, 24(3): 287–93.

Chapter 20

Urban Development and Sound Level in Ichalkaranji City, Maharashtra

C.T. Pawar and M.V. Joshi***

**Department of Geography, Shivaji University, Kolhapur – 416 004, Maharashtra, India*

***Jaywant Mahavidyalaya, Ichalkaranji – 416 115, Maharashtra, India*

ABSTRACT

The environment is an integrated system in which all its elements act and react in such a way that a balance is naturally maintained. Man is a user of the environment for his developmental activities and always disrupts this natural system and creates a background for environmental degradation. In view of this present study alms to analysis the impact of urban development on sound level of Ichalkaranji city which is known as little Manchester of Maharashtra. For the present investigation empirical data regarding noise pollution for four selected spots have been collected through intensive field work. Ambient sound levels in various spots of Ichalkaranji city were monitored and compared with that of standards provided by schedule III, of Environmental Protection Rules, 1986. The processed data has been shown with the help of tables and columnar diagrams. The analysis reveals that the sound level in the city is found beyond the standards at all the spots. Further and attempt is made to evaluate the causes of noise pollution and some remedies are suggested to minimise it.

Introduction

The man's environment is the basis of human existence and survival. The rapid technological developments no doubt have helped civilization but at the same time led to environmental degradation. This in turn has threatened human survival. Noise is one of the constituents of over all environmental pollution. It has been established that excessive noise is not only adversely effecting the health of human beings but is also a health hazard to all living beings. Even the non-living things are not left unaffected by high intensity of noise (Trivedi, 1999). The common man is made aware of the hazards of noise pollution. Low and society should join hands in making environment healthy. Keeping these facts in view, an effort has been made to assess the impact of sound level at various spots in Ichalkaranji city. Ichalkaranji city region is one of the important industrial centres of Maharashtra, where the development of powerlooms and weaving industry has brought overall prosperity to the city. Being an industrial centre, it is a major city place on the boundary of Maharashtra and Karnataka states. Ichalkaranji city is located in south Maharasthra, in Hatkanangale tahsil of Kolhapur district and due to weaving industry it is called as, "Little Manchester of Maharashtra". It is situated on the bank of river Panchaganga, a tributary of river Krishna. The civic administration of Ichalkaranji is governed by Municipality. The city located on 16°46′ North latitude and 74°25′ East longitude at an altitude of 542 meters above MSL, occupies 18.13 sq.kms. area.

Database and Methodology

Here an attempt is made to study the sound level of Ichalkaranji city for a weeks average, for which empirical data regarding noise pollution for four selected spots have been collected in each category of area, through intensive field work. Ambient sound levels in various spots of Ichalkaranji city were monitored and compared with that of standards provided by Schedule III of Environmental Protection Rules, 1986 (Table 20.1). Sound Level Meter Cygnet System was used for monitoring the sound level. Sound levels are measured in decibels. One decibel is the threshold of hearing. Data has been represented with the help of tables and columnar diagrams. Population and industrial figures were collected from the records of Ichalkaranji Municipal Council, Ichalkaranji.

Table 20.1: Ambient Air Quality Standards in Respect of Sound Area

Area Code	*Category of Area*	*Limits in dB (A)*	
		Day Time (6 a.m. to 9 p.m.)	*Night time (9 p.m. to 6 a.m.)*
A	Industrial area	75	70
B	Commercial area	67	55
C	Residential area	55	45
D	Silence zone	50	40

Source: Environmental Protection Rules, 1986, Schedule III, Tripathy, 1999.

Results and Discussion

Ichalkaranji city is the industrial and urbanized centre and no exception to environmental problems of various types. Population of Ichalkaranji has phenomenally increased in the last four decades due to rapid industrialization, which has resulted in the rise of immigrant population of the Ichalkaranji city (Table 20.2). Apart from powerlooms, 13 spinning mills, 173 sizing units, 190 winding units, 11 main processing houses, 81 hand processing units, 70 dyeing and printing units, 10 foundaries, 1 sugar mill and various supported engineering industrial units are in operation in the city. The studies made by Pawar and Joshi (2002 and 2003) indicate that the increased urbanization and industrialization of Ichalkaranji city have resulted in deterioration of water quality. Similarly the problem of solid waste disposal has also created environmental hazards.

Table 20.2: Ichalkaranji City: Growth of Powerlooms and Population

Sl.No.	Year	Number of Powerlooms	Population	Growth Rate of Population (%)
1.	1951	2,000	27,423	–
2.	1961	8,000	50,978	+ 85.90
3.	1971	15,000	87,731	+ 72.10
4.	1981	33,000	1,33,704	+ 52.40
5.	1991	90,000	2,14,835	+ 60.68
6.	2001	1,28,000	2,57,600	+19.90

Source: Record of Ichalkaranji Municipal Council, Ichalkaranji.

The analysis reveals that the sources of noise pollution in the city are mainly traffic junctions, industrial and commercial areas with various sounds. Sound levers are checked at various intervals in different areas such as industrial area, commercial area, traffic junctions etc. and average decibel values are given in the Table 20.3 and presented in Figure 20.1.

Table 20.3: Ichalkaranji City: Monitoring Locations and Ambient Sound Level in Different Areas

Sl.No.	Category of Area	Location	Period	Average Sound Level (dB)
1.	Industrial	Near Shahu Putala Industrial Estate	Day	90
			Night	85
2.	Commercial	Janata Bank	Day	80
			Night	65
3.	Residential	Rajwada	Day	58
			Night	53
4.	Silence	Dr. Ambedkar Hospital	Day	55
			Night	45

Source: Compiled by Authors.

The residential areas of Rajwada has recorded more noise level 58 dB (Figure 20.1A) than the prescribed standard limits. Similarly, in the commercial areas such as Janata Bank sound levels are higher than norms. Janata Bank area records maximum noise level at 80 dB during day time and 65 dB (Figure 20.1B) during night time. Similarly in the sensitive area at Dr. Babasaheb Ambedkar Hospital, which has court and Govindrao High school in the vicinity, the noise levels are higher with 55 dB during day timc and 45 dB during night time. Various industrial sectors also show higher noise levels because of processes causing impact, vibration or reciprocation movements, friction and turbulence in air or gas streams. In short, sound levels in Ichalkaranji city are at higher side in some residential, commercial, silence zone. Noise causes several undesirable effect. Damage to hearing and loss of sleep are only two of the obvious results (Saxena, 1999). Noises produce irritability and a feeling of fatigue and may reduce a worker's efficiency. The noise is one of the major cause of stress and anxiety, also impact on blood pressure and is a cause of other cardiovascular diseases.

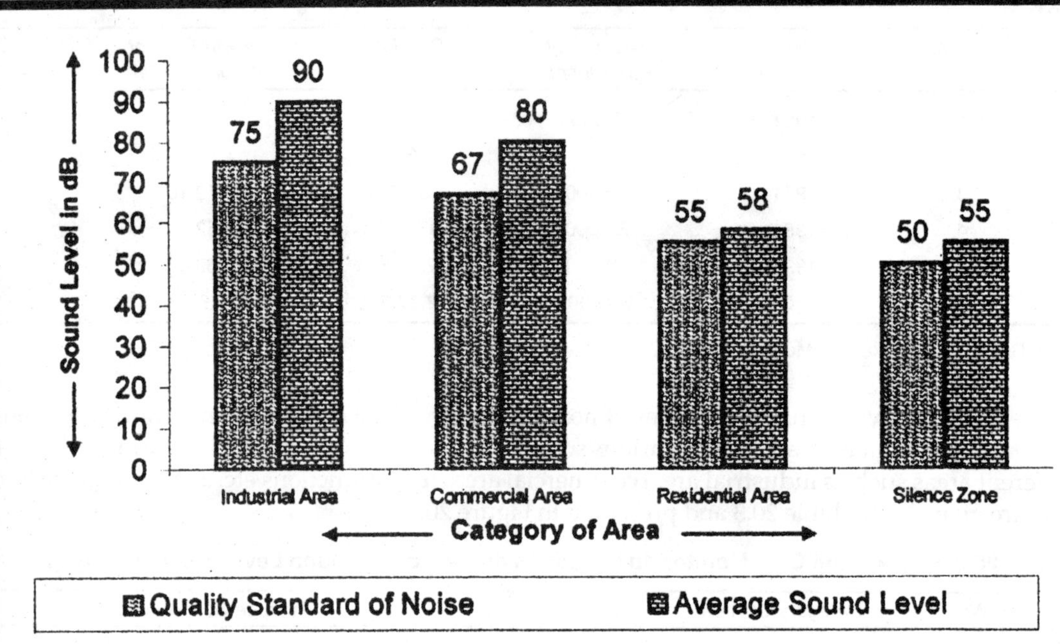

Figure 20.1A: Ichalkaranji City: Ambient Sound Level in Different Areas During Day Time

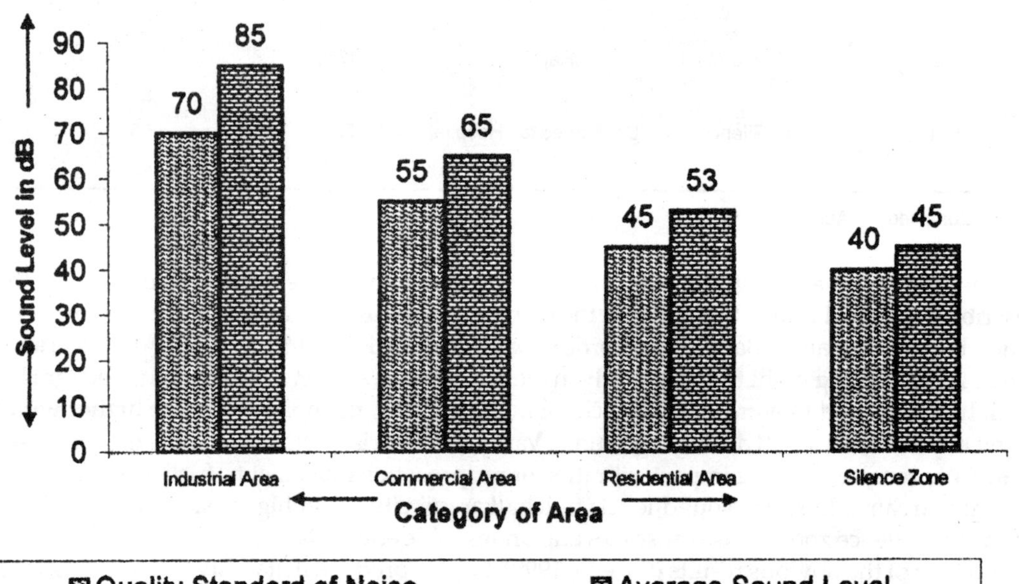

Figure 20.1B: Ichalkaranji City: Ambient Sound Level in Different Areas During Night Time

Recommendations

Since noise pollution is created by man, it can be controlled by adopting certain measures. Some of the measures are as follows.

1. Noise producing industries should be located away from residential areas.
2. To avoid traffic jamming, there is a need of over bridge near Janata Bank.
3. Use of earplugs or similar devices be made compulsory for laborers working in industries, which record higher noise levels.
4. Old machines often create more noise, therefore, all the machines should be well maintained and replaced if necessary.
5. Dense tree plantations around or nearby areas of noise can also helps to reduce its intensity.
6. To create awareness among the people regarding noise pollution, the efforts like organizing workshops, seminars and demonstrations need be adopted.

Conclusions

The foregoing analysis reveals that increased urbanization have resulted sound levels in Ichalkaranji city are at higher side in some residential, commercial and silence zone. Noise is any sound that is not wanted. It is one of the more common forms of atmospheric pollution. So the some recommendations made need be implemented rigorously.

References

Katyal, T. and Satake, M. (1998). *Environmental Pollution*. Anmol Publications, New Delhi, pp. 300.

Pawar, C.T. and Joshi, M.V. (2002). Impact of urbanization and industrialization on water quality: A case study of Ichalkaranji city of Maharashtra. *Nature, Environment & Pollution Technology*, 1(4) 351–355.

Pawar, C.T. and Joshi, M.V. (2003): Solid waste and its environmental impact in Ichalkaranji city, Maharashtra. *Nature, Environment & Pollution Technology*, 2(2) 237–240.

Rao, K.V. and Padmaja, P. (1999). Ambient noise level monitoring in Gwalior at various zones. *Journal of Environment and Pollution*, 6 (2&3): 211–214.

Saxena, H.M. (1999). *Environmental Geography*. Rawat Publication, Jaipur, pp. 307.

Tripathy, D.P. (1999). *Noise Pollution*. A.P.H. Publishing Corporation, New Delhi, pp. 200.

Trivedi, P.R. (1999). *Encyclopaedia of World Environment*. A.P.H. Publishing Corporation, New Delhi, Vol. IV and V.

Chapter 21
Studies on Limnological Characteristics of Guruvayanakere Pond Near Belthangady, S.K. District

B.A. Kumara Hegde, G. Suresha*, K. Ramadas* and B. Yashovarma***

**Department of Botany, Technical Resource Center, Sri Dharmasthala Manjunatheswara College, Ujire – 574 240, South Canara District, Karnataka, India*
***Principal, S.D.M. College, Ujire, South Canara District, Karnataka, India*

ABSTRACT

The present study deals with study of limnological characteristics of Guruvayanakere pond. The water samples from the pond was collected and analyzed for physico-chemical characteristics and planktonic composition. The pond is polluted as it contains high values of BOD. Phosphates and dissolved organic matter. 69 species of phytoplanktons belonging to 37 genera were observed. Seven zooplanktons and seven Macrophytes were recorded.

Keywords: Guruvayanakere, Limnology, Phytoplanktons, Zooplanktons, Macrophytes.

Introduction

The fresh water's a vital concern for mankind since it is directly linked with human welfare. The surface water bodies are most important sources of water for human activities. Unfortunately the surface water bodies have put under severe environmental stress and being threatened as a consequences of developmental activities and domestic uses.

Limnology is the science of inland water bodies particularly ponds and lakes. The central aspect of limnology is the biogenic material balance of natural waters.

Guruvayanakere is a small town situated by the side of Ujire–Mangalore road. The pond is situated at Guruvayanakere and it is annual receives water from adjacent paddy, fields. The total area of the pond is 79 acres of which spreads over an area of 52 acres with an average depth of 7 ft. The pond receives agricultural run-off from the paddy fields. Domestic activities like washing of clothes and animals, disposal of sewage from Guruvayanakere are contributing towards the pollution of the pond. A perusal of available literature has revealed that there is no scientific work carried out on the limnological aspects of Guruvayanakere pond. Thus with this background the present work was taken to explore the limnological characteristics of Guruvayanakere pond.

Materials and Methods

The study was conducted from June 2004 to October 2004. The water samples for the physico-chemical and planktonic composition were brought to the laboratory. The parameters such as air temperature, water temperature and pH were analyzed on spot of collection. The standard methods of APHA (1992), Saxena (1990); Trivedy and Goel (1986) were followed for physico-chemical characteristics. For studying planktonic composition monographs of Welch (1969), Krishnan Pillai (1987) and Presscot (1982) were used.

The results of Physico-chemical characteristics of Guruvayanakere pond are presented in Table 21.1. The checklist of phytoplanktons and macrophytes are appended in Table 21.2 and 21.3 respectively.

Table 21.1: Physico-chemical Parameters of Guruvayanakere Pond

Sl.No.	*Physico-chemical Parameters*	*Values*
1.	Air temperature °C	31.2
2.	Water temperature °C	29.4
3.	pH	8.26
4.	Turbidity NTU	6.8
5.	Electrical conductivity μ mhols cm^{-1}	248.0
6.	Total solids mg/L^{-1}	1436.0
7.	Total dissolved solids mg/L^{-1}	518.0
8.	Calcium as $CaCO_3$ mg/L^{-1}	24.0
9.	Magnesium mg/L^{-1}	11.0
10.	Total hardness mg/L^{-1}	166.0
11.	Phosphates mg/L^{-1}	2.82
12.	Sulphates mg/L^{-1}	64.6
13.	Sodium ppm	61.0
14.	Potassium ppm	29.0
15.	Dissolved oxygen mg/L^{-1}	5.82
16.	Free CO_2	2.18
17.	Biochemical oxygen demand 5 day 20°C mg/L^{-1}	88.6
18.	Dissolved organic matter mg/L^{-1}	21.8
19.	Total nitrogen mg/L^{-1}	1.86

Table 21.2: List of Phytoplankton

Sl.No.	Phytoplankton	Sl.No.	Phytoplankton
Chlorophyceae			
1.	*Cosmarium reticuloformes*	2.	*Cosmarium contractum*
3.	*Cosmarium subtumidum*	4.	*Cosmarium granatum*
5.	*Scenedesmus platydiscus*	6.	*Scenedesmus quadricada*
7.	*Scenedesmus dimorphus*	8.	*Scenedesmus opoliensis*
9.	*Scenedesmus protruberans*	10.	*Scenedesmus accuminatus*
11.	*Kirchinella lunaris*	12.	*Kirchinella limnetica*
13.	*Oedogonium* sp.	14.	*Coelastrum contractum*
15.	*Coelastrum microsporum*	16.	*Coelastrum* sp.
17.	*Paediastrum simplex*	18.	*Paediastrum duplex*
19.	*Paediastrum tetras*	20.	*Tetradon minimum*
21.	*Tetradon caudatum*	22.	*Tetradon trilobium*
23.	*Tetradon trigonum*	24.	*Tetrado muticum*
25.	*Tetrado protiformae*	26.	*Spirogyra* sp.
27.	*Selenastrum* sp.	28.	*Actinastrum hautzschii*
29.	*Schoroderia indica*	30.	*Volvox aurens*
31.	*Volvox globator*	32.	*Volvox* sp.
Bacillariophyceae			
1.	*Melosira granulata*	2.	*Melosira* sp.
3.	*Synedra* sp.	4.	*Navicula* sp.
5.	*Nitzschia*	6.	*Cymbella* sp.
7.	*Cymbella turgida*	8.	*Gamphonema lanceolatus*
9.	*Cyclotell stelligera*	10.	*Cocconies placentula*
Desmids			
1.	*Staurunies phoenicentron*	2.	*Euglypha* sp.
3.	*Geminela crenlatocollis*		
Cyanophyceae			
1.	*Oscillatoria limosa*	2.	*Oscillatoria princeps*
3.	*Oscillatoria* sp.	4.	*Anabaena spiroides*
5.	*Anabaena* sp.	6.	*Phormidium* sp.
7.	*Synechocystis* sp.	8.	*Spirulina nodesti*
9.	*Chlamydomonas* sp.	10.	*Gamphosphaeria* sp.
11.	*Schoroderia setigera*	12.	*Asterionella* sp.
13.	*Staurastrum* sp.	14.	*Diplonies subvalis*
Euglenophyceae			
1.	*Euglena minuta*	2.	*Euglena acus*
3.	*Euglena gracilis*	4.	*Euglena elongata*

Contd...

Table 21.2–Contd...

Sl.No.	Phytoplankton	Sl.No.	Phytoplankton
5.	*Euglena proxima*	6.	*Phacus curvicauda*
7.	*Phacus anacoelus*	8.	*Phacus longicauda*
9.	*Phacus* sp.	10.	*Trachelomonas* sp.
Zooplanktons			
1.	*Sporodimium* sp.	2.	*Elphidium* sp.
3.	*Keratella* sp.	4.	*Paramoecium caudatum*
5.	*Euplotes* sp.	6.	*Peranema* sp.
7.	*Uronema* sp.		

Table 21.3: List of Macrophytes

Sl.No.	Macrophytes	Sl.No.	Macrophytes
1.	*Eriocaulon* sp.	2.	*Eichhornea crassipes*
3.	*Monochhornea* sp.	4.	*Cyperus* sp.
5.	*Jussea* sp.	6.	*Ipomoea aquatica*
7.	*Colocassia* sp.		

Discussion

In the present study the results of the physicochemical analysis has revealed that the pond is polluted due to the anthropogenic activities such as washing of clothes and animals and discharge of the sewage, as it possess high values of BOD, phosphates and total nitrogen. The remaining parameters comply with the standards. In total 69 species of phytoplanktons belonging to 37 genera were recorded in which the Chlorophycean members were found to be dominant (32 Species). Bairagi and Goswami (1992) while working on ecology of water blooms in some ponds of north east India reported that the higher concentrations of nutrients such as dissolved organic matter, Phosphates, sulphates and total nitrogen favours the abundance of the Chlorococcales. The results of the present work are in consonance with that of the above researchers. The water is alkaline in nature (pH–8.26). The pond receives continuous inflow of domestic sewage and favours a good number of blue greens and Euglenoids. The presence of Zooplanktons represented by Rotifers and Copepods indicate that the water in monsoon season is in septic condition and this may be due to the enrichment of nutrients and alkalinity of water. Ravikumar and Puttaiah obtained similar results in their work on the ecological investigations on the lakes of Hassan dist. The results of our study are in essential agreement with that of Ravikumar and Puttaiah. In the present study it is concluded that the Guruvayanakere pond is polluted from the human interferences. The discharge of sewage and the agriculture runoff should be prevented to keep the pond unpolluted and conserve it for future generation.

Acknowledgements

The authors are grateful to Honorable President S.D.M.E. trust, Dr. D. Veerendra Heggade for his encouragement and facilities provided to us to take-up this work.

References

APHA, AWWA and WPCF. 1992. *Methods for the Examination of Water and Wastewater*, 19th ed. Washington D.C., U.S.A.

Bairagi, S.P. and Goswami, M.M. 1994, Ecology of water blooms in some ponds of N.E. India. *J. Environment and Ecology*, 12(3) 568–571.

Krishnan Pillai, N. 1986. *Introduction to Planktonology*. Himalaya Publishing House, Bombay.

Mishra and Saxena, D.N. 1992. *Aquatic Ecology*, 1st ed. Ashish Publishing House, New Delhi.

Prescott, G.W. 1982. *Algae of Western Great Lakes Area*. Otto, Koeltz Science Publishing House, W. Germany.

Ravikumar, B.S. and E.T. Puttaiah. 1996. Ecological investigations on the lakes, of Hassan dist. (Karnataka): Biological index of pollution. *Geobios*, 15(1): 13–16.

Swarnalatha, N. and Narasinga Rao, A. 1993. Ecological investigation of two lentic environment with special reference to Cyanobacteria and water pollution. *Ind. J. Microbiology and Ecology*, 3: 41–48.

Trivedy, R.K., Goel, P.K. and Trisal, L. 1998. *Practical Methods in Ecology and Environmental Science*. Enviromedia Publishing House, Karad (India).

Chapter 22

Diel Variation in Waterfowl During Winter at Sirpur Tank, Indore

Manjeet Malhotra, M.M. Prakash* and K. Pawar***

**Department of Zoology, Govt. Holkar Science College, Indore, M.P.*
***Department of Botony, Govt. Holkar Science College, Indore, M.P.*

ABSTRACT

The present study describes the diel variation in waterfowl at Sirpur Tank in the month of March (winter). The survey revealed that 32 species of waterfowl belonging to 10 families and 7 orders were present during the study month. Ducks and Geese in the morning hours followed by Gallinules and Coots, while the contribution of group grebe was lowest. In general the total number of waterfowl and species observed in the morning was more than at noon and in the evening hours. However, species like *Phalacrocorax fuscicollis, Egretta intermedia, Anas acuta* were not present in the evening hours and species like *Phalacrocorax carbo, Anhinga rufa, Ardea alba, Ardea cinerea, Porphyrio purpurea, Tadorna ferruginea, Anas sterepera, Porphyrio porphyrio, Vanellus malabaricus* were found only in the morning hours. It is interesting to note that the shore birds were recorded in the maximum number in the evening.

Keywords: Waterfowls, Winter count, Sirpur tank, Diel variation.

Introduction

Waterfowl play a very important role in wetland ecosystem, because they are conspicuous elements of consumer levels in such ecosystem. They are also considered as an important indicator of changes in the aquatic environment. Bezzel, 1974; Karlsson *et al.*, 1976; Nilsson and Nilsson, 1976; Eriksson, 1984 and Koskimies, 1987. Considerable work has been done on the characteristics, distribution, migration as well as the general ecology of birds found in the Indian subcontinent (Baker, 1931; Ali, 1968; Ali and Ripley, 1983; Vijayan, 1986 and Singh and Roy, 1989). Population trends of water birds have been studied by Barman *et al.* (1995) and Prakash (1997), while seasonal population density has been studied by Sharma and Belsare (1997) and Kumar and Bohra (2002). Literature available reveals that study of diel variation in wintering waterfowl is wanting. Therefore, an attempt has been made to study the diel variations of waterfowl at Sirpur Tank, Indore, (M.P.).

Material and Methods

The waterbody selected for the present study is known as Sirpur Tank. It is situated in Indore on the left side of Indore-Dhar Road. Sirpur tank is a shallow and tropical waterbody, covering 1,17000 ha. water spread area with 4.5 km. shore line (Sharma and Belsare, 1997). The waterfowl were identified with the help of binocular, consulting Woodcocks (1983) and Ali and Ripley's (1983) books. The count was made in the month of March 2004 weekly in the morning (6 to 7 hrs), noon (12 to 13 hrs), evening (18 to 19 hrs) and at night (24 to 1 hrs). Status of waterfowl given in text and table is taken from Ali's book (2002).

Results and Discussion

Results obtained are summarised in Tables 22.1–22.5 and presented through Bar diagrams, Pie diagrams and Line diagrams (Figures 22.1–22.11).

Prakash (1997) reported 13 species of waterbirds during winter at Bahadur Sagar, Jhabua (M.P.). Barman *et al.*, (1995) described 62 species of waterbirds belonging to 13 families from Beel Wildlife Sanctuary. Joyti *et al.* (2001) observed 24 species of waterbirds in Gharana wetland reserve. Kumar and Bohra (2002) recorded 103 species of birds belonging to 43 families and 13 orders from Udhuwa lake. In the present investigation 32 species belonging to 10 families and 7 orders were observed (Table 22.1) in the Sirpur tank during March 2004. However, Sharma and Belsere (1997) in his annual study reported total 59 species of waterfowl from the same tank belonging to 13 families. This showed that wintering avifauna of Sirpur Tank included less species and familes. In authors' opinion this may be due to the departure of many migratory species of waterfowl from this water body before March. Waterfowl reported in the present investigation included 13 species of winter visitor or migratory status. The rest were of resident or local migratory nature.

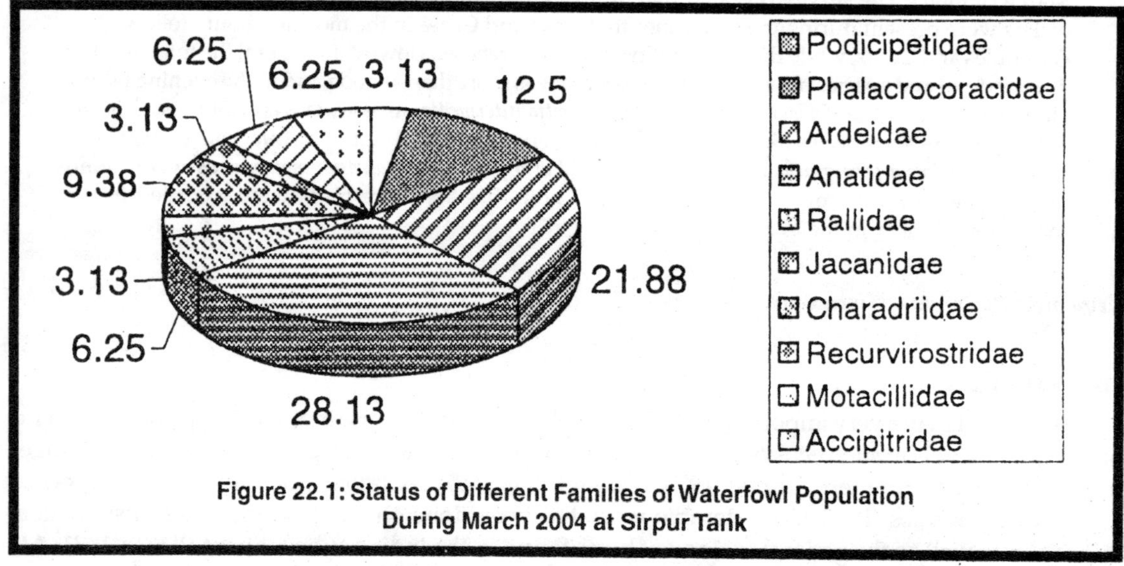

Figure 22.1: Status of Different Families of Waterfowl Population During March 2004 at Sirpur Tank

Table 22.1: Waterfowls of Sirpur Tank, Indore

Sl.No.	Species	Common Name	Local Name	Family	Order	Status
1.	*Podiceps ruficollis*	Little grebe or Dabchick	Pandubbi	Podicipetidac	Podicipediformes	M
2.	*Phalacrocorax fuscicollis*	Indian Shag	Pankawwa	Phalacrocoraciedae	Pelecaniformes	R
3.	*Phalacrocorax niger*	Little Cormorant	Pankawwa	Phalacrocoraciedae	Pelecaniformes	R
4.	*Phalacrocorax carbo*	Large Cormorant	Pankawwa	Phalacrocoraciedae	Pelecaniformes	R
5.	*Anhinga rufa*	Darter or Snake Bird	Banbi	Phalacrocoraciedae	Pelecaniformes	R, L, M
6.	*Egretta intermedia*	Median Egret	Patangkha Bagula	Ardeidae	Ciconiformes	R
7.	*Egretta garzetta*	Little Egret	Karchia Bagula	Ardeidae	Ciconiformes	R
8.	*Bubulcus ibis*	Cattle Egret	Gai Bagula	Ardeidae	Ciconiformes	R
9.	*Ardea alba*	Large Egret	Bada Bagula	Ardeidae	Ciconiformes	R, N
10.	*Ardeola grayii*	Pond Heron	Andha Bagula	Ardeidae	Ciconiformes	R
11.	*Ardea cinerea*	Grey Heron	Anjan	Ardeidae	Ciconiformes	R
12	*Ardea purpurea*	Purple Heron	Lal Anjan	Ardeidae	Ciconiformes	R, L, M
13.	*Nettapus caromandelianus*	Cotton Teal	Girria, Silli	Anatidae	Anseriformes	R, L, M
14.	*Dendrocygna juvanica*	Lessor whistling teal	Seelhi	Anatidae	Anseriformes	R, L, M
15.	*Tadorna ferruginea*	Ruddy Shelduck	Surkhab	Anatidae	Anseriformes	M
16.	*Anas acuta*	Pin Tail	Seenkh	Anatidae	Anseriformes	M
17.	*Anas poecilorhyncha*	Spot billed duck	Garm, Pai	Anatidae	Anseriformes	R, N
18.	*Anus strepera*	Gadwall	Myla, Bhuar	Anatidae	Anseriformes	M
19.	*Anas penelope*	Wigeon	Chhota Lalsir	Anatidae	Anseriformes	M
20.	*Anas querquedula*	Garganey	Palari	Anatidae	Anseriformes	M
21.	*Netta rufina*	Red crested pochard	Lal Chonch	Anatidae	Anseriformes	M
22.	*Porphyrio porphyrio*	Indian purple moorhen	Kalim	Rallidae	Gruiformes	R, L, M
23.	*Fulica atra*	Common Coot	Dasari Dasarni	Rallidae	Gruiformes	M
24.	*Hydrophasianus chirurgus*	Pheasant tailed jacana	Pihuya	Jacanidae	Charadriformes	R
25.	*Limosa limosa*	Black Tailed Godwit	Gudera	Charadriidae	Charadriformes	M
26.	*Tringa glareola*	Spotted Sandpiper	Chupka	Charadriidae	Charadriformes	M
27.	*Tringa hypoleucos*	Common Sandpiper	Merwa	Charadriidae	Charadriformes	M
28.	*Himantopus hemantopus*	Blackwinged Still	Gazpaon	Recurvirostridae	Charadriformes	R
29.	*Venellus indicus*	Red Wattled Lapwing	Titeeri	Motacillidae	Passeriformes	R
30.	*Vanellus malabaricus*	Yellow Wattled Lapwing	Zirdi	Motacillidae	Passeriformes	R
31.	*Charadrius dubius*	Little Ringed Plover	Merwa	Accipitridae	Falconiformes	M
32.	*Calidris minuta*	Little Stint	Chhota Pankawwa	Accipitridae	Falconiformes	M

Abbreviation: M: Migrant; RM: Resident migrant; R": Resident.

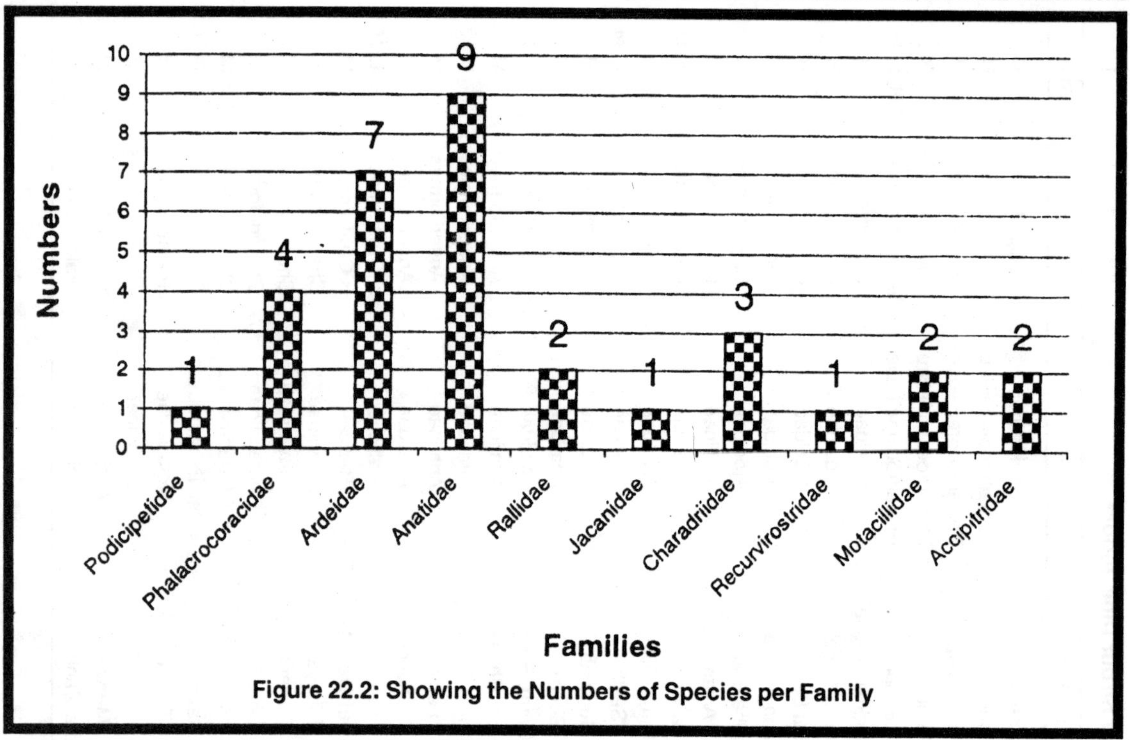

Figure 22.2: Showing the Numbers of Species per Family

Table 22.2: Species-wise Dominance of Family at Sirpur Tank, Indore

Sl.No.	*Family*	*No. of Species/Family*	*% Dominance*
1.	Podicipetidae	1	3.13
2.	Phalacrocoracidae	4	12.50
3.	Ardeidae	7	21.88
4.	Anatidae	9	28.13
5.	Rallidae	2	6.25
6.	Jacanidae	1	3.13
7.	Charadriidae	3	9.38
8.	Recurvirostridae	1	3.13
9.	Motacillidae	2	6.25
10.	Accipitridae	2	6.25
	Total	**32**	**100**

The most dominant family in the present investigation was Anatidae, which consists of 9 species. Its percentage contribution was 28.12, which is highest in the total observed families (Table 22.2 and Figure 22.2). Braman *et al.* (1995) also reported Anatidae as dominant family in his observed the basis of percentage contribution and number of species the ladder of family dominence can be represented as:

Table 22.3: Mean Diel Winter Count of Waterfowl at Sirpur Tank (Indore)

Sl.No.	Species	Time (Hours)			
		6 to 7 (Morning)	12 to 13 (Noon)	18 to 19 (Evening)	24 to 1 (Night)
(I)	**Grebes**				
1.	*P. ruficollis*	25	00	15	00
(II)	**Cormorant and Darter**				
2.	*P. fuscicollis*	50	25	00	00
3.	*P. niger*	50	30	10	00
4.	*P. corbo*	10	00	00	00
5.	*A. rufa*	2	00	00	00
(III)	**Herons and Egrets**				
6.	*E. intermedia*	10	5	00	00
7.	*E. garzetta*	55	10	50	00
8.	*B. ibis*	10	10	15	00
9.	*A. alba*	2	00	00	00
10.	*A. grayii*	20	5	25	00
11.	*A. cinerea*	6	00	00	00
12.	*A. purpurea*	4	00	00	00
(IV)	**Geese and Ducks**				
13.	*N. caromandelianus*	500	00	50	00
14.	*D. javanica*	150	00	15	00
15.	*T. ferruginea*	10	00	00	00
16.	*A. acuta*	25	40	00	00
17.	*A. poecilorhymcha*	10	00	50	00
18.	*A. strepera*	80	00	00	00
18.	*A. penelope*	200	100	25	00
19.	*A. querquedula*	100	80	35	00
21.	*N. rufina*	80	50	40	00
(V)	**Gallinules and Coot**				
22.	*P. porphyrio*	4	00	00	00
23.	*F. atra*	1000	250	50	00
(VI)	**Jacanans**				
24.	*H. chirurgus*	40	00	20	00
(VII)	**Shore Birds**				
25.	*L. limosa*	15	00	10	00
26.	*T. glareola*	100	30	50	00
27.	*T. hypoleucas*	100	40	60	00
28.	*H. hementopus*	200	50	500	00
29.	*V. indicus*	50	15	40	00
30.	*V. malabaricus*	2	00	00	00
31.	*C. dubius*	4	00	2	00
32.	*C. minuta*	60	00	30	00
	Total No. of Species	**32**	**15**	**20**	**00**
	Total No. of Birds	**2974**	**740**	**1092**	**00**

Table 22.4: Group-wise Diel Variation of Waterfowl at Sirpur Tank, Indore

Sl.No.	Group Name	Count at Hours		
		6 to 7 (Morning)	12 to 12 (Noon)	18 to 19 (Evening)
1.	Grebe	25	00	15
2.	Cormorants and Darter	112	55	10
3.	Herons and Egrets	107	30	90
4.	Geese and Ducks	1155	270	215
5.	Gallinules and Coot	1004	250	50
6.	Jaconas	40	00	20
7.	Shore Birds	531	135	692
	Total	**2974**	**740**	**1,092**

Note: Column of 24 to 1 hours is deleted because of absent of waterfowl.

Table 22.5: Deil Variation in Percent Composition of Various Groups of Waterfowl at Sirpur Tank, Indore

Sl.No.	Group Name	Percent Contribution at Hours		
		6 to 7 (Morning)	12 to 13 (Noon)	18 to 19 (Evening)
1.	Grebe	0.84	00	1.37
2.	Cormorants and Darter	3.77	7.43	0.92
3.	Herons and Egrets	3.60	4.05	8.24
4.	Geese and Ducks	38.84	36.49	19.69
5.	Gallinules and Coot	33.76	33.38	4.58
6.	Jacanas	1.34	00	1.83
7.	Shore Birds	17.85	18.24	63.37

Note: Column of 24 to 1 hours is deleted because of absent of waterfowl.

Anatidae > Ardeidae > Phalacrocoracidae > Charadridae > Rallidae, Motacillidae, Accipitridae > Podicipetidae, Jacanidae, Recurvivostridae

Diel winter counts of waterfowl are presented in the Table 22.3. This shows that total counted species are higher in the morning and lower at noon (Figure 22.10). However, in the evening their positions were in between them. In the night no species were observed. A perusal of Table 22.4 revealed that Geese and Ducks (1155) were maximum in number followed by Gallinules and Coots (1004), while Grebes (25) were recorded in the minimum number.

Species-wise diel winter count are presented in Table 22.3 and Figures 22.3–22.9, which shows that in the morning 32 species were present. In noon their numbers was 15 which further rase in the evening and reached up to 20. In the morning total 2974 waterfowl were counted. The total number reduced to 740 at noon. However, in the evening their number again increased and reached upto 1092 (Table 22.4). The birds *P. carbo, A. rufa, A. alba, A. cinerea, P. purpurea, T. ferruginea, A. sterepera, P. porphyrio, V. malabaricus* were observed only in the morning hours. The birds *P. ruficollis, E. intermedia, A. acuta* were not found in the evening. However, the rest of the species recorded were present at all times

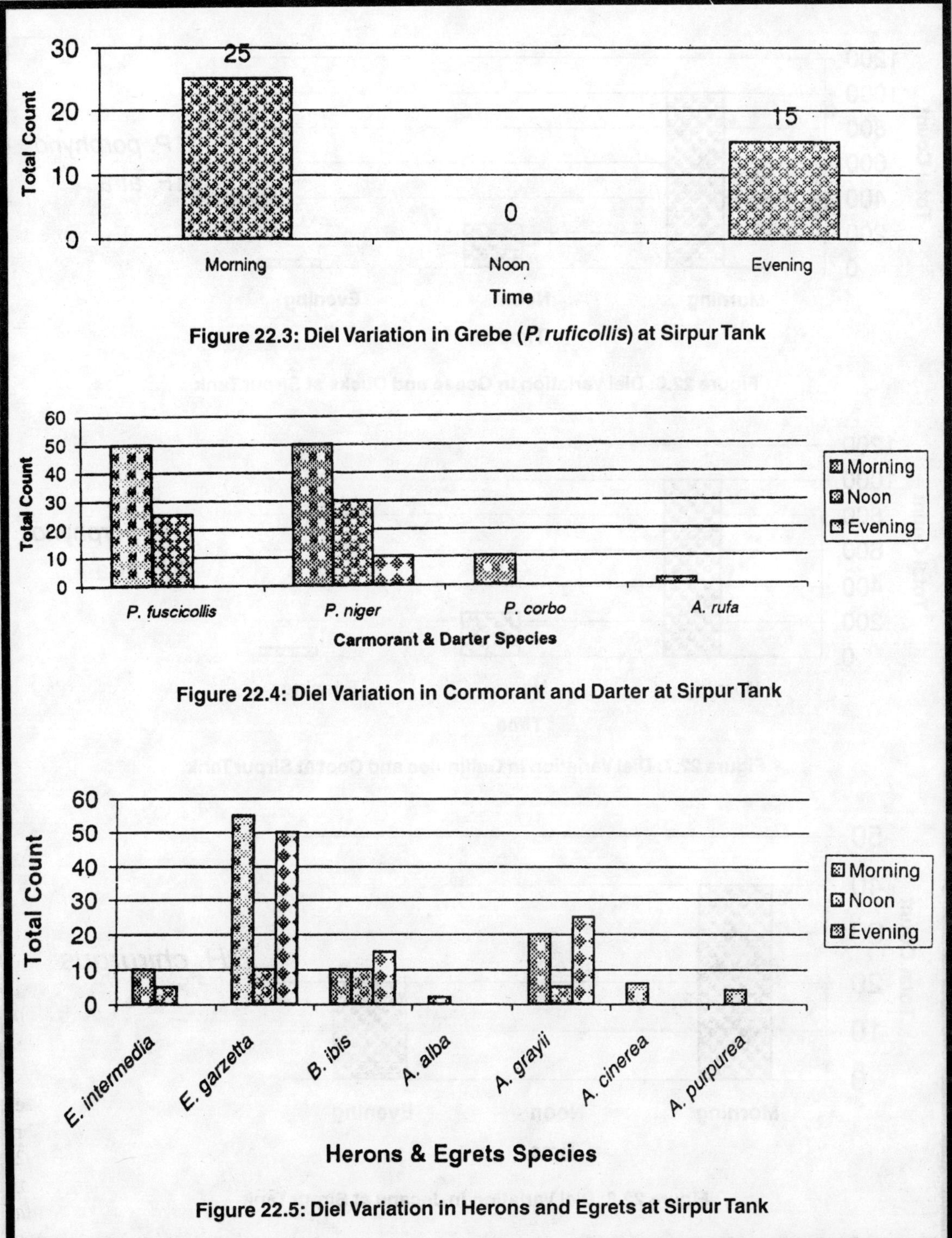

Figure 22.3: Diel Variation in Grebe (*P. ruficollis*) at Sirpur Tank

Figure 22.4: Diel Variation in Cormorant and Darter at Sirpur Tank

Figure 22.5: Diel Variation in Herons and Egrets at Sirpur Tank

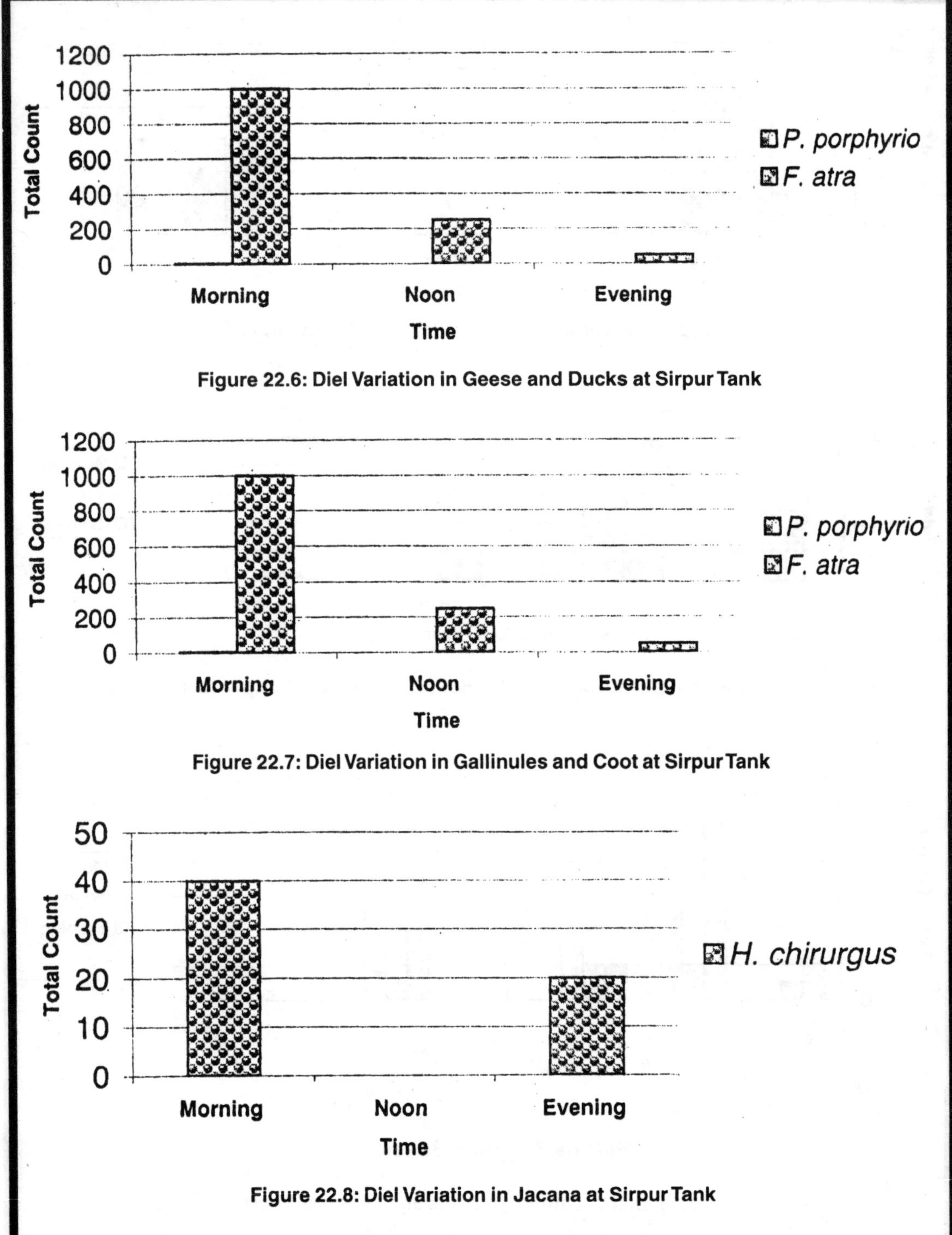

Figure 22.6: Diel Variation in Geese and Ducks at Sirpur Tank

Figure 22.7: Diel Variation in Gallinules and Coot at Sirpur Tank

Figure 22.8: Diel Variation in Jacana at Sirpur Tank

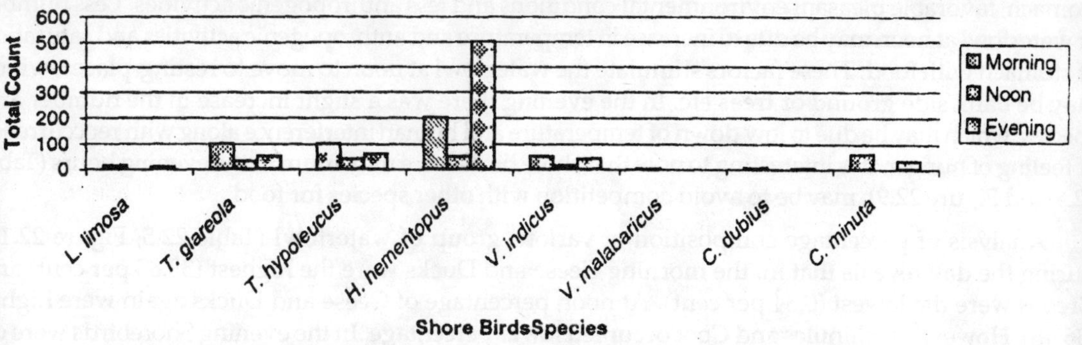

Total Count
600
500
400
300
200
100
0
Morning
Noon
Evening
L. limosa
T. glareola
T. hypoleucus
H. hementopus
V. indicus
V. malabaricus
C. dubius
C. minuta
Shore BirdsSpecies

Figure 22.9: Diel Variation in Shore Birds at Sirpur Tank

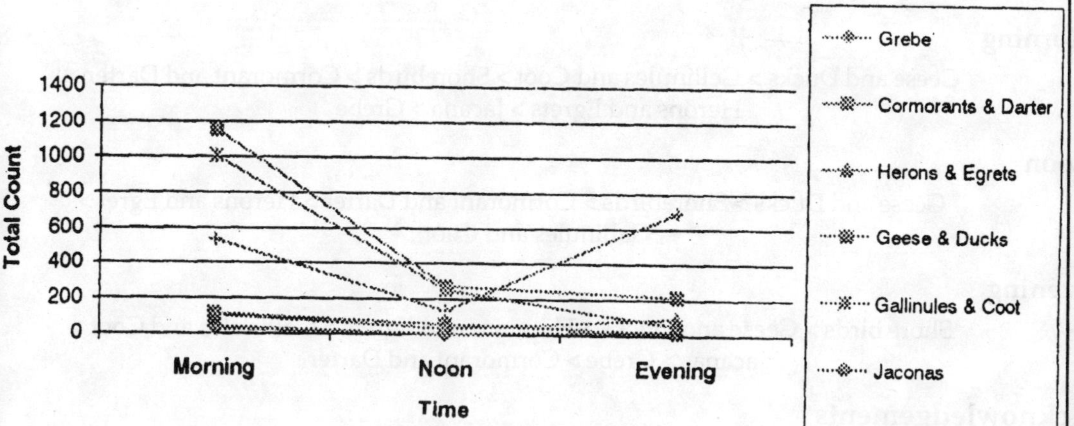

Total Count
1400
1200
1000
800
600
400
200
0
Morning
Noon
Evening
Time
Grebe
Cormorants & Darter
Herons & Egrets
Geese & Ducks
Gallinules & Coot
Jaconas
Shore Birds

Figure 22.10: Diel Variation in Waterfowl Group at Sirpur Tank (March, 2004)

100%
80%
60%
40%
20%
0%
Morning
Noon
Evening
Time
Shore birds
Jacanas
Gallinules & Coot
Geese & Ducks
Herons & Egrets
Cormorants & Darter
Greebe

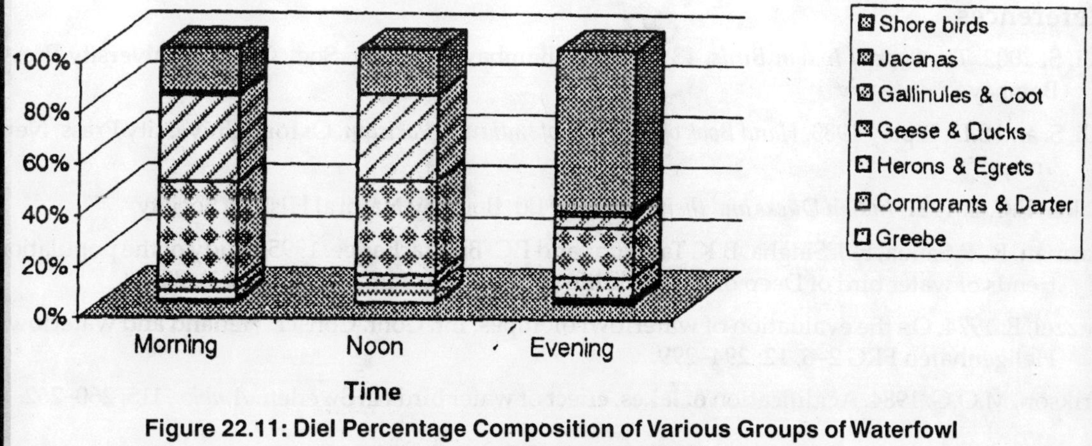

Figure 22.11: Diel Percentage Composition of Various Groups of Waterfowl at Sirpur Tank (March, 2004)

except night. Presence of more birds in number and species in the morning may be due to their empty stomach, favorable pleasant environmental conditions and less anthropogenic activities. Less number of waterfowl at noon may be due to increase in temperature and anthropogenic activities and saturation of stomach with food. These factors stimulate the waterfowl at noon to move to resting-places, which may be bank side ground or trees etc. In the evening there was a slight increase in the number and species, which may be due to low down of temperature and human interference along with reoccurrence of feeling of hunger. It is interesting to note that shore birds were maximum in the evening hours (Table 22.4 and Figure 22.9), may be to avoid competition with other species for food.

Analysis of percentage composition of various group of waterfowl (Table 22.5, Figure 22.11) during the day reveals that in. the morning Geese and Ducks were the highest (38.83 per cent) and Greebs were the lowest (0.84 per cent). At noon percentage of Geese and Ducks again were higher (36.48). However, Gallinules and Coot occupied lower percentage. In the evening Shorebirds were on the highest position (63.36 per cent) and Greebs were on the lowest position (1.37 per cent). The order of dominance of groups can be presented as:

Morning

Geese and Ducks > Gellinules and Coot > Shorebirds > Cormorant and Darter > Herons and Egrets > Jacana > Grebe

Noon

Geese and Ducks > Shorebirds > Cormorant and Darter > Herons and Egret > Gallinules and Coot

Evening

Shore birds > Geese and Ducks > Hermons and Egrets > Gallinules and Coot > Jacanas > Grebe > Cormorant and Darter

Acknowledgements

The authors are thankful to Dr. N.K. Dhakad, Principal Holker Science College, Indore and Dr. Ashok Sharma, The Head of the Zoology Department, Holker Science College, Indore for providing necessary facilities and healthy research environment.

References

Ali, S. 2002. *The Book of Indian Birds*, 13th edition. Bombay Nat. Hist. Soc., Oxford University Press, Bombay.

Ali, S. and S.D. Ripley 1983. *Hand Book of the Birds of India and Pakistan*. Oxford University Press, New York.

Baker, E.C.G. 1921. *Indian Ducks and their Allies*, 2nd Ed. Bombay Natural History Society.

Barman, R., P. Saikia, H.J. Singha, B.K. Talukdar and P.C. Bhattacharjee. 1995. Study of the population trends of water bird of Deep or Beel wildlife sanctuary, Assam. *Pavo*, 33(1&2): 25–40.

Bezzel, E. 1974. On the evaluation of waterfowl biotopes. Int. Conf. Conser. Wetland and waterfowl, Heligenhafen FRG 2–6, 12: 294–299.

Erikson, M.O.G. 1984. Acidification of lakes, effect of water birds in Sweden. *Ambio.*, 115: 260–262.

Jyoti, M.K., Sharma, K.K., Sharma, Surinder P. and Singh P. 2001. Biodiversity of Gharana wetland reserve with special reference to water birds. Abstract in Nat. Sem. on Biodiversity Conservation at J.U. Gwalior (Nov. 26–28, 2001), pp. 16.

Karlsson, J., Lindergreen, A. and Rudebeek, G. 1976. Drastiska for andringer Och flagelfauna; krankesjon Och Biorkesakrasion. 1973–1976 (In Swedish). *Anser*, 15: 165–184.

Koskimies, P. 1987. Monitoring of finnish bird fauna. Bird as environmental indicator (In finnish with English summary) Pub. Ministry Environment rer. A. 49: 1–255.

Kumar, A. and Bohra, C.J. 2002. Faunistic composition, communities structure, population dynamics and ethology of the birds of Udhuwa lake (Santhal Pargana) and the strategies for the conservation in new millennium. In: *Ecology and Conservation of Lakes, Reservoir and River* Vol. II, (Ed.) A. Kumar). ABD Publisher, Jaipur, p. 411–434.

Nilsson, S.G. and Nilsson, I.N. 1976. Valuation of South Swedish wetland for conservation with the proposal of a new method for valuation of wetland as breeding habitat for birds (In Swedish with English summary) Fauna. *Ack Flora*, 71: 136–144.

Prakash, M.M. 1997. Some Salient features of a wintering habitat of aquatic birds. In: *Modern Trends of Research in Zoology*, (Ed) M.M. Prakash, RBSA Publisher, Jaipur, p. 212–223.

Sharma, S. and Belsare, D.K. 1997. A check-list of waterfowl of Sirpur Lake. *Ind. J. Z. Spect.*, 8(2): 39–42.

Singh, J.P. and Roy, S.P. 1989. Faunistic composition community structure and behavior of birds of the Karwar lake, Begusaraia, Bihar. *Biol. Bull., India*, 11: 1–8.

Vijayan, V.S. 1986. On conserving the bird fauna of Indian wetland. Proc. Indian Acad. Sci. (Animal Science/Plant Science). Supplement. Nov., p. 99–101.

Woodcock, M. 1983. *Collins Hand Guide to the Birds of the Indian Sub-continent*. William Collins Sons and Co. Ltd. London.

Chapter 23

Removal of Heavy Metals from Electroplating Industrial Effluent Using Plants–Phytoremediation

S. Anitha, V. Mahesh and C. Sheela Sasikumar

P.G. Department of Biochemistry, D.G Vaishnav College, Chennai, India

ABSTRACT

Toxic metal contamination of soil, aqueous waste streams and groundwater causes major environmental and human health problems. The most commonly used methods for dealing with heavy metal pollution are still extremely costly. Phytoremediation is emerging as an innovative tool with greater potential for achieving sustainable development to decontaminate metal polluted air, soil and water. Plants are the ideal agents for soil and water remediation because of their unique genetic, biochemical and physiological properties. The present study is an attempt to investigate the ability of easily available and inexpensive plant bark to adsorb heavy metals from electroplating industrial effluent.

Keywords: *Adsorption, Heavy metal, Plant bark, Electroplating effluent, Phytoremediation.*

Introduction

The current concern for increased soil contamination due to both industrial and domestic pollution depends on several economic and health considerations: in many cases contaminated sites cannot be safely reused unless prior remediation measures are implemented (Hagemeyer, 1999), long-term effects may transform toxic compounds into non-predictable and even more toxic hazardous substances that may leak into water sources; and the volume of domestic and industrial waste is increasing continuously (Ghosh *et al.*, 2000).

Heavy metals are highly toxic because of their non-biodegradable nature and is found generated especially 90 per cent in electroplating industries alone (Longman, 1996). Plants can be compared to solar driven pumps, that extract and concentrate the elements from the environment (atleast 5 per cent of its dry weight). All plants extract necessary nutrients, including metals, from their soil and water environments (Mofa, 1995). Some plants, called hyperaccumulators, have the ability to store large amounts of metals, even some metals that do not appear to be required for plant functioning (Williams, 2000).

In the present study, an attempt has been made to record the phytoremediation capacity of *Tamarindus indicus* bark and *Pongamia glabra* bark. The influence of contact time on the adsorption of heavy metal like Cr, Ni, Zn, Pb in the electroplating effluent has been investigated (Singh *et al.*, 1994: Martin-Dupont *et al.*, 2002).

Materials and Methods

Tamarind and pongamia barks collected were dried in room temperature and ground. 500mg of bark powder was taken in conical flasks containing 100ml of effluent. The setup was agitated in a mechanical shaker for 30, 60, 90, 120 minutes at room temperature. Estimation of heavy metals were done following the standard procedures (Basett *et al.*, 1986).

Results and Discussion

Plants that hyper accumulate metals have tremendous potential for application in remediation of metals in the environment (Azadpour *et al.*, 1996). The control values were found to be 50, 40, 15 and 150 mg for Zn, Ni, Pb, Cr respectively.

Pongamia barks exhibited some appreciable affinity for the binding of Pb ions from the solution by 22 per cent to 56 per cent during the treatment. It is observed during the investigation that Cr concentration has decreased from 150 mg to 112.5mg level in 60min and the per cent reduction is 25. It can also be seen from the graph that there is a further decline in Cr steadily with increase in contact time up to 120 min. (40 per cent).

Table 23.1: Effect of Tamarind (*Tamarindus indicus*) Bark on Heavy Metal Concentration in Electroplating Effluent

Time	*Heavy Metals (mg/l)*			
	Zinc	*Nickel*	*Lead*	*Chromium*
0	50	40	15	150
30	31.66	35	13.33	130
60	29.99	35	11.66	112.5
90	25	30	9.99	105
120	23.33	30	9.99	101

It is interesting to note that removal efficiency of Zn by tamarind bark is higher at 30 min. time (37 per cent) compared to Pongamia (33 per cent). But as time passes (l20min) removal efficiency of Pongamia is more (77 per cent). The graph shows the faster adsorption of Ni ions from the effluent especially in first 30 min. (13 per cent). Whereas Pongamia could adsorb only 38 per cent even after 120 min. Plants can take up heavy metals by their roots or even via their stems and leaves and accumulate them in their organs (Cheng, 2003).

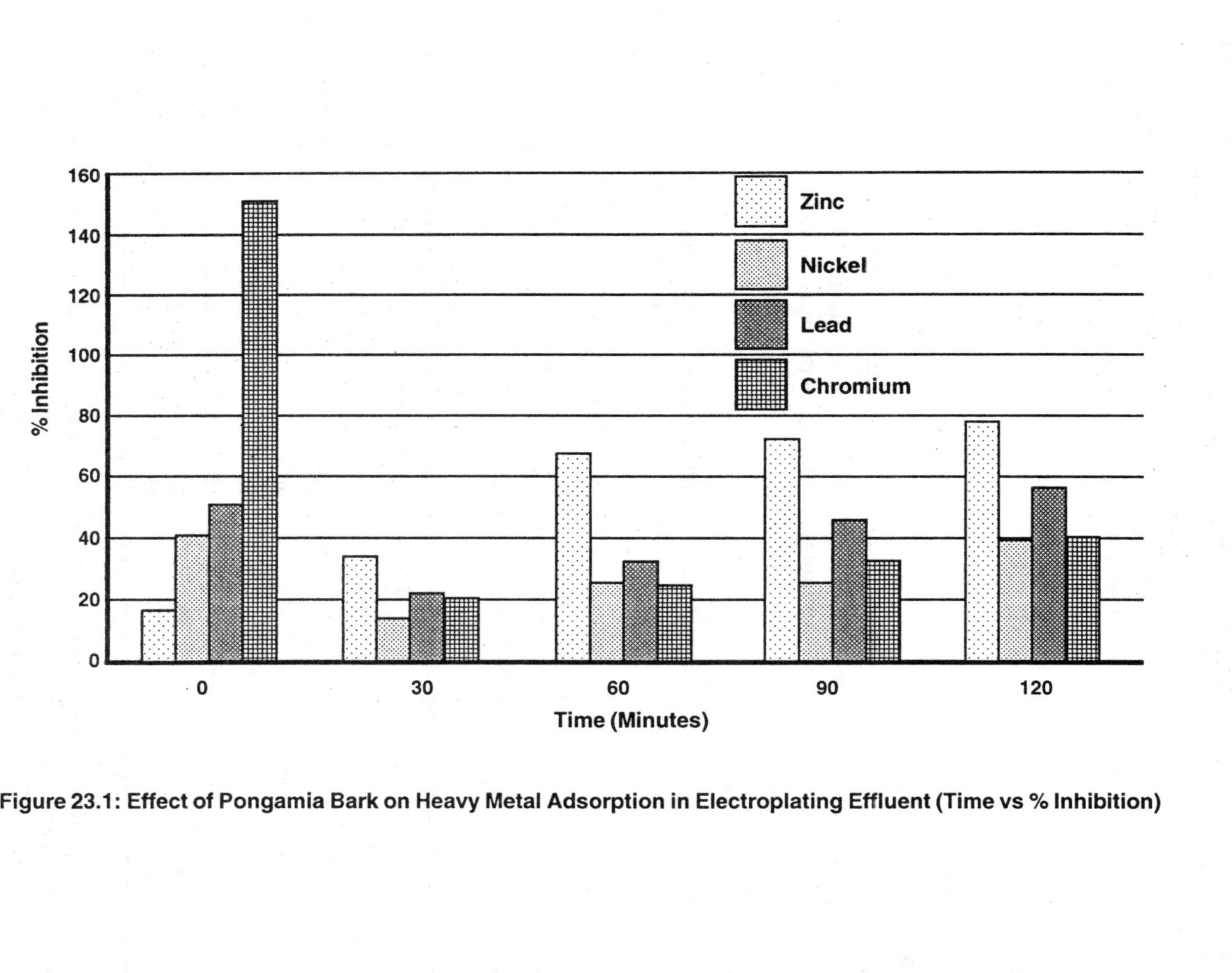

Figure 23.1: Effect of Pongamia Bark on Heavy Metal Adsorption in Electroplating Effluent (Time vs % Inhibition)

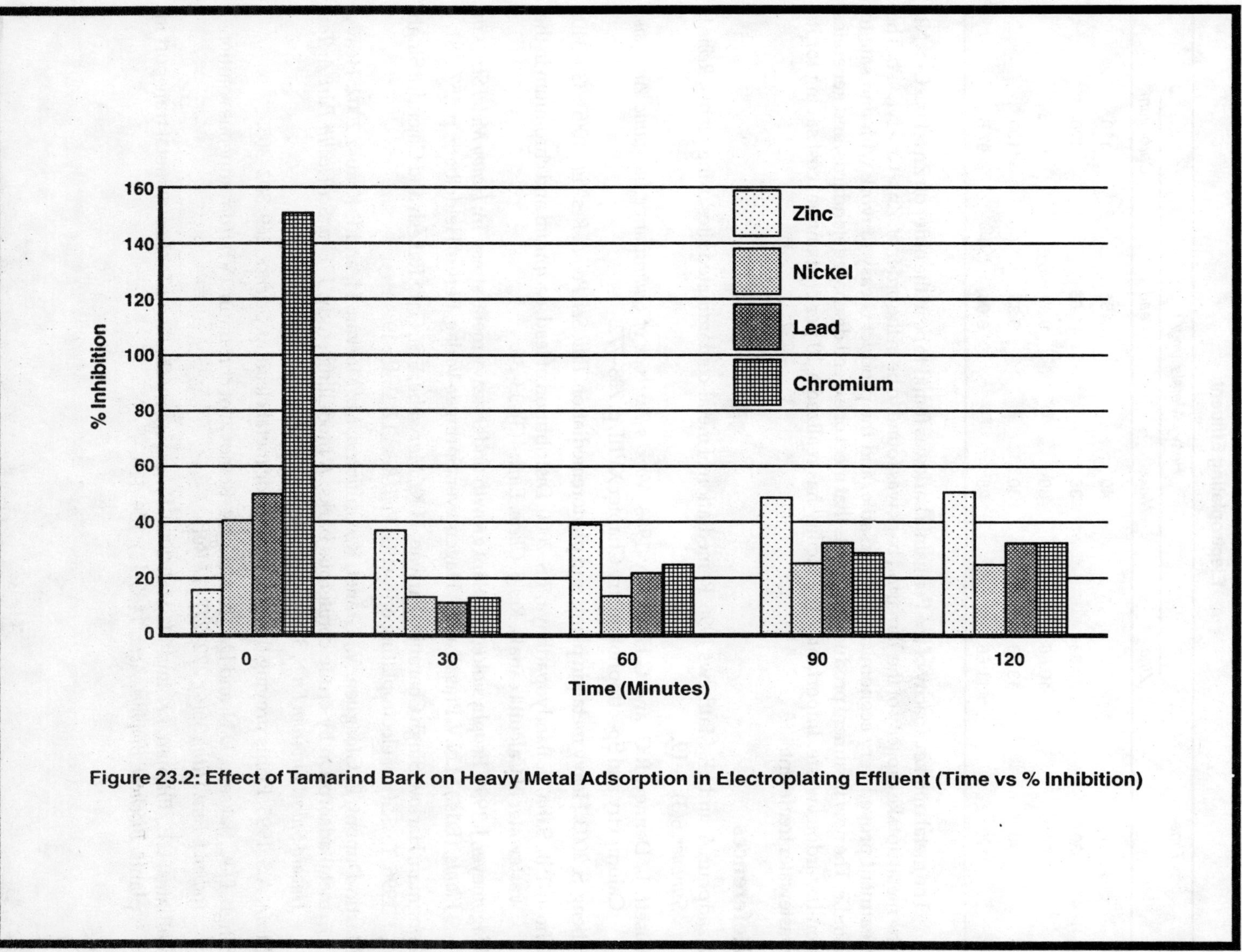

Figure 23.2: Effect of Tamarind Bark on Heavy Metal Adsorption in Electroplating Effluent (Time vs % Inhibition)

Table 23.2: Effect of Pongamia (*Ponganlia glabra*) Bark on Heavy Metal Concentration in Electroplating Effluent

Time	*Heavy Metals (mg/l)*			
	Zinc	*Nickel*	*Lead*	*Chromium*
0	50	40	15	150
30	33.33	35	11.66	120
60	16.66	30	9.99	112.5
90	13.33	30	8.33	100
120	11.66	25	6.66	97.5

The metal uptake capacity of the Pongamia bark was found to be in the order of: Zn > Pb > Cr > Ni. The metal uptake capacity of the Tamarind bark was found to be in the order of: Zn > Cr > Ni > Pb. The treatment proved very economical and versatile, and the product is easy to work with and safe to handle. The new treatment produces effluents that meet or exceed discharge standards, and generates non-hazardous waste. Importantly, the facility has realized a 10 per cent annual cost saving for its wastewater treatment.

References

Azadpour, A. and J.E. Matthews 1996. Remediation of metal-contaminated sites using plants. *Remed. Summer.* 6(3): 1–19.

Basett, J., Denney, R.C. and G.H. Jeffery 1986. *Vogel's Textbook of Quantitative Inorganic Analysis.* Calorimetry and Spectrophotometry Chapter XVIII, p. 738–772.

Cheng, S. 2003. Heavy metals in plants and phytoremediation. *Env. Sci. Pollut. Res. Int.,* 10(5): 335–340.

Ghosh, P.B., Saha, T., Bandyopadhaya, T.S. 2000. Distribution of lead, cadmium and chromium in the wastewater of Calcutta canals. *Res. J. Chem. Env.,* 4(3): 33–36.

Hagemeyer, J. 1999. Ecophysiology of plant growth under heavy metal stress. In: *Heavy Metal Stress in Plants,* (Eds.) M.N.V. Prasad and J. Hagemeyer. Springer-Verlag, Berlin, Heidelberg, p. 157–181.

Longman, Harlow, Singh Charanjit, Parwan, H.K., Merwaha, S.S., Garg Rakesh and Gajendra Singh 1996. Toxicity of electroplating effluents. *Poll. Res.,* 12(1): 15–19.

Martin-Dupont, F., Gloaguen, V., Granet, R., Guilloton, M., Morvan, H. and P. Krausz 2002. Heavy metal adsorption by crude coniferous barks: A modelling study. *J. Env. Sci Health Part A: Tox Hazard Subst. Environ. Eng.,* 37(6): 1063–1073.

Mofa, A.S. 1995. Plants proving their worth in toxic metal cleanup. *Science,* 269: 302–305.

Singh, D.K., Saksena, D.N. and D.P. Tiwari 1994. Removal of chromium (VI) from aqueous solutions. *Indian J. Env. Hlth.,* 36(4): 272–277 (17 Ref).

Williams, L.E., Pittman, J.K. and J.L. Hall 2000. Emerging mechanisms for heavy metal transport in plants. *Biochem. Biophys. Acta,* 1465(12): 104–126.

Chapter 24
Effects of DAP Toxicity on Certain Hematological Parameters of *Acridotheres tristis*

Anant Kumar, R.K. Yadav** and B.P. Akela****

**Lecturer, Department of Zoology, Adarsh College, Ghailarh-Jiwachpur, Madhepura*
***Principal, Adarsh Inter College, Ghailarh-Jiwachpur, Madhepura*
****P.G. Department of Zoology, L.N. Mithila University, Darbhanga*

ABSTRACT

The effect of diammonium phosphate, a chemical fertilizer led to significant decrease in haemoglobin content and packed cells volume of *Acridotheres tristis*. Fertilizer toxicity caused neutrophilia, lymphocytopaenia, monocytopaenia, eosinophilia upto 1000 ppm but thereafter reverse trend was observed in 1500 ppm DAP treatment. The fall in the haemoglobin content was due to hypochromatic microcytic anaemia which in turn was due to decreased utilization of haemoglobin synthesis on the application of DAP toxicity.

Keywords: Hematology, D.A.P. toxicity, A. tristis.

Introduction

The hematology is a growing field of research in animal biology and the importance of hematological research has considerably increased due to the impact of environmental pollution to indicate the level of harmful effect on animals. It has been a dependable diagnostic tool in the treatment of diseases but studies on avian hematology is very scarce. The haematological parameters vary according to change in age, sex, season and administration of toxicants. The variations in the hematological parameters due to toxicants have been worked out by Breuer and Baldwin (1995), Maqueet *et al.*, (1993), Agrawal and Srivastava (1980), Banerjee and Verma (1992), Banerjee (1986), Das and Konar (1990), Gilbert (1968), Quli and Banerjee (1994), Rani (1999), Sharma and Saxena (1997). A number of references are available on the age and body weight related hematological variations in birds but scanty literature is available on hematological changes in birds on the effect of an organic fertilizer diammonium phosphate (DAP). Hence, in the present paper an attempt has been made to investigate the changes in differential count of leucocytes, haemoglobin content and packed cells volume (P.C.V.) in the blood of Acridotheres tristis on the effect of DAP toxicity.

Materials and Methods

The common Indian Myna, *Acridotheres tristis* (Lenn) was used as experimental animal. It was collected from the local catches and were brought to the laboratory. The LC_{50} of DAP for myna was 2000 ppm. Thus experiments were set at three sublethal Concentration of DAP *i.e.,* at 500 ppm 1000 ppm and 1500 ppm. The toxic efffect of DAP was observed by using contaminated food. The experiments were also held in both sexes of birds. The birds were anaesthesized and blood was collected form the left brachial vein form wing under side and was mixed with sodium as an citrate anticoagulant. The haemoglobin content was estimated by Sahli's acid haematin method (Darmady and Davenport, 1954), the differential count of blood was made with Leishman's method (Bell, 1959) and P.C.V. was determined by microcentrifuge method (Srivastava and Das, 1995).

Results and Discussion

The haemoglobin content in Control birds was higher in males (14.0±0.460 gm per cent) than that of females (13.57±0.188 gm per cent). It was found that DAP toxicity caused a gradual and significant fall in the level of haemoglobin up to 1000 ppm treatment but thereafter, significant eleration in the haemoglobin content was found in both sexes of bird (Table 24.1). The P.C.V. percentage was also found greater in male birds (27.46±0.364 per cent) than that of females (24.911±0.853 per cent). In male and female the trend of variation was found similar.

Table 24.1: Showing the Effect of DAP Toxicity on Blood Parameters of *Acridotheres tristie*

Parameters	*Sex*	*Control*	*DAP Toxicity (in ppm)*		
			500	*1000*	*1500*
Neutrophils (%)	Male	58.7±0.633	60.6±0.513	63.4±0.569	59.9±0.498
	Female	57.3±0.513	59.3±0.470	62.3±0.490	57.1±0.573
Lymphocyte (%)	Male	33.3±0.800	25.8±0.946	28.3±1.294	30.8±0.834
	Female	32.6±0.695	27.7±0.566	24.2±0.629	34.3±0.678
Monocytes (%)	Male	4.6±0.209	3.6±0.154	2.8±0.189	4.7±0.356
	Female	5.6±0.379	4.6±0.289	3.9±0.221	4.9±0.221
Eosinophils (%)	Male	2.6±0.209	8.7±0.678	9.8±0.562	3.2±0.613
	Female	3.5±0.291	7.3±0.448	8.4±0.473	2.6±0.209
Basophils (%)	Male	0.7±0.202	1.1±0.242	1.2±0.189	1.0±0.244
	Female	0.7±0.141	1.1±0.282	1.2±0.126	1.0±0.141
Haemoglobin (in gm %)	Male	14.0±0.460	13.04±0.219	11.78±0.059	13.06±0.163
	Female	13.57±0.188	13.23±0.144	11.6±0.134	12.82±0.113
P.C.V. (%)	Male	27.46±0.364	23.19±0.261	20.13±0.272	25.3±0.196
	Female	24.911±0.853	21.9±0.246	18.28±0.262	24.61±0.256

In birds the neutrophils were maximum in number. Lymphocytes were next to neutrophils. The basophils were minimum in number. The percentage of neutrophils, lymphocytes, monocytes, eosinophils and basophils in control male birds were 58.7±0.633 per cent, 33.3±0.80 per cent, 4.6±0.209 per cent, 2.6±0.209 per cent and 0.7±0.202 per cent respectively (Table 24.1). DAP toxicity (500 ppm) caused nutrophilia, lymphocytopaenia, monocytopaenia and eosinophilia in both sexes of birds. However, no significant effect of DAP was seen on the number of basophils. Same trend continued upto 1000 ppm DAP treatment but 1500 ppm treatment showed a recovery state.

Wintrobe (1967) stated that stress caused adrenocorticosteroid hyperactivity through pituitary ACTH followed by neutrophils and eosinopaenia. He opined that nutrophilia is accounted for, in part at least by mobilization of cells from the marginal grapulocytes pool of the blood. Villa sensor *et al.* (1948) has already suggested that the adrenal stimulation by pituitary ACTH or a variety of stresses are responsible for eosinopaenia. Neutrophilia is due to abnormal increase of neutrophils in DAP intoxicated birds upto 1000 ppm concentration. Lymphopaenia and monocytopaenia in *Acridotheres tristis* was due to fertilizer treatment. Eosinophilia recorded in the present study was due to increased secretion of adrenaline byadrenal medulla.

The significant fall in the haemoglobin content in *Acridotheres tristis* was due to anaemia. DAP treated bird may develop hypochromatic microcytic anaemia due to reduction of R.B.C. and haemoglobin percentage values which was attributed to the deficiency of iron and its decreased utilization for haemoglobin synthesis (Ranganatha and Ramamurthi, 1979). The decrease in its concentration also might be due to disturbance in metabolic activities of haemopoietic organs caused by diammonium phosphate intoxication. The decrease in values of P.C.V. was due to decrease in the number of erythrocytes.

References

Agrawal, S.J. and Srivastava, A.K. 1980. Haematological responses in a fresh water fish to experimental manganese poisoning. *Toxicology*, 17: 97–100.

Banerjee, V. 1986. Effect of heavy metal poisoning on peripheral haemogram in *Heteropneustes fossilis* (Bloch) 1. Mercuric sulphate (LC_{50}) Camp. *Physiol. Ecol.*, 2(4): 173–176.

Banerjee, V. and Verma, G.K. 1992. Effect of three pesticides on erythrocyte and Leucocyte morphology in *Channa punctatus* (Bloch). *Proc. Natl. Symp.*, p. 33–38.

Bell, D.J. 1957. The Distribution of glucose between the plasma water and the erythrocyte water in hen's blood. *J. Exp. Physiol.*, 42: 410–416.

Breuer, K., Lill, A. and Baldwin, J. 1995. Haematological and body mass change of small passerines over wintering in South-eastern Australia. *Australian J. Zoology*, 43(1): 31–38.

Darmady, E.M. and Davenport, S.G.T. 1954. *Haematological Technique for Medical Laboratory Technicians and Medical Students*. J and A. Churchill Ltd., London, p. 27–46.

Das, P.K.M.K. and Konar, S.K. 1990. Influence of mixture of light Iranian crude oil and cryptoanionic detergent on aquatic ecosystem. *Environ. Ecol.*, 8(3): 834–841.

Gilbert, A.B. 1968. The relationship between the erythrocyte sedimentation rate and packed cells volume in the domesticated fowl. *Brit. Poult. Sci.*, 9(3): 297–299.

Maqueet, M.A., Yadgirkar, G., Ahmad, S.R. and Sarma, J. 1993. Study of haematological and biochemical response in avian mycoplasmosis to tisamutin. *J. Vet and Animal Sci.*, 23(2):73–77.

Quli, S.M.S. and Banerjee, M. 1994. Qualitative study of polymorphonuclear leucocytes in Bank myna, *Acridotheres ginginanus*. 2. *Environ. Ecol.*, 12(4): 832–834.

Rani, K. 1999. Haematological dynamics in *Gallus domesticus* with response to sex and breeding. *Ph.D. Thesis*, L.N. Mithila University, Darbhanga.

Ranganatha, P. and Ramamurthi, R. 1979. Haematological studies in *Sarotherodon mossembica* exposed to lethal (LC_{50}) 48 h concentration of sumithion and sevelon. *Curr. Sci.*, 48: 19.

Sharma, L.L. and Saxena, P.N. 1997. Carbaryl induced haematological changes in *Columba libia* (Camelin). *J. Environ. Biol.*, 18(1): 17–22.

Srivastava, B.K. and Das, N.L. 1995. *A Manual of Practical Physiology*. Vijay Bhagat, Scientific Book Company, Ashok Raj Path, Patna, p. 65–67, 72–74.

Villasensor, J.B., Rath, C.E. and Finch, C.A. 1948. The blood picture in Addison 's disease. *Blood*, 3: 769–773.

Wintrobe, M.M. (1967). *Clinical Haematology*, 6th Edn. Lea and Febiger, Philadelphia, p. 23, 94, 425.

Chapter 25

Comparison of Mosquito Fauna in Srivilliputhur Town and Krishnankovil Village, Tamil Nadu*

K. Karuppasamy & T. Sooravan

Post Graduate and Research Department of Zoology, Ayya Nadar Janaki Ammal College (Autonomous), Sivakasi – 626 124, Tamil Nadu

ABSTRACT

Systematic survey of mosquito fauna of Srivilliputhur town and Krishnankovil village was carried out from October 2000 to December 2001. Nineteen species of mosquito fauna belonging to five genera, such as *Armigeres, Culex, Aedes, Anopheles* and *Mansonia* were recorded in both study areas. Eleven mosquito species were recorded in Srivilliputhur and Krishnankovil. One species, *Aedes vittatus* was present in Krishnankovil alone. In both study areas three mosquito species, such as, *Armigeres subalbatus, Culex quinquefasciatus* and *Aedes aegyptii* were recorded throughout the study period. Vector mosquitoes for malaria, filariasis, dengue fever and encephalities were recorded in the study areas.

***Keywords:** Mosquitoes, Vector.*

Introduction

Mosquitoes are familiar insects as they cause discomfort in the sound sleep of man. They are found in damp and marshy localities but are abundant in tropics and subtropics. They are nocturnal in feeding habitat (Kotpal, 1976). Mosquitoes are the vectors of various diseases, such as, malaria,

* Presented in the National Conference on Zoology, helt at Vivekananda College, Agastheeswaram, Kanyakumari on 21–23 December, 2003

filariasis, dengue fever, yellow fever, encephalitis etc. Mosquito survey provides valuable information as occurrence, distribution, prevalence and species composition of various mosquitoes in an area which assumes significance due to their public health importance (Prakash *et al.*, 1998). Mosquitoes are considered as serious vectors of many dreadful diseases both in urban and rural areas. The problem is more severe in the rural areas especially those are associated with a well irrigated agro ecosystem. In these areas, poor sanitation facilities, severe agricultural practices and lack of medical facilities enhance the diversity and density of the population of mosquitoes and vector born diseases. For successful implementation of vector management programmes, adequate knowledge about the species composition, density and feeding behavior is essential (Pandian *et al.*, 1997).

Materials and Methods

Survey of mosquito fauna were carried out in Srivilliputhur and Krishnankovil for a period of 15 months from October 2000 to December 2001. The study area Srivilliputhur is a town located in Western region of Virudhunagar district about 10 km away from the Western ghat. The town bearing newly formed colonies in the peripheral region, having stagnation of domestic sewages around the town. Krishnankovil village is located in the foot hill of Western ghat located in the National highway 208 which is surrounded by agricultural fields. There are number of breeding sites available for the breeding of mosquito throughout the year. Biting collections of mosquitoes from 18 hour to 6 hour was carried out in outdoor habitat in Srivilliputhur and Krishnankovil to investigate the species diversity. The collected mosquitoes were killed using diethyl ether packed in separate container depending upon the hour of collection. The collected mosquitoes were identified using the dichotomous key in the Regional Entomological Research Station at Virudhunagar.

Result

In the present study 19 species of mosquito fauna belonging to 5 genera, such as, *Armigeres, Culex, Aedes, Anopheles* and *Mansonia* were recorded in the study areas. 11 mosquito species were recorded in Srivilliputhur and Krishnankovil. 7 species were found only in Srivilliputhur. The mosquito species *Aedes vittatus* was present in Krishnankovil alone. 3 species such as *Armigeres subalbatus, Culex quinquefasciatus* and *Aedes aegyptii* were recorded through the study period (Table 25.1). Vector mosquitoes such as *Culex quinquefasciatus, Culex tritaeniorynchus, Aedes aegyptii, Mansonia uniformis, Mansonia annulifera* and *Anopheles stephensi* were recorded in the study areas.

Discussion

Mosquitoes are vectors proliferating in large number and transmitting several diseases. To control the vector born disease, it is essential to know the diversity of mosquito species of a particular place. 19 species of mosquito belongs to 5 genera were identified in the study area. The species diversity in the present study may be due to the availability of different types of mosquitogenic habitat and mismanagement of ecosystem and application of agricultural practices.

Among the five genera recorded in the study area *Culex* is predominant genus with 7 species recorded in Srivilliputhur and 6 species in Krishnankovil. The *Anopheles* genus was represented by 6 species in Srivilliputhur and 2 species in Krishnankovil (Table 25.1). The diversity of *Culex* and *Anopheles* species are due to the presence of different breeding habitat, such as sewage stagnant pools, freshwater pools, paddy fields and irrigated canals around the study areas. Similar results were reported by Pandian (1997) in the urban, sub urban and rural areas associated with agro ecosystem. Two species in the genus *Mansonia* were recorded in Srivilliputhur town and one species was represented in Krishnankovil village (Table 25.1). The *Mansonia* species breed in the water containing aquatic vegetation. There are peranial and temporary ponds with aquatic vegetation in Srivilliputhur and Krishnankovil respectively this may be the reason for variation in the number of species in the study area. Similar results were reported by Khamre and Kaliwal (1998) from Daman; Kulkarni and Naik (1989) are reported in Goa. *Armigreres* genus was represented by one species recorded in both study areas (Table 25.1). The *Aedes* genus represented by 3 species (Table 25.1). The mosquito *Aedes vittatus* was recorded in Krishnankovil alone. This species breeds in the rock holes of the hill areas. There are 18 species were recorded in the Srivilliputhur town, only 12 species were found in the Krishnankovil. This is due to the availability of breeding sites. Srivilliputhur town having many stagnant pools of sewage and freshwater body, are the ideal sites for breeding of mosquito. Therefore more species were recorded in the town area than the rural village. In the study area 6 species of vector mosquitoes, such as *Culex quinquefasciatus, Culex tritaeniorynchus, Aedes aegyptii, Mansonia uniformis, Mansonia annulifera* and *Anopheles stephensi* were recorded (Table 25.1). The present study suggested that if any carrier of vector born pathogen may present in the study area, there is a chance for out break of vector born disease. Therefore it is essential to control the mosquito population. By controlling the mosquito population, the vector born disease may be controlled or prevented the transmission of vector born diseases.

Table 13.1: Showing the Mosquito Species Recorded in Srivilliputhur Town and Krishnankovil Village

Sl.No.	*Mosquito Species*	*Srivillipthur*	*Krishnakovil*
1.	*Armigeres subalbatus*	+	+
2.	*Culex quinquefasciatus**	+	+
3.	*Culex tritaeniorynchus**	+	+
4.	*Culex bitaeniorynchus*	+	+
5.	*Culex gelidus*	+	–
6.	*Culex epidesmus*	+	+
7.	*Culex vishnui*	+	+
8.	*Culex infula*	+	+
9.	*Aedes aegyptii**	+	+
10.	*Aedes albopictus*	+	–
11.	*Aedes vittatus*	–	+
12.	*Anopheles subpictus*	+	+
13.	*Anopheles sinensis*	+	+
14.	*Anopheles stephensi**	+	–
15.	*Anopheles annularis*	+	–
16.	*Anopheles barbirostris*	+	–
17.	*Anopheles tessellates*	+	–
18.	*Mansonia uniformis**	+	–
19.	*Mansonia annulifera**	+	+
	Total	**18**	**12**

Notes: *: Vector mosquitoes; +: Present; –: Absent.

References

Khamre, J.S., and Kaliwal 1988. Mosquitoes of Daman. *Ind. J. Malarial.*, 25(2): 109–111.

Kotpal, R.L., S.K. Agarwal and R.P. Khetarpal 1976. *Modern Textbook of Zoology: Invertebrates.* Rastogi Publications, Meerut, India, pp. 471.

Kulkarni, S.M., and P.S. Naik 1989. Breeding habitats of mosquitoes in Goa. *Ind. J. Malarial.*, 26(1): 41–44.

Pandian, R.S. 1997. Habitat selection by the urban and suburban mosquitoes. *J. Env. & Poll.*, 4(1): 45–47.

Pandian, R.S., R. Rethinavelu and A.C. Manoharan 1997. Species diversity and feeding behaviour of mosquitoes in a rural area associated with an agro ecosystem: A case study. *Proceedings of the 2nd Symposium on Vectors and Vector Borne Diseases*, p. 207–211.

Prakash, A., D.R. Bhattacharyya, P.K. Mohapatra and J. Mahanta 1998. Mosquito fauna of a broken forest ecosystem of district Dibrugarh with updated systematic list of mosquitoes recorded in Assam, India. *Entomon*, 23(4): 291–298.

Chapter 26

Biodegradation of Tannery Effluent Using *Pseudomonas aeruginosa* and *Bacillus subtilis*

P. Mythili, K.T.P.B. Ponneelan**, R. Suchitra** and M. Poonkothai****

**P.G. and Research Department of Environmental Science,*
P.S.G. College of Arts and Science, Coimbatore –641 041
***Department of Microbiology, Kandaswamy Kandar's College, Velur – 638 182*
****Department of Biochemistry, Avinashilingam Deemed University,*
Coimbatore – 641 043, Tamil Nadu, India

ABSTRACT

The present study was carried out to find out the degradation efficiency of microorganisms isolated from tannery effluent. Tannery effluent was collected and analysed for its physico-chemical characteristics. *Pseudomonas aeruginosa* and *Bacillus subitlis* were isolated from tannery effluent using biochemical tests. Different dilutions of the effluent (25 per cent, 50 per cent, 75 per cent and 100 per cent) were prepared and inoculated with the above bacterial cultures. After incubation for 0, 24, 48 and 72 hrs the biodegradation rate of the effluent was analysed. The results showed that among the bacterial cultures *Pseudomonas aeruginosa* showed maximum degradation efficiency.

Keywords: Biodegradation, Bacteria, Tannery effluent.

Introduction

Biological treatment system is potentially a simple, low cost, self sustaining option for amelioration of wastewater (Brix, 1987 and Schlerup, 1989). The tannery effluent is a major source of water pollution in the country and a threat to environment due to indiscriminate use of chromate (Karhadkan, *et al.*, 1987). Tannery effluents containing large amounts of waters, especially tannins are toxic to plants, animals, and soil microorganisms. Tannins also cause deleterious effect on soil microorganisms degrade tannins and utilize them as carbon source. The industrial effluent that contains toxic metal ions and a variety of organic compounds which are harmful to different forms of life. Among them the notable are chromium and sulphide in the form of basic chromium sulphate and sodium sulfide (Balasubramaniam and Pugalendhi, 1998). pH, DO, BOD, COD, TDS, TSS, Bicarbonate, Choloride, total protein, total tannin and total phenol were found to be higher in tannery of effluent. This study was aimed to reduce 'the level of BOD, COD, TSS and TDS of the tannery effluent using bacterial isolates.

Materials and Methods

Collection of Effluent

The tannery effluent was collected freshly from Bay Leather Exports (P) Ltd Erode, Tamil Nadu, India, from the disposal site.

Isolation and Identification of Microorganisms

The serial dilution technique was adopted to isolate the microbial strains from the tannery effluent. 1 ml of the effluent was serially diluted and different dilutions were plated on Nutrient agar (Seeley and Demark, 1981). The plates were incubated at 37°C for 24 hrs and well grown colonies were picked and further purified by streaking. The purified bacterial cultures were identified based on Gram staining, motility test and biochemical characteristics (Kannan, 2002).

Biodegradation Study and Physiochemical Analysis of Effluent

The effluent was diluted, (25 per cent, 50 per cent, 75 per cent, 100 per cent) using distilled water and 1ml of the bacterial culture was inoculated into flasks. The flasks were incubated at 37°C for 0, 24, 48 and 72 hrs. The untreated and bacterially degraded tannery effluent were analysed after 0, 24, 48 and 72 hrs of incubation. The analysis of wastewater for various parameters were done as per standard methods (APHA, 1992).

Results and Discussion

Pseudomonas aeruginosa and *Bacillus subtilis* was isolated from the tannery effluent and their biochemical characters were presented in Table 26.3. The pH, DO (Dissolved Oxygen), BOD (Biological Oxygen Demand), COD (Chemical Oxygen Demand), TDS (Total Dissolved Solids), TSS (Total Suspended Solids), Bicarbonate and chloride was found to be higher in the raw tannery effluent. The conventional waste water treatment methods are unable to remove the bicarbonate and chloride from tannery effluent. However biodegradation using activated microorganisms can be versatile, inexpensive and can potentially transform a toxic material in the harmless product (Ramchandra, 2001). The pH at 0th hr for the dilutions ranges from 11.5 and it is partially reduced to 8.6 in *Pseudomonas aeruginosa* where as in *Bacillus subtilis* it was reduced to 9.0 from 11.0. Similarly the BOD was found to be decreased from 56.0 to 50.0 in case of *Pseudomonas aeruginosa* and 59.0 to 30 in case of *Bacillus subtilis*. The same is observed in the case of all the parameter and Tables 26.1 and 26.2 depicts the rate of degradation at different intervals of time. Thus among the two bacteria, *Pseudomonas aeruginosa* was found to be predominant in degradation of tannery effluent.

Table 26.1: Biodegradation of Tannery Effluent Using *Pseudomonas aeruginosa*

Parameters	*25%*				*50%*				*75%*				*100%*			
	0	*24*	*48*	*72*	*0*	*24*	*48*	*72*	*0*	*24*	*48*	*72*	*0*	*24*	*48*	*72*
pH	11.5	10.0	10.2	9.8	11.0	9.8	9.4	9.3	10.9	9.5	9.2	9.0	10.5	9.0	8.8	8.6
DO	5.28	2.88	2.56	2.52	5.75	3.84	3.72	2.84	4.85	4.16	4.00	3.12	4.22	5.60	5.40	3.51
BOD	56	47	43	40	68	53	46	48	75	55	54	55	86	66	60	50
COD	190	160	150	142	280	240	214	158	315	256	235	169	415	288	255	180
TDS	5215	5090	4105	4098	5.10	5460	5235	4234	5812	5868	5678	4568	6182	5902	5690	5126
TSS	7618	5213	5123	5012	7752	6617	5224	5246	8068	6872	5843	5464	8237	7470	6140	5701
BC	210	195	180	179	235	205	200	198	250	225	215	200	300	280	260	205
Cl^-	28.00	27.12	25.147	24.98	28.50	31.09	30.02	26.18	32.85	35.50	32.90	27.25	35.52	25.70	35.80	27.84

Table 26.2: Biodegradation of Tannery Effluent Using *Bacillus subtilis*

Parameters	*25%*				*50%*				*75%*				*100%*			
	0	*24*	*48*	*72*	*0*	*24*	*48*	*72*	*0*	*24*	*48*	*72*	*0*	*24*	*48*	*72*
pH	11.0	10.5	9.5	9.0	10.8	9.3	9.1	8.7	10.1	8.5	8.7	8.4	9.9	9.5	9.2	9.0
DO	6,11	4.64	4.20	4.01	7.55	5.60	5.82	4.98	7.02	6.72	6.10	5.60	6.85	7.68	6.82	5.95
BOD	59	54	51	47	66,	59	48	39	78	62	43	35	90	70	34	30
COD	180	160	156	148	255	224	148	129	290	304	136	100	310	320	120	110
TDS	4125	4761	4430	4120	5210	4832	1638	4468	5760	5725	58.15	5630	6l.85	5896	6018	6182
TSS	7660	7418	7124	7028	8010	7506	7418	7412	8095	7654	8014	8068	8315	7660	8208	8124
BC	220	160	150	142	253	223	198	165	280	240	227	198	295	299	268	210
Cl^-	27.15	26.98	22.18	20.96	28.60	32.66	24.96	23.82.	35.02	35.07	30.45	26.32	42.50	38.45	32.12	30.66

Table 26.3: Characteristics of Cultures Isolated from the Tannery Effluent

Parameters	*Pseudomonas aeruginosa*	*Bacillus subtilis*
Gram staining	–	+
Motility	+	–
Catalase	+	–
Oxidase	+	+
Indole	–	–
Methyl red	–	–
VP	–	+/–
H_2S	+	–
Citrate	+	–
Urease	+	–
Glucose	A^+	–
Lactose	–	A^+
Maltose	–	–
Sucrose	–	A^+
Starch hydrolysis	–	–

+: Positive; –: Negative; A^+: Acid Production.

Acknowledgement

The authors are thankful to the management of Kandaswamy Kandar's College, Velur, for providing us the facilities to carry out the present work.

References

APHA, AWWA and WPCF. 1992. *Standard Method for the Examination of Water and Wastewater*, 18th ed. American Public Health Association, Washington, D.C. USA.

Balasubramaniam, S. and Pugalenthi, V. 1998. Determination of major pollutants in tannery effluents. *Indian J. Env. Prot.*, 19: 15–17.

Brix and Schlerup, H. 1989. The use of aquatic macrophytes in water pollution control. *AMBIO*, 18: 100–107.

Kannan, N. 2002. *Laboratory Manual in General Microbiology*. Palani Paramount Publications, 2:74–150.

Karhadkar, P.P., Audic, J.M. Paup, G.M. and Khanna, P. 1987. Sulphide and sulfate inhibition of methanogenesis. *Water Res.*, 21: 1061–1066.

Sekaran, G., Chitra, K. and Mariappan, M. 1993. Removal of sulphide in tannery effluents by catalytic oxidation. *Indian J. Env. Prot.*, 19: 19–24.

Saravanan, G. and Saravana, A. 1998. Decolorization of tannery effluent by *Flavobacterium* sp. Tkl. *Indian J. Env. Prot.*, 19: 19–24.

Chapter 27
Pesticide Disrupted Ovarian Physiology and Biochemistry of *Channa striata*

G.M. Natarajan, D. Kalavathi, N. Parthi, P. Esther Joice, P. Palanisamy, G. Sasikala and N. Bhuvaneshwari*

*Post Graduate and Research Department of Zoology, Government Arts College (Autonomous) Coimbatore – 641 018, *Dublin-9, Ireland*

ABSTRACT

Metasystox exposure induced significant changes in the histophysiology of the ovary, blood glucose content and glycogen concentration in the air-breathing fish, *Channa striata*. The results are discussed.

Keywords: Channa striata, Metasystox, Blood glucose.

Introduction

Oxydementon–methyl (Demeton) is a locally used pesticide against cotton, paddy and vegetable crop pests. Lethal and sublethal conentration of metasystox exposure causes many change in respiration, oxidative enzymes, and histopathological changes in the air–breathing fish *Channa striata* (Natarajan *et al.*, 1998). But no information is available on the effect of Oxydementon–methyl on the reproduction of any freshwater fish. In the present work, an attempt has been made to study the effect of sublethal concentration of metasystox on some aspects of reproduction in *Channa striata*. The test fish, *Channa striata* is indigenously available in large quantities and is extensively cultured in ponds,

* Corresponding Author.

lakes, direct water bodies and paddy fields where they may have chances of encounting this pesticides attack.

Material and Methods

Specimens of snake headed fish, *Channa striata* used in the present study (10–35 g) was obtained from commercial sources locally. The fish were acclimized to the laboratory conditions in glass aquaria for at least two weeks. The fish were fed on boiled eggs adlibitum on alternate days. Water was changed once a week. During acclimation, the fish were kept under a 12 hour light–12 hour dark cycle (L : D 12 : 12). In the laboratory, all specimens were measured made weighted, sexed and allocated a maturity stages. These allocations were made initially macroscopically and were later verified by examination of smears of small sections from each gonad.

Technical grade metasystox of 95 per cent purity was obtained from Bayer LTD, Bombay. LC 50/ 48 hr calculated by the probit method was 5 mg/l for metasystox and hence 1/3 of the LC 50 /48 hr concentration (1.7 mg/l) was chosen for sublethal exposures. At this concentration, the fish survived even after prolonged periods of exposure. Two batches containing all stages of ovarian development were taken for experimentation. Group I served as control Group II was exposed to a sublethal dose of metasystox of 1.7 mg/l for 30 days. Water was changed once a week and predetermined quantity of pesticides was added after changing the water.

Blood from control and metasystox exposed fishes were obtained by serving the caudal peduncle and blood glucose was determined after Nelson–Somogyi's method (as given by Oser, 1975). Ovarian pieces from control and metasystox exposed fishes were minced and placed in hot 30010 KOH. After precipitation, the ovarian glycogen was estimated by the method of Montgomery (1975) as given by Oser (1975). The test described by Fischer (1950) was employed to calculate the statistical significance between control and experimental values.

Results

Following metasystox exposure, marked changes were seen in the histophysiology of the ovary of *Channa straita*. 30 days exposure increased to immature and atretic oocytes (Table 27.1): A progressive reduction in maturing ($P < 0.001$) and matured oocytes ($P < 0.001$) is visible at this exposure. A corresponding insignificant increase in oogonia ($P < 0.001$), immature ($P < 0.05$) and atretic oocyte ($P < 0.005$) is an important observation in worth mentioning. Histologically, the oocytes exhibited prominent vacuolization and extensive disintegration. The cytoplasm showed many "inclusion bodies and vacuolization." The nuclei appeared necrotic. The diameter of the oocytes reduced significantly at 30 days exposure. Further, it reduced the mean body weight ($P < 0.05$), the mean body length ($P < 0.01$) and the gonadosomatic index ($P < 0.01$).

Table 27.1: Sublethal Concentration of Oxydemeton–methyl on the Percentage Occurrence of Different Stage of Oocytes of *C. straita*

Stages	*Control*	*Oxydemeton–methyl, Exposed for 30 Days*
Oogonia	700±018	11.20±0.42*
Immature	21.00±0.63	41.30±1.70*
Maturing	27.00±1.00	2.60±0.46*
Mature	38.00±0.80	1.40±0.21*
Artetic	8.00±0.56	43.5±0.34*

All values are $\overline{X} \pm$ SD of six individual determinations; *$P < 0.001$; ** $P < 0.05$.

Metasystox exposure lowered the blood glucose content at all stages of development of the ovary (Table 27.2) stage III (–22.12 per cent; P < 0.05) and stage V (–21.01 per cent; P < 0.01) registered a decrease of more than 20 per cent.

Table 27.2: Sublethal Concentration of Oxydemeton–methyl Exposure (30 days) on the Blood Glucose (mg/100 ml) and Ovarian Glycogen (mg/g) Contents *C. striata*

Maturity Stage of Ovary	Blood Glucose		Ovarian Glycogen	
	Control	Experimental	Control	Experimental
I	71.4±21.3	60.2±19.6 – 18.60 P < 0.05	1.38±0.02	1.10±0.01 – 25.45 P < 0.01
II	63.6±20.6	56.4±17.5 – 12.76 P < 0.01	2.71±0.04	2.16±0.03 – 25.46 P < 0.05
III	76.2±21.3	62.4±18.9 – 22.12 P < 0.05	3.54±0.01	2.60±0.01 – 36.15 P < 0.05
IV	61.3±17.6	56.2±16.4 – 9.07 P < 0.01	6.40±0.03	4.20±0.02 – 52.38 P < 0.01
V	50.1±18.8	41.4±13.2 – 21.01 P < 0.01	1.64±0.02	1.02±0.03 – 60.78 P < 0.05

Each value is an average ($\overline{X}$ ± SD) of six individual determinations.

Pesticide treatment consistently depleted the ovarian glycogen content at all stages of maturity of the ovary. Maximum reduction (-60.78 per cent; P <0.05) is observed at V stage. This is followed by a 52.4 per cent reduction (P< 0.01) at the IV stage. The III maturity stage (–36.15 per cent; P < 0.05) showed a significant reduction. Both the (–25.45 per cent; P < 0.01) and the II maturity stages (–25.46 per cent; P < 0.05) maintained steady and uniform reduction in glycogen content.

Discussion

Following metasystox exposure, a number of dramatic changes were introduced in the ovarian morphology and physiology. The number of immature and atretic oocytes increase enormously. At the same time, the population of matured and maturing oocytes decreased. The matured oocytes were either shrunken, deformed, lysed or vacuolated. Similar deformities and architectural changes were reported for a number of pollutants (Singh, Lal and Yadav, 1998). Cytoplasmic clumping, hyperplasia of germinal epithelium and involution of some ova, decreased frequency of oocyte maturation and fragmentation and karyolysis of ova are some of the most frequently recorded changes in fishes during pesticide treatment (Ramalingam, 1998). An increase in the population of atretic and immature oocytes will definitely reduce the spawning actively which will ultimately lead to the decline in the number of oocyte stage II and III accompanied with a remarkable decrease in gonadosomatic index (GSI), assation of vitellogenesis and inactivation of steroidogenesis in gravid *Mystus vittatus* after 12 weeks treatment with malathion. Pesticide also inhibits LH induced *in vitro* ovulation and causes atresia of the oocytes in *O. latipes* (Hirose, 1975).

Generally, in teleost fishes, ripening gonads tend to accumulate glycogen. This glycogen is utilized for the maturation and copper development of the oocytes. Since pesticide exposure is a physical stress, the depletion is more probably likely to be switched to gear to give the energy giving substances like ATP. In teleost fishes, depletion of ovarian glycogen is under the control of hormones (Singh, Lal and Yadav, 1998). Naturally, in the pesticide exposed fishes, the hormones play an important role in the rise and fall of glycogen in the ovary.

Among the various blood parameters used to monitor the pesticide "stress" blood glucose play a significant place (Ramalingam, 1988). A tendency to increase the blood glucose level is shown for DDT and malathion exposure in *Sarotherodon mossambicus* (Ramalingam, 1998). In the present investigation, the blood glucose level continue to decrease at all stages of exposure. This hypoglycemic condition might have developed either for rapid energy yielding purpose or stored as glycogen some where for further purpose. Since blood glucose level in under the control of endocrine organs, the present observations indicate a hormonal imbalance at the cellular level with warrants further studies.

Acknowledgements

Authors are thankful to Government of Tamil Nadu and Bharathiar University for all the facilities.

References

Fisher, R.A. 1950. *Statistical Methods for Research Works*, 11th edn. Oliver and Boyd, London.

Haider, S. and Upadhayay. H. 1986. Ovarian damage in *Mystus vittatus* after exposure to an organophosphorus pesticides. *Bull. Environ. Contam. Toxicol.*, 21: 270–275.

Hirose, K. 1975. Reproduction in medaka, *Oryzias latipes* exposed to sublethal concentration of Y benzene hexachloride (BHC). *Bull. Tokal. Reg. Fish. Res. Lab.*, 81: 139.

Montgomery, R. 1957. The determination of glycogen. *Arch. Biochem. Biophys.*, 67: 378–386.

Natarajan, G.M., Mallikaraj, D., Vijayalakshmi, P., Rani, R.P. Parthi, N. 1998. Impact of pesticides on the Haematoenzymological changes in fish. *In: Pesticides, Man and Biosphere*, (Eds.) Shukla, O.P., Omkar and Kulshretha, A.K. APH Pub. Corp., New Delhi, p. 337–350.

Oser, B.L. 1975. *Hawk's Physiological Chemistry*. Mc-Graw Hill Book Co, New York.

Ramalingam, K. 1998. Toxic effects of DDT, malathion and mercury on the tissue carbohydrate metabolism of *Sartherodon mossambicus*. *Proc. Indian. Acad. Sci. Anim. Sci.*, 97: 443–448.

Chapter 28

Effect of Metal Poisoning on Total Body Ascorbic Acid in *Sphaerodema rusticum* (Belostomatidae: Hemiptera)

S. Mumtazuddin and S. Ehteshamuddin***

**University Department of Chemistry, **University Department of Zoology B.R.A. Bihar University, Muzaffarpur – 842 001, India*

ABSTRACT

Although many metals are required for normal physiological functions of animals in very low concentrations, altered physiological functions result when one or more of these reach sufficiently high concentrations in body cells. In the present investigation the effect of three metal compounds, *viz.*, copper sulphate, cobalt nitrate and Mohr's salt on the total body ascorbic acid of an aquatic insect *Sphaerodema rusticum* has been studied. It was observed that both the male and the female insects showed decline in the total body ascorbic acid concentrations on the treatment with all these three metal compounds. It is assumed that metal ions in high concentrations act as inhibitors to different enzyme systems responsible for ascorbic acid syntheses.

Keywords: Sphaerodema rusticum, Total body ascorbic acid, Metal poisoning.

Introduction

Many metals are required for normal physiological function of mammals, but only in very low concentrations. These include Copper, Iron, Zinc, Manganese, Cobalt, Selenium and Chromium (Harper,

Rodwell and Mayer, 1977). It is assumed that most of these also probably have similar functions in fish and other aquatic animals. However, altered physiological functions result when one or more of these reach sufficiently high concentrations in body cells both in aquatic animals and mammals. Metals enter in the aquatic environment by a variety of industrial effluents and old mines. Acid precipitation also causes leaching of metals from surrounding soils (Norton, 1982). In fact, aquatic environment is the ultimate sink for the pollutant and any compound either produced on the industrial scale or present in terrestrial environment is likely to reach there, sooner or later.

The study of physiological changes as means for understanding the responses of aquatic animals, like insects, fish, etc. to the stress of various types of pollutants is of great help in studying the nature and the extent of pollution. Eishler *et al.* prepared bibliographies on the biological effects of metals in aquatic environment (Eisler, Rossoll and Gaboury, 1979).

The present investigation is an attempt to study the effects of three metal compounds, *viz.*, copper sulphate, cobalt nitrate and Mohr's salt on the total body ascorbic acid of an aquatic insect *Sphaerodema rusticum.*

In the present work, the investigation of body ascorbic acid has been undertaken. Although, ascorbic acid is not used in large amounts in cells, its lack is accompanied by characteristic deficiency symptoms. Since all organisms cannot synthesise ascorbic acid, the organisms rely on external dietary source. In insects, specially the aquatic ones, no literature is available on the ascorbic acid concentration of tissues in normal and various treated conditions.

Experimental

The insects (*S. rusticum*) were treated separately in different batches by the following metal compounds in different concentrations.

Copper Sulphate ($CuSO_4 . 5H_2O$)

The LC_{50} of the compound for the insects was worked out to be one molar concentration (1 M) and as such ten different concentrations of the compound were taken from 0.1 M to 1 M. In each concentration of the solution, 30 to 50 insects were kept and biochemical assays were performed after 24 hours of each treatment.

Cobalt nitrate [$Co(NO_3)_2$]

The LC_{50} for the insects in this case was found to be 0.055 M and 30 to 50 insects were placed in nine different concentrations ranging from 0.047 M to 0.055 M and the biochemical assays were performed in each case after 24 hours of treatment.

Mohr's Salt [$FeSO_4 . (NH_4)2SO_4 . 6H_2O$]

The LC_{50} of Mohr's salt for the insects was found to be 0.1 M. Hence, ten different concentrations of the compound ranging from 0.01 M to 0.1 M were taken and in each 30 to 50 insects were placed for 24 hours for biochemical assays.

Quantitative estimation of total body ascorbic acid of treated and control (untreated) insects was done by 2, 4-D.N.P.H. method (Roe, 1961), using Erma colorimeter. Mean and S.D. for each set of reading was calculated. Students-t-test, where necessary, was applied in order to find level of significance between two sets of readings.

Result and Discussion

Total Body Ascorbic Acid Concentration of 'Control' Insects

Total body ascorbic acid concentration was found to be low in untreated male and female insects 2.87±0.2 S.D. mg/gm of body weight in male insect and 3.57±0.4 S.D. mg/gm of body weight in female insect.

Total Body Ascorbic Acid Concentration of the 'Treated' Insects

Treatment with Copper Sulphate

When exposed to 0.1 M concentration of copper sulphate, there was an insignificant ($P > 0.05$) decline in the total body ascorbic acid concentration of male insect but a significant ($P < 0.05$) decline in the total body ascorbic acid concentration of the female insect. But on treatment with 0.2 M concentration of $CuSO_4$ onwards body ascorbic acid concentration was found to be zero in both male and female insects. The results are described in Table 28.1.

Table 28.1: Body Ascorbic Acid Concentration in Insects on Treatment with Copper Sulphate

Concentration of Copper Sulphate (molar)	*Body Ascorbic Acid Concentration (mg/gm body weight*±S.D.)*	
	Male	*Female*
0.1	2.5±1.5	2.42±1.8
0.2	Nil	Nil
0.3	Nil	Nil
0.4	Nil	Nil
0.5	Nil	Nil
0.6	Nil	Nil
0.7	Nil	Nil
0.8	Nil	Nil
0.9	Nil	Nil
1.0**	Nil	Nil

* Mean wt. of insect:
Male insect: 250 mg ± 9 S.D.
Female insect: 265 mg ± 10 S.D.

** LC_{50}
Body ascorbic acid in 'control':
Male insect: 2.87 ± 0.2 S.D.
Female insect: 3.57 ± 0.4 S.D.

Treatment with Cobalt Nitrate

When exposed to 0.047 M concentration of cobalt nitrate, the total body ascorbic acid concentration of male insect showed a significant ($P < 0.01$) fall, whereas the ascorbic acid was not traceable from the body of female insect. The total loss of ascorbic acid from the body of the male insect was also noted from 0.048 M concentration of cobalt nitrate onwards. The results are described in Table 28.2.

Table 28.2: Body Ascorbic Acid Concentration in Insects on Treatment with Cobalt Nitrate

Concentration of Cobalt Nitrate (molar)	*Body Ascorbic Acid Concentration (mg/gm body weight*±S.D.)*	
	Male	*Female*
0.047	0.76±0.04	Nil
0.048	Nil	Nil
0.049	Nil	Nil
0.050	Nil	Nil
0.051	Nil	Nil
0.052	Nil	Nil
0.053	Nil	Nil
0.054	Nil	Nil
0.055**	Nil	Nil

* Mean wt. of insect:
Male insect: 250 mg ± 9 S.D.
Female insect: 265 mg ± 10 S.D.

** LC_{50}
Body ascorbic acid in 'control':
Male insect: 2.87 ± 0.2 S.D.
Female insect: 3.57 ± 0.4 S.D.

Treatment with Mohr's Salt

Total body ascorbic acid concentration showed a significant decline ($P < 0.01$) when exposed to 0.01 M concentration of Mohr's salt in both male and female insects. When exposed to 0.02 M concentration of the compound the total body ascorbic acid showed an even more significant ($P < 0.001$) decline in both male and female insects. On treatment with 0.03 M onwards the total body ascorbic acid concentration vanished completely. The results are described in Table 28.3.

Hence, on treatment with all the three metal compounds in a very low concentration a total loss of body ascorbic acid was found. This is suggestive of the possible role of the enzyme L-gulonolactone oxidase, responsible for the synthesis of ascorbic acid from glucose. The results also indicate that the insect *S. rusticum* possibly synthesises its own ascorbic acid for metabolic requirement from glucose. Dietary requirements for the ascorbic acid can be ruled out as the insects were sufficiently provided with food prior to their 'treatment' and in each case the readings were taken after twenty hours exposure to the chemicals and such a short period can not be the cause for sharp decline or complete loss of body ascorbic acid in them.

It may thus be concluded that all the three metal compounds pollute the aquatic environment and even in low concentrations affect the physiology of the insect by inhibiting the synthesis of essential biochemical components like proteins (Mumtazuddin and Ehteshamuddin, 2000) and ascorbic acid, thus affecting the production and growth of the insect. As the insect is an important constituent of the aquatic ecosystem in maintaining a normal functional life of pond, lake, etc., the decline in the growth and production of the insect would ultimately affect the normal life of a pond. Moreover, similar adverse effect on the physiology of fishes and other aquatic animals is also expected under the influence of these chemicals.

Table 28.3: Body Ascorbic Acid Concentration in Insects on Treatment with Mohr's Salt

Concentration of Mohr's Salt (molar)	*Body Ascorbic Acid Concentration (mg/gm body weight*±S.D.)*	
	Male	*Female*
0.01	1.53±0.1	1.53±0.2
0.02	0.76±0.08	0.90±0.06
0.03	Nil	Nil
0.04	Nil	Nil
0.05	Nil	Nil
0.06	Nil	Nil
0.07	Nil	Nil
0.08	Nil	Nil
0.09	Nil	Nil
0.10**	Nil	Nil

* Mean wt. of insect:
Male insect: 250 mg ± 9 S.D.
Female insect: 265 mg ± 10 S.D.

** LC_{50}
Body ascorbic acid in 'control':
Male insect: 2.87 ± 0.2 S.D.
Female insect: 3.57 ± 0.4 S.D.

References

Eisler, R., Rossoll, R.M. and G.A. Gaboury. (1979). Fourth annotated bibliography on biological effects of metals in aquatic environments. U.S. Environmental Protection Agency Res. Rep. EPA-600/3-79084, Washington, D.C.

Harper, H.A., V. Rodwell and P.A. Mayer. (1977). *Review of Physiological Chemistry*, 16th edition. Lange, Los Altos, California.

Mumtazuddin, S. and S. Ehteshamuddin (2000). Effect of metal poisoning on total body protein in *Sphaerodema rusticum* (Belostomatidae; Hemiptera). *Res. J. Chem. Environ.*, 4(4): 49–51.

Norton, S.A. (1982). The effect of acidification on the chemistry of ground and surface water. In: *Acid Rain Fisheries*, (Ed.) R.E. Johnson. American Fisheries Society, Betbesda, Md., pp. 93.

Roe, J.H. (1961). In: *Standard Methods of Clinical Chemistry*, (Ed.) David Seligson. Academic press, New York, 3: 35.

Chapter 29

Microbial Contamination in Drinking Water: Cause, Detection and Remedy

M.K. Bhutra and Ambica Soni*

Department of Chemistry, J.N.V. University, Jodhpur (Rajasthan)

ABSTRACT

Most microorganisms in water derive from air, soil, living and decaying plants or animals, and fecal excrements of humen and other warm blooded animals. Many of these organisms come from unknown sources and have no sanitary significance because they are widely distributed in the natural environment and have no suspected association with human or animal wastes. This discussion will consider only those disease producing bacteria, which have been associated with recognized outbreaks in India and abroad. The paper also discusses merits and demerits of existing microbiological analytical techniques and highlights importance of rapid analytical techniques in effective implementation of microbiological quality assurance. Some remedial measures for control of bacterial population in water have also been discussed.

Keywords: *Microorganisms, Coliforms, Quality assurance, Waterborne disease outbreak.*

Introduction

Water is in fact a vehicle for the transfer of a wide range of diseases of microbial origin. The widespread prevalence of waterborne diseases continues to pose a very serious public health issue. Microorganisms get into water from air, soil, sewage, organic wastes, dead plants and animals etc. The disease producers most frequently found are:

---

* *Corresponding Author.*

1. *Salmonella,* which causes typhoid fever and other gastrointestinal disorders.
2. *Shigella,* which are causative agents for diarrhoeal diseases.
3. *Entamoeba histolytica,* which produces dysentery, and
4. *Enteric viruses,* like infectious hepatitis, polioviruses, coxsackie, ECHO and adenoviruses which are associated with enteric and other complex diseases.

Waterborne Disease Outbreaks

Frequent outbreaks of waterborne diseases are being reported from different parts of the world due to absence of proper sanitation and rapid monitoring facilities. Some of them, which drew the attention of health workers, are cited below.

The outbreak of cholera, attributed to the Broad street pump in London in 1854 and was probably the first time that water was implicated by epidemiological methods, in transmitting a specific disease. Cholera, an illness caused by ingestion of the bacterium *Vibrio cholerae,* is characterized by intense diarrhoea, which results rapidly to massive fluid depletion and death in a very large percentage of untreated patients. Thousands of Indians continue to die each year even at present due to this disease. The epidemic of typhoid fever that occurs in Lausanne, Switzerland in 1872 was the first case to attract general attention. Since then more than 275 outbreaks have been observed throughout the world. In America the major well-documented epidemic of typhoid was attributed contaminated municipal water supply of Keene in December 1959. The Riverside outbreak of salmonellosis in May 1965 was attributed to the community water supply. The basis of this conclusion was the epidemiological investigation that eliminated all other likely possibilities except water. An outbreak of bacillary dysentery occurred in Newton, Kan in 1942, when sewage entered the water distribution system through frost-proof hydrants and water closet valves. An outbreak of water borne gastroenteritis also occurred in California, USA in August 1965. The city obtained its water from deep wells. There was considerable evidence that the outbreak was a result of sewage contamination of the city's water supply through one of the wells.

Apparently the transmission of amoebiasis has been traced directly to drinking water on only few occasions. In the first epidemic in 1933 and 1934 at two Chicago hotels three sources of direct and fecal contamination of drinking water were found, *viz.,* back siphonage form toilets, draining from a defective sewer over an open water cooler and direct cross connections between sever lines and water supply. Another outbreak occurred under similar circumstances in the Mantetsu apartment building in Tokyo. All these epidemics followed fecal contamination of drinking water in the distribution systems.

Infectious hepatitis is currently the only viral disease for which a waterborne route of infection is generally accepted, although person-to-person contact is considered the most frequent type of transmission. In 1971 Indo-Pak war more casualties at western border of Rajasthan (India) were due to spread of this infectious, disease for which root cause was considered contaminated water.

The status of water supply in India, till today, is far from satisfactory. Frequent outbreaks of waterborne diseases are being reported from different parts of country in absence of proper sanitation and rapid monitoring facilities. It is reported by WHO that about 15 million children are said to die every year in the world because of waterborne disease and 50 per cent of world hospital beds in underdeveloped and developing countries are filled with people suffering from waterborne disease.

Some Facts about Water-Related Diseases

1. Each year more than ten million people die from water-related disease.
2. The leading cause of child death in the world is diarrhoea.
3. UNICEF estimates that in 1993 alone, 3.8 million children under the age of five died from diarrhoea resulting from ingesting waterborne pathogens.
4. Of the 37 major diseases in developing countries, 21 are water and sanitation related.
5. The World Health Organization (WHO) estimates that 80 per cent of all sickness in the world is attributed to unsafe water and sanitation.
6. WHO also estimates that each year, throughout the world, children under five suffer 1.5 billion episodes of diarrhoea, approximately 4 million of which are fatal.
7. If no action is taken to address unmet basic human need for water, as many as 135 million people will die from water related diseases by 2020.
8. Even if the explicit Millennium Goals announced by the United Nations in 2000 are achieved, between 34 and 76 million people will perish from water-related diseases by 2020.
9. No single type of intervention has greater overall impact upon the national development and public health than does the provision of safe drinking water and the proper disposal of human excreta.
10. Human health improvements are influenced not only by the use of clean water, but also by personal hygiene habits and the use of sanitation facilities.
11. Water, sanitation and hygiene interventions have been shown to reduce sickness from diarrhoea on average between a quarter and a third.
12. Six to nine million people are estimated to be blind from trachoma. The population at risk for this disease is 500 million. This disease can be reduced 25 per cent by provision of adequate quantities of water.
13. In reducing the toll of sickness and death from diarrhoea, the supply of adequate quantities of water is usually more important than improving its quality. This is because the organisms that cause diarrhoea can be spread through many routes besides drinking water, and increased quantities of water can improve household and personal hygiene to prevent these.
14. About one-third of the population of the developing world are infected with intestinal worms that can be controlled through better water, hygiene, and sanitation. These parasites can lead to malnutrition, anaemia, and retarded growth.
15. 200 million people in the world are infected with schistosomiasis, of which 20 million suffer severe consequences. This disease can be reduced 77 per cent from well-designed water and sanitation interventions.
16. At any given time, half the people in developing countries are suffering from water-related diseases.
17. More people die each year from unsafe water from all forms of violence, including war.

Microbiological Analytical Techniques

The objective of the bacteriological examination of water is usually to estimate the hazards due to fecal pollution, and therefore, the probability of disease producing organisms being present. Because tests for the detection of pathogens are expensive and time consuming, so normally occurring bacteria in the intestines of warm blooded animals have been used as indicators of fecal pollution; and, indirectly of disease producing potential. The coliform group and fecal coliforms have been used at various times-in making these estimations.

Coliforms

The coliform group (total coliforms) as described in USPHS Drinking Water Standards includes organisms that differ in biochemical and serological characteristics and in natural sources and habitats. This group of bacteria is used world wide as an indicator of the microbiological quality of the water (WHO, 1993). The coliforms include all aerobic and facultative anaerobic, gram negative, non-spore forming, rod shaped bacilli, which ferment lactose within 48 hours at 35±2°c and which are native to the intestinal tract of mammals. They are members of the family Enterobacteriaceae and like the other members can be cultured from mammalian feces. The group mainly consists of several genera of bacteria belonging to this family, mostly *Escherichia, Klebsiella, Enterobacter, Citrobacter*, and *Serratia*. Among these *Escherichia* (*E. coli*) is of fecal origin while others are frequently found on vegetation. The detection of coliforms in tap water is supposed to indicate to utilities and health authorities that a breach has occurred in treatment or in the distribution system, allowing microbial (fecal) contamination of water to occur and thereby rendering it inappropriate for drinking.

Fecal Coliforms (Thermo Tolerant Coliforms)

Studies reveal that fecal coliform analysis is a more definitive test for recent fecal pollution than the total coliforms. The word fecal coliform primarily defines *E. coli* and occasionally *Klebssiella*. It is relatively harmless bacterial specie and is almost always present in water that contains enteric pathogens. Thus, because they are relatively easy to isolate and because they normally survive longer than the disease producing organisms, fecal coliforms are a useful indicator of possible presence of enteric pathogens and viruses. In most cases water that is free from fecal coliforms is considered free of disease producing bacteria. So nowadays estimation of fecal coliforms is preferred over total coliforms.

Fecal coliforms are distinguished from total coliforms by having the capability of fermenting lactose even at elevated temperature like 44.5±0.2°C. At this temperature due to heat shock the non-fecal coliforms do not grow. In sewage, the fecal coliforms may constitute 30–40 per cent of total coliforms.

Bacteriological Standards

For Water Entering the Distribution System in Piped Supply after Chlorination or Disinfection

Fecal coliform–zero per 100 ml of the sample.

Total coliforms–zero per 100 ml of the sample.

A sample of water that does not conform to this standard, calls for immediate investigation into the efficiency of the purification process and the method of sampling.

For Untreated Water Maximum Allowable Limit is:

Fecal coliforms–zero per 100 ml of the sample.

Total coliforms–8–10 per 100 ml of the sample.

If the number exceeds, the supply should be disinfected.

Conventional Method for Detection of Fecal Coliforms

Multiple Tube Fermentation Technique (MTFT Technique)

It is a 3-stage techniques, which can be used for estimation of total coliforms as well as fecal coliforms. In this procedure 3 sets of tubes (each set comprising of 5 test tubes) were taken. Water (under test) in 10 ml, 1 ml and 0.1 ml amount is inoculated into lactose broth tubes. The tubes are incubated at 37°C for maximum 48 hours and coliform organisms may be identified by their production of gas from the lactose broth. By referring to a MPN table (Table 29.1) a statistical range of the number of coliforms may be determined by observing how many broth tubes showed gas formation and turbidity.

Table 29.1: MPN Index and 95 per cent Confidence Limit for Various Combinations of Positive Results when 5 Tubes are Used per Dilution (1.0 ml, 1.0 ml and 0.1 ml)

Combination of Positives	*MPN Index/ 100 ml*	*95% Confidence Limits*		*Combination of Positives*	*MPN Index/ 100 ml*	*95% Confidence Limits*	
		Lower	*Upper*			*Lower*	*Upper*
0–0–0	< 2	–	–	4–2–0	22	9.0	56
0–0–1	2	1.0	10	4–2–1	26	12	65
0–1–1	2	1.0	10	4–3–0	27	12	67
0–2–0	4	1.0	13	4–3–1	33	15	77
				4–4–0	34	16	80
1–0–0	2	1.0	11	5–0–0	23	9.0	86
1–0–1	4	1.0	16	5–0–1	30	1.0	110
1–1–0	4	1.0	15	5–0–2	40	20	140
1–1–1	6	2.0	18	5–1–0	30	10	120
1–2–0	6	2.0	18	5–1–1	50	20	150
				5–1–2	60	30	180
2–0–0	4	1.0	17	5–2–0	50	20	170
2–0–1	7	2.0	20	5–2–1	70	30	210
2–1–0	7	2.0	21	5–2–2	90	40	250
2–1–1	9	3.0	24	5–3–0	80	30	250
2–2–0	9	3.0	25	5–3–1	110	40	300
2–3–0	12	5.0	29	5–3–2	140	60	360
3–0–0	8	3.0	24	5–3–3	170	80	410
3–0–1	11	4.0	29	5–4–0	130	50	390
3–1–0	11	4.0	29	5–4–1	170	70	480
3–1–1	14	6.0	35	5–4–2	220	100	580
3–2–0	14	6.0	35	5–4–3	280	120	690
3–2–1	17	7.0	40	5–4–4	350	160	820
4–0–0	13	5.0	38	5–5–0	240	1.00	940
4–0–1	17	7.0	45	5–5–1	300	100	1300
4–1–0	17	7.0	46	5–5–2,	500	200	2000
4–1–1	21	9.0	55	5–5–3	900	300	2900
4–1–2	26	12	63	5–5–4	1600	600	5300
				5–5–5	1600		

This technique is applicable to the analysis of even saline of brackish water but method utilizes five days till completion test, so there is a risk that the water will be used by many consumers before analysis is completed. Moreover different culture media are required during different stages, which are of perishable nature.

Results of the examination are reported in terms of most probable number (MPN) and for unusual combinations (not cited in the table) following formula, which is suggested by Thomas, can be used:

$$\text{MPN}/100\text{ml} = \frac{\text{No. of Positive Tubes} \times 100}{\sqrt{\text{ml of Sample in negative tubes} \times \text{ml sample in all tubes}}}$$

Membrane Filter Technique (MF Technique)

The MF technique is highly reproducible and yield numerical results more rapidly than the MTFT technique. The method employs a filtration unit. About 100 ml water sample is filtered through a membrane filter and the filter pad is then transferred to a plate of bacteriological medium. Bacteria trapped in the filter will form colonies, which may be counted. For total coliform and fecal coliform different culture media are used. Total number of bacteria can be calculated by following statistical formula:

$$\text{MPN}/100\text{ml} = \frac{\text{Colonies counted} \times 100}{\text{ml of sample filtered}}$$

The test may be carried out in field also. But the test is quite expensive because both filtration unit as well as media are imported and the media used are unstable too.

Standard Plate Count (SPC)

Water sample is diluted in sterile buffer. Measured amounts are pipetted into petri dishes. Agar medium is added and plates incubated. Colony counts are made and multiplied by the reciprocal dilution factor to have total bacteria per ml water. This method is similar to that used for milk. This method requires skilled operators and the media used are unstable too.

Need for Rapid Technique

Enumeration and isolation of organisms using conventional method involves steps like enrichment, isolation on selective agar medium, biotyping and serotyping. Although these techniques are serving well over the years but there are some inherent limitations with them. One of the major limitation is the long time period for the completion of test and another is that media used for growth of bacteria are of perishable nature. So there is a strong need to develop methods for rapid detection of indicator organisms as well as pathogens (using chemically defined culture media which are generally non perishable and cheap) in order to assess quality of water without compromising on the sensitivity and accuracy of test methods.

Treatment for Removal of Organisms

Water and sewage harbor a variety of organisms responsible for diseases in man. It is thus necessary to remove or rather interrupt the disease cycle in nature through treatment of water and sewage. Water purification involves 3 basic steps.

Sedimentation

The objective is to remove bulky objects such as leaves, sand and, gravel particles and the materials that have run off form soil. It is done in large reservoirs or in restricted area of settling tank. Aluminium sulphate (alums) or iron sulphate may be added to hasten sedimentation process. After mixing, chemicals form jelly like masses of coagulated materials called flocs. They fall through the water and cling to organic particles and microorganisms, dragging a major portion to the bottom sediment in a process called flocculation.

Filtration

The process removes the remaining microorganisms from the water. Although different types of filter materials are available, but mostly a layer of sand and gravel is utilized to trap the microorganisms.

Disinfection

It is the last step to ensure 100 per cent killing of microorganisms. Mostly bleaching powders are used. Studies reveal that enough chlorine is added which should, not only satisfy chlorine demand of water but should leave some residual free chlorine to ensure complete disinfection.

Now-a-days chloramination, ozonation, UV irradiation and use of hydrogen peroxide is preferred over chlorination, because their bactericidal power is high and long lasting. The efficacy of silver as a bactericidal against *E. coli* has also been established.

Conclusion

Waterborne diseases are major health problem in many countries. Fecal contamination of drinking water is one of the main cause of these disease outbreaks as many pathogens are carried through human feces. At present, the quality of drinking water is ensured by testing the water sample for total coliforms and thermo tolerant fecal coliforms or *E. coli*. However when applied to remote areas the standard methods have various limitations. The distance to laboratory facilities and the high cost of analysis restrict water quality assessment in remote areas and in developing countries. Under such circumstances as on-site bacteriological test is considered to be highly essential for routine water quality testing.

The combination of a more effective rapid detection method and an analytical tool integrating a variety of parameters will greatly help water utilities in identifying the causes of water quality problems and consequently, allow them to apply corrective measures to eliminate coliform occurrence from their distribution system.

References

Ahmed, F. and M.E. Sikder. 1997. Performance of pond sand filter in saline problem areas. *J. IPHE*, p. 10–16.

Anand, S.K., 2002. Role of rapid analytical techniques in effective implementation of microbiological quality assurance. *Dairy Microbiology*, p. 10–15.

Battling Waterborne ills in a Sea of 950 Million, February 17, 1997. The Washington Post.

Brion, G.M. *et al.* 2000. New approach to use of total coliform test for watershed management. *Water Sci. & Technol.*, 42: 1–65.

Chalmers, R.M. *et al.* 2000. Waterborne *Escherichia coli* 0157. *Jour. Applied Microbiol.*, 88: 124S.

Chan, M.S. 1997. The global burden of intestinal nematode infections: Fifty years on. *Parasitology Today*, 13(11): 438–443.

Chatterjee, D. 1986. Some rapid but sensitive methods for routine determination of drinking water quality. *J. IPHE*, 1: 29–35.

Craun, G.F. and Calderon, R.L. 1999. *Waterborne Disease Outbreaks: Their Causes, Problems and Challenges to the Treatment Barriers*, 15th edn. Waterborne Pathogens, AWWA Manual M48, AWW A, Denver.

Craun, G.F. and Calderon, R.L. 2001. Waterborne disease outbreaks caused by distribution system deficiencies. *J. AWWA*, 93: 9–64.

Craun, G.F. *et al.* 1997. Coliform bacteria and waterborne disease outbreaks. *J. AWWA*, 89: 3–96.

Easterbrook, G. September 11, 1994. Forget PCBs. Radon. Alar. The World's Greatest Environmental Dangers Are Dung Smoke and Dirty Water, The New York Times Magazine.

Edberg, S.C. *et al.* 2000. *Escherichia coli*: The best biological drinking water indicator for public health protection. *J. Applied Microbiol.*, 88: 106S.

Esrey, S.A., J.B. Potash, Roberts, L. and Shiff. C. 1991. Effects of improved water supply and sanitation on ascariasis, diarrhoea, dracunculiasis, hookworm infection, schistosomiasis, and trachoma, WHO, 69(5): 609–621.

Furtado, C. *et al.* 1998. Outbreaks of waterborne diseases in England 1992–95. *Epidemiol. Infect.*, 121: 1–109.

Gale, P. 1996. Coliforms in the drinking water supply: What information do the 0/100 ml Samples Provide? *J. Water SRT-Aqua.*, 45: 4–155.

Gatel, D. *et al.* 1995. Forecasting the Number of Coliform–positive Sample. Proc., AWW A WQTC, New Orleans.

Geldreich, E.E. 1996, Pathogenic agents in freshwater resources. *Hydrological Processes*, 10: 2–315.

Gleick, Peter H., August 15,2002. "Dirty Water: Estimated Deaths from Water-Related Diseases 2000–2020".

Greenberg, A.E. and Ongerth, H.J. 1966. Salmonellosis in riverside California. *J. AWWA*, 58: 145.

Healy, W.A and Grossman, R.P. 1961. Waterborne typhoid fever epidemic at Keene, New Hampshire. *J. NEWWA*, 75: 38.

Henry, F. *et al.* 1990. Environmental sanitation, food and water contamination and diarrhoea in rural Bangladesh. *Epidemiol. Infect.*, 104: 253–259.

Kinnman, C.H. and Beelman, F.C. 1944. An epidemic of 3000 cases of bacillary dysentery involving a war Industry and members of armed forces. *Am. J. Public Health*, 34: 948.

Lenore, S.C., Arnold, E.D. and Andrew, D.E. 1998. *Standard Methods for Examination of Water and Waste Water*, 20th edition.

Michael, A.B., Lance Adling and Michael, P.L. 2003. The efficacy of silver as a bactericidal agent: Advantages, limitations and considerations for future use. *Aqua*, 522: 6.

Momba, M.N.B., Cloeta, T.E. and Kaflf, R. 2000. Influence of disinfection processes on the microbial quality of potable ground water in laboratory-scale system model. *Aqua*, 49: 1.

Mosley, J.W 1959. Waterborne infectious hepatitis. *New Eng. J. Med.*, 261: 703.

Parry-Jones, S. and P. Kolsky, WELL Technical Brief. Some global statistics for water and sanitation related diseases.

Power and Daginwala. 1999. *General Microbiology*, Vol. 2. Himalaya Publishing House.

Rodier, J. 1975. *Analysis of Water*. John Wiley and Sons, New York.

Rodriquez, M.J., Milot, J. and Serodes, J.B. 2003. Predicting trihalomethane formation in chlorinated waters using multivariate regression and neutral networks. *Aqua*, 52: 3.

Ross, A.I. and Gillespie, E.H. 1952. An outbreak of waterborne gastroenteritis and some dysentery. *Gr. Brit. Ministry of Health, Bull.*, 11: 36.

Sharma, P.D. 1994. *Microbiology*, 1st Ed. Vivek Rastogi and Company Educational Publishers, Meerut.

Water and Sanitation for Health project, sponsored by the US Agency for International Development. 1993. Lessons Learned in Water, Sanitation and Health.

World Health Organization and Harvard University. 1996. *Global Burden of Disease*.

World Health Organization Fact Sheet, November 1996. Water and Sanitation, Water-related diseases include not only those contracted by ingesting unsafe water but also those that result from improper sanitation.

World Health Organization. 1992. Our Planet. Our Health: Report of the WHO Commission on Health and Environment.

Chapter 30

Analysis and Seasonal Comparative Study of Amanishah Nallah and Neighbouring Ground Water Source in Sanganer Town, Jaipur

*Dinesh Kumar, Hari Singh Shivran, Mahavir Prasad and R.V. Singh**

Department of Chemistry, University of Rajasthan, Jaipur – 302 004 (India)
E-mail: kudiwal@datainfosys.net

ABSTRACT

Increasing urbanization has led to the vast problem of disposal of municipal solid waste and industrial untreated effluents. Industrialization has its inevitable effect on pollution of air, water and soil based on the type of industry, nature of raw materials used, the manufacturing processes involved, data on the effect of industrial effluents from a paper factory, an automobile industry, a texthe factory and on soil properties and water qualities. Water pollution not only affects the human health but the total ecosystem including the flora and funa. Physico-chemical monitoring of Sanganer Nallah and surrounding tube wells and hand pumps was carried out in post monsoon in the months of October 2003 and January 2004. A lot of sewage industrial effluent's are being discharged into Amanishah nallah. The analysis was carried out for the parameters namely temperature, pH, D.O., B.O.D., Cl^-, NO_3^-, total alkalinity, total solids, total dissolved solids, total suspended solids, total hardness, calcium hardness, magnesium hardness, carbonate hardness, non-carbonate hardness, fluorides, odour and colour in the water samples. The quality parameters were compared with the standards laid by WHO and ICMR for drinking water quality. It is significant to note that the level of fluoride and total hardness concentration is high in tube well samples nearby to the nallah.

---

* Corresponding Author.

Therefore, serious attentions are required for the safe drinking water for which some remedial measures have been suggested in this paper. Comparative studies have also been carried out for the results of both months in the present article.

Keywords: *Physico-chemical parameter, Industrial effluent, Water quality, Amanishah Nallah.*

Introduction

According to the American College Dictionary, pollution is defined as: "to make foul or unclean; dirty." Water pollution occurs when a body of water is adversely affected due to the addition of large amounts of materials to the water. When it is unfit for its intended use, water is considered polluted. Two types of water pollutants exist; point source and non-point source. Point sources of pollution occur when harmful substances are emitted directly into a body of water (Gannon, Osmond, Humenik, Gale and Spooner, 1996). A non-point source delivers pollutants indirectly through environmental changes. An example of this type of water pollution is when fertilizer from a field is carried out into a stream by rain, in the form of run-off which in turn affects aquatic life (Winter, Britton and Krough, 1992). The technology exists for point sources of pollution to be monitored and regulated, although political factors may complicate matters. Non-point sources are much more difficult to control. Pollution arising from non-point sources accounts for a majority of the contaminants in the streams and lakes (Onsdorff, 1996; Terry, 1996).

Causes of Pollution

There are many causes of pollution including, sewage and fertilizers which contain nutrients such as nitrates and. phosphates. In excess levels, nutrients over stimulate the growth of aquatic plants and algae. Excessive growth of these types of organisms consequently clogs our waterways, use up dissolved oxygen as they decompose, and block light to deeper waters (Shivran, 2000). This, in turn, proves very harmful to aquatic organisms as it affects the respiration ability of fish and other invertebrates that reside in water. Pollution is also caused when silt and other suspended solids, such as soil, wash off plowed fields, construction and logging sites, urban areas, and eroded river banks during the rains. Under natural conditions, lakes, rivers, and other water bodies undergo eutrophication, an aging process that slowly fills in the water body with sediments and organic matters. When these sediments enter into various bodies of water, fish respiration becomes impaired, plant productivity and water depth become reduced, and aquatic organisms and their environments become suffocated. Pollution in the form of organic material enters waterways in many different forms as sewage, as leaves and grass clippings, or as runoff from livestock feedlots and pastures. When natural bacteria and protozoan in the water break down this organic material, they begin to use up the oxygen dissolved in the water (Richman, 1997). Many types offish and bottom-dwelling animals cannot survive when levels of dissolved oxygen drop below two to five parts per million. When this occurs, it kills aquatic organisms in large numbers which leads to disruptions in the food chain.

Materials and Methods

Nine samples were collected in two litres polythene bottles from different stations at Amanishah nallah and neighbouring tube wells from nallah. The bottles were cleaned earlier by soaking these with dilute chromic acid which was washed well with distilled water. At the sampling site temperature, pH and D.O. were measured using the potable water analysis kit and odour by smelling and colour by sight. The other parameters were measured by standard methods (APHA, WPCF 1995). For the determination of other parameters, samples were preserved by adding appropriate reagents and brought to the laboratory for detailed chemical analysis. pH was determined using digital pH meter; TDS, TS and TSS by gravimetric method and BOD by azide modification, total alkality and hardness by titrimetric, chloride by argentometric titration, nitrates and fluoride by the selective electrode method.

Results and Discussion

The analysed physico-chemical characteristics of water samples are summarized in Tables 30.1 and 30.2. The impact of industrial and domestic water on the water quality of drinking water samples indicated that most of the parameters were high in summer due to the low volume of water. In monsoon period however, dilution of nallah water occurs due to turn off from the land and accordingly some parameters showed lower values.

Table 30.1: The Physico-chemical Parameters of Various Sampling Stations During October 2003

Sl.No.	*Parameters*	*Site 1 Nallah*	*Site 2 O.T.W.*	*Site 3 Nallah*	*Site 4 H.P.*	*Site 5 Nallah*	*Site 6 B.W.*	*Site 7 Nallah*	*Site 8 B.W.*	*Site 9 Nallah*
1.	Temperature °C	292	30.0	29.1	30.0	29.6	292	28.9	29.1	28.8
2.	Odour	Foul smell	Odour-less	Bad smell	Odour-less	Bad smell	Odour-less	Bad smell	Odour-less	Bad smell
3.	Colour	Brown	Colour-less	Red	Slight brown	Brown	Colour-less	Grey	Colour-less	Grey
4.	pH	7.8	7.0	7.6	8.0	7.4	7.6	7.5	8.1	8.0
5.	D.O.	Nil	5.1	Nil	1.8	Nil	5.4	Nil	5.1	Nil
6.	B.O.D.	100	Nil	40	50	80	Nil	70	Nil	62
7.	Total alkalinity	520	370	600	390	590	400	620	530	570
8.	Total hardness	630	470	680	110	630	220	620	110	590
9.	Calcium hardness	350	250	320	60	490	120	370	70	330
10.	Magnesium hardness	280	220	360	50	140	100	250	40	260
11.	Carbonate hardness	520	370	600	110	590	220	620	110	570
12.	Non carbonate hardness	110	120	80	Nil	40	Nil	Nil	Nil	20
13.	Total solids (TS)	1610	1350	1750	1040	1790	880	1760	1140	1660
14.	Total dissolve solids (TDS)	1400	910	1470	840	1540	700	1540	1050	1470
15.	Total suspended solids (TSS)	210	440	280	200	250	180	220	90	190
16.	Chloride	24	340	320	100	350	70	390	180	350
17.	Nitrates	48	32	50	0.40	50	36	60	40	44
18.	Fluorides	0.34	033	0.35	0.61	0.38	0.32	0.37	1.6	0.51

O.T.W.: Open Tube Well; H.P.: Hand Pump; B.W.: Bore Well.

Note: All the values are in mg/L except, pH, colour, odour, and temperature (°C).

Table 30.2: The Physico-chemical Parameters of Various Sampling Stations During January 2004

Sl.No.	Parameters	Site 1 Nallah	Site 2 O.T.W.	Site 3 Nallah	Site 4 H.P.	Site 5 Nallah	Site 6 B.W.	Site 7 Nallah	Site 8 B.W.	Site 9 Nallah
1.	Temperature °C	28.0	282	28.1	28.1	28.0	27.9	27.8	28.0	28.1
2.	Odour	Foul smell	Odour-less	Bad smell	Odour-less	Bad smell	Odour-less	Bad smell	Odour-less	Bad smell
3.	Colour	Brown	Colour-less	Red	Brown	Grey	Colour-less	Blue	Colour-less	Red
4.	pH	6.7	7.6	7.4	8.3	7.6	7.7	7.6	8.1	7.8
5.	D.O.	Nil	7.3	Nil	3.4	Nil	7.4	Nil	10.1	Nil
6.	B.OD.	18	Nil	70	4	40	6	40	Nil	65
7.	Total alkalinity	400	700	480	250	610	300	590	570	430
8.	Total hardness	600	630	530	150	550	290	650	150	560
9.	Calcium hardness	170	110	400	30	170	140	530	120	390
10.	Magnesium hardness	430	520	130	120	380	150	120	30	230
11.	Carbonate hardness	400	630	480	150	550	290	590	170	430
12.	Non carbonate hardness	200	70	50	100	60	10	60	400	130
13.	Total solid (TS)	1400	1880	2205	1690	1600	900	1730	1100	1200
14.	Total dissolve solids (TDS)	960	1130	1650	1130	1420	600	1390	910	690
15.	Total suspended solids (TSS)	440	750	555	560	180	300	340	190	510
16.	Chloride	350	330	280	100	290	130	330	170	230
17.	Nitrates	15	90	160	31	24	80	23	121	21
18.	Fluorides	0.32	0.33	0.32	0.68	0.31	0.45	0.34	1.58	0.34

O.T.W.: Open Tube Well; H.P.: Hand Pump; B.W.: Bore Well.

Note: All the values are in mg/L except, pH, colour, odour, and temperature (°C).

Temperature

The variation in temperature in different months of the year followed the seasonal variation in the atmospheric temperature.

pH

The pH ranged from 7.0 to 8.1 and 6.7 to 8.3 in October 2003 and January 2004, respectively in the analyzed water samples. In every station the water remained alkaline. The pH has direct correlation with salinity and temperature. Higher pH at station 4 in the month of January 2004 may be due to the consumption of CO_2 by phytoplantation.

Biochemical Oxygen Demand (BOD)

BOD ranged from 0 to 100 mg/L in the month of October 2003 and from 0 to 65 mg/L in the month of January 2004. However, B.O.D. in the ground water samples were nil.

Total Alkalinity

The present study has revealed that water of the entire experimental area was alkaline, being more towards the source of effluent and loss away from the source. Maximum alkalinity (700 mg/L) had been found in Bore well water sample in the month of January. Abebisi (1980) suggested that water having 60 mg/L alkalinity added hardness to water. The results obtained in the present experiments clearly suggested that the effluent as well as the effluent affected crops field water were considerably hard and hence toxic to crop plants.

Total Hardness (TH)

Total hardness of all the water samples is within the highest desirable limit according to WHO, ICMR and BIS. While nallah water samples in both the months of October and January are having high range of total hardness showing the high concentration of carbonate, bicarbonate and sulphate ions in these samples.

Calcium and Magnesium Hardness

An amount of calcium and magnesium hardness varies from 60 to 490 mg/L and 40 to 360 mg/L, respectively in the month of October 2003. However, in January 2004 the amount varies from 30 to 530 mg/L and 30 to 520 mg/L of calcium and magnesium hardness, respectively. This type of hard water is useful for the growth of children due to the presence of calcium but hard water cause excessive consumption of the soap used for the cleaning purposes.

Carbonate and Non-carbonate Hardness

Carbonate and non-carbonate hardness found in October 2003 ranged from 110 to 620 mg/L and 0 to 120, respectively while in January 2004, these ranged from 150 to 630 mg/L and 10 to 400 mg/L, respectively.

Total Solid (TS)

Total solids ranged from 900 to 2205 mg/L in the month of January 2004 while in October 2003 results are below than the present analysis. Total solids determination is used to assess the suitability of potential supply of water for various uses.

Total Dissolve Solid (TDS)

TDS varies from 600 to 1650 mg/L and these values are lower than the values obtained in October 2003. Many dissolved substances are undesirable in water. Dissolved minerals, gases and organic constituency may produce aesthetically displeasing colour, taste and odour.

Total Suspended Solid (TSS)

TSS varies from 180 to 750 mg/L. These values are higher than October 2003 results. Biologically active (live) suspended solids may include disease causing organism as well as organisms such as toxin producing strains of algae.

Chlorides

Chlorides concentration varies from 100 to 350 mg/L. The present analysis and comparative studies during both the months indicated more variation only at two places. Chlorides associated with sodium exerts salty taste, when its concentration is more than 250 mg/L.

Fluorides

Fluorides ranged from 0.31 to 1.58 mg/L. All water samples bear low fluoride concentration. While site 8 (Borewell) water samples fluoride concentration found high *i.e.* 1.58 mg/L. Presence of large amounts of fluoride is associated with dental and skeletal fluorosis (WHO, 1996) if it is greater than 1.5 mg/L.

Nitrates

Nitrates vary from 15 to 160 mg/L. All water samples bear low NO_3^- concentration in October 2003. In the month of January, 2004, the amount of NO_3^- vary by fluctuation. Many ground waters have significant quantities of nitrates due to leaching of the nitrate with the percolating water. Ground water can also contaminated by sewage, septic tank effluents and other wastage rich in nitrate. Therefore, nitrate poisoning has been referred to as the 'blue baby' diseases. The scientific term is 'Methemoglobinemia'. This disease occurs in infants, only when the concentration of nitrates exceeds 45 mg/L in drinking water.

Conclusion

The present studies have shown occurrence of high fluoride content in the ground water of Bore Well (Site 8) in both the seasons. The author suggested to the villagers to eliminate the problem with the help of defluoridation techniques. The high concentration of nitrates in the ground water of open tube well (Site 2), Bore Well (Site 6) and Bore Well (Site 8) in the January 2004 is according to the permissible limits prescribed by WHO. As a suggestion, for the removal of high nitrate concentration biological denitrification technique may be used. The observed values of the BOD and DO of sampling Site 4 (H.P.) in the month of October 2003 have been found above the permissible limits due to percolation of rain and nallah water.

Acknowledgement

The authors are thankful to CSIR, New Delhi for financial assistance.

References

APHA (American Public Health Association). (1995). American Water Works Association and Water Pollution Control Federation, *Standard Methods for the Examination of Water and Waste Water*, 19th edition, New York, U.S.A.

Gannon, R.W., Osmond, D.L., Humenik, F.J., Gale, J.A. and J. Spooner (1996). Agricultural water quality. *Water Res. Bull.*, 32: 437.

Onsdorff, K.A. (1996). Water pollution. *Environ. Law Pract.*, 4: 14.

Richman, M.IND. (1997). Water pollution. *Waste Water*, 5: 24.

Shivran, H.S. (2000). *Studies on Water Quality Parameters: A Dissertation Report*. MDS University, Ajmer.

Terry, L.A. (1996). Water pollution. *Environ. Law Pract.*, 4: 19.

W.H.O. (1996). *Guideline for Drinking Water Quality*, 2nd edn., Vol. 2, Health Criteria and Other Supporting Information, WHO, Geneva.

Winter, J., P. Britton, O. Krough. (1992). *Analytical Quality Control Gems/Water Operational Guide*, 3rd edn. UNEP/WHO/UNESCO Programme on Global Water Quality Monitoring and Assessment, pp. 19.

Chapter 31

Status of Drinking Water Quality Awareness and its Impact on Student Health: A Study of Schools of Buldana District

S.V. Agarkar and B.S. Thombre***

**Head, Department of Chemistry, Anuradha Engineering College, Chikhli – 443 201, Dist. Buldana*
***Government B. Ed. College, Buldana*

ABSTRACT

Educational Institutes are leaders in socio-economic development of country. They plays key role in all way development of individuals personality so that every individual can contribute effectively in the progress of the development. This study focuses on status of environmental awareness specially drinking water quality awareness among students of schools of Buldana District. The study tries to find out the impact of water quality on student health and ultimately on educational progress.

Introduction

In addition to regular curricular activities, educational institutes organizes various programmes for overall development of student personality. These activities helps to impart knowledge about social ethics, social problems, environment and its components, natural resources and their management, etc. Nowadays environmental education has become essential component of all walks of life. The rapid industrialisation, population explosion and illiteracy has generated various environmental problems. One of the most important aspect of life is non-availability of safe drinking water. The scarcity of drinking water has forced the population to use whatever the quality of water available in the region. This has seriously affected human health. To avoid this problem water quality awareness through environmental education at school level is essential.

Students are the leaders of next generation. They should be exposed to all the problems and make them capable to face and solve the problem. The students should be well aware about impact of drinking water quality on health and ultimately educational progress. To do so educating our students on the aspects of water quality, quantity requirement, pollution, resources management and their importance in the day to day life and its relation to human health. Schools should impart knowledge and information through co-curricular activities. In the present study attempts were made to study status of awareness and activities organised by the schools.

Objectives

1. To study the drinking water quality awareness among the school students.
2. To study the status of co-curricular activities organised by schools to create environmental awareness.
3. To study the status of quality of drinking water available at schools.
4. To study the impact of water quality on student health and ultimately educational progress.
5. To find out the causes of health problems and suggest remedial measures to the students of schools.

Hypotheses

1. There is significant relationship between quality of drinking water, students health and educational progress.
2. School organizes activities on water quality awareness and its impact on human health is relatively less than activities focusing on cultural and entertainment components.

Area

The study was conducted in the Buldana District of Maharashtra State.

Samples

For this study altogether 100 samples were collected from ten different schools of Buldana District. The names of villages and city along with number of samples collected are shown in Table 31.1.

Table 31.1: Number of Samples Collected from Schools of Buldana District of Maharashtra

Location of School	*Number of Samples Collected*
Chikhli	10
Divthana	10
Sakegaon	10
Buldana	10
Chikhli	10
Buldana	10
Kelvad	10
Sawargaon	10
Deulgaon	10
Raipur	**10**
Total Samples	**100**

Tools

In this study the investigators used the random sampling technique and the following tools for date collection.

1. Questionnaire
2. Interview
3. Observation

A questionnaire was prepared by the investigators to collect data from the students, teachers and headmaster of schools selected for the study. The items in the questionnaire were both open as well as close ended. The questionnaire was designated to get information about present status of drinking water awareness level, water storage facilities, health problems associated and its impact on educational progress.

Interviews of headmasters were conducted to know about administrative involvement in monitoring of drinking water quality and its impact on students educational progress. The status of drinking water supply, quality, quantity storage and utility were observed by the investigators while visiting the schools of the which are under study.

Method

The study was based on a survey. The quality of drinking water samples were analysed as per standard procedure (APHA, 1995; Trivedy and Goel, 1986 and Manivasakam, 1996).

Limitations of the Study

This study has been limited to ten schools of Buldana District. The study was confined to the students studying in VIII class and not the entire secondary stage. Ten students from each schools were selected randomly for the study. The head master and one science teacher teaching to VIII class selected for getting more information.

Findings and Discussion

Following are the findings of the study presented objective-wise:

1. Present status of drinking water quality awareness well aware. In all the schools, the percentage students well aware of water quality is less than 40 per cent. The school-wise percentage of students awareness is expressed in the Table 31.2.

Programmes Organised by Schools

In most of the schools, there were no separate schedule for organizing activities concerned with drinking water quality awareness. These programmes are organised as a part of environmental education. The activities like exhibitions of posters, lectures, competitions, etc. were organised in the schools.

Status of Quality of Drinking Water Availability at Schools

In all the schools there were separate arrangement for storage of drinking water, but no separate trained expert staff member is appointed to monitor the water supply and quality aspects. The filtration and disinfections of water by chlorination are the methods of purification at each school. The water quality testing laboratory facilities are not available in any of the school.

Table 31.2: Drinking Water Quality Awareness Among School Students of Buldana District

Names of the Schools Villages	*No. of Samples Collected*	*Percentage of Students Awarded*
S1	10	20
S2	10	20
S3	10	10
S4	10	30
S5	10	20
S6	10	20
S7	10	30
S8	10	30
S9	10	30
S10	10	30
Total Samples	**100**	

The physico-chemical and microbiological analysis of drinking water samples from each schools reveals that in all the schools physico-chemical parameters are within permissible limit but microbiological parameters exceeds the permissible limit as per various standards (WHO 1977). The results of analysis are shown in Table 31.3.

Table 31.3: Physico-chemical and Microbiological Analysis of Drinking Water Samples from Schools of Buldana District (Pre monsoon mean values)

Parameters	*Schools of Buldana District*									
	S1	*S2*	*S3*	*S4*	*S5*	*S6*	*S7*	*S8*	*S9*	*S10*
pH	7.12	7.46	7.37	7.40	7.12	7.11	7.40	7.52	7.22	7.17
Turbidity	2.9	3.5	2.7	4.3	5.7	4.2	4.1	4.2	5.7	3.9
Electrical Conductivity	0.62	0.72	0.6.)	0.97	0.37	0.65	0.55	0.71	0.62	0.55
Total Solids	175	139	162	178	191	177	167	190	134	127
Total hardness	200	195	175	155	176	167	187	191	131	154
Dissolved Oxygen	7.5	6.9	6.2	6.8	7.0	7.3	7.3	6.9	6.8	7.1
Chloride	97	82	110	150	171	181	162	112	99	101
Fluoride	0.122	0.09	0.17	0.99	0.19	0.72	0.82	0.71	0.75	0.92
Iron	0.12	0.015	0.15	0.75	0.37	0.101	0.33	0.71	0.12	0.09
Nitrate	0.09	0.007	0.062	0.0079	0.0068	0.0079	0.0019	0.0067	0.0082	0.090
Sulphate	0.17	1.2	1.74	2.20	1.79	1.22	1.71	2.10	1.62	1.51
Free CO_2	Nil	Nil	Nil	Nil	Nil	Nil	Nil	Nil	Nil	Nil
Oil and Grease	Nil	Nil	Nil	Nil	Nil	Nil	Nil	Nil	Nil	Nil
MPN of coliforms	350	200	300	225	375	350	425	375	375	250

All the values are in mg/lit. except pH, Turbidity, Electrical conductivity.

Turbidity: NTU, Electrical conductivity: mmhos/cm.

Impact of Water Quality on Health

Students of forty per cent of the schools were facing problems of water borne diseases like dysenteries, diarrhoea specially in monsoon season. The students suffering from water borne diseases remains absent in the school during treatment. As result of this they unable to complete home work, fail in unit tests and ultimately it affect annual result.

Causes of Health and Remedial Measures

The students of the schools were suffering for water borne diseases specially in rainy season. It observed that these are mainly caused due to microbiological parameters exceeding permissible limit as per various standards. To avoid this regular water quality monitoring cleanliness and disinfections of water by chlorine is recommended.

Conclusion and Suggestions

The teachers should be trained regularly about drinking water quality and its impact on health. The schools should organize various programmes, the list of some of the important programmes which can be organised in the schools are mentioned below:

1. Guest lectures of expert
2. Posters exhibition
3. Slide show/film show
4. Visit to water treatment plant
5. Various competition
6. Project work
7. Health check up camps.

These activities will create water quality awareness among student effectively and the problem of health hazards and impact on educational progress can be avoided.

In conclusion, it can be suggested that to irradiate the that problem of health hazards due to water quality and its impact on educational progress of the student training the teachers, appointment separate staff for monitoring the water quality and organisation of various programmes are needed.

References

Agarkar S.V. (1998). Physico-chemical aspects of ground water quality in Chikhli town of Buldana District. *Poll. Res.*, 17(3): 291–292.

APHA. (1995). *Standard Methods for Examination of Water and Waste Water*, 19th edn. American Public Health Association, American Water Works Association and Water Pollution Control Federation, Washington, DC.

Manivasakam, N. (1996). *Physico-chemical Examination of Water, Sewage and Industrial Effluents*, 3rd edn.

Trivedy, R.K. and Goel P.K. (1986). *Chemical and Biological Methods for Water Pollution Studies*. Environmental Publications, Karad.

WHO (1977). Regional Publication, South East Asia Series No. 14, WHO, New Delhi.

Chapter 32

Assessment of Pollution Soils Near Aluminium Industry (Industrial Estate) Warangal, Andhra Pradesh

B. Lalitha Kumari and M.A. Singara Charya

Department of Botany, Kakatiya University, Warangal – 506 009, Andhra Pradesh

ABSTRACT

An investigation of polluted soils amended with aluminium industry was under taken during 1995–96. Various physical, chemical and biological characteristics were assessed and discussed to understand the pollution in the soils. High pH values were recorded in January, and the mean value of bulk density was 1.24 gm/cm^3. The maximum alkalinity was recorded in March with a range of variation in between 0.01 to 0.62 mg/g. No significant quantitative changes were seen in nitrate, nitrates and ammonia contents. The variation for fungi was $0.7 \times 1.10 \times 10^5$ in the polluted or $0.5–2.3 \times 10^5$ in control soils. The soil enzymes varied differently with different seasonal variations.

Introduction

Aluminium industry is located in industrial estate Warangal, Andhra Pradesh. It was established in the year 1980 and concentrating on various activities on manufacturing, aluminium materials which includes, coasting, rolling and. spinning and washing materials. During the process the heavy amounts of costic soda and HNO_3 are used and they appear equally in high amounts in the effluents too. The effluents discharged from the industry without any treatment were applied on soils causing a major threat to the environment. The soil quality is being degraded and not used for any cultivation purpose.

Materials and Methods

The soils from the sewage irrigated soils adjacent to sewage canals near aluminium industry were selected and surface soils were collected for analysis. Soils of Kakatiya university campus which were unpolluted and physically similar in nature, served as control. The soils was air dried and crushed to parts in 2 mm screen. The physical, chemical and biological characteristics were analysed as methods suggested by Trivedi *et al.* (1987). The data is presented in Tables 32.1–32.4.

Results and Discussion

The physical properties of the soils from July 1995 to June 1996 was presented in Table 32.1. From the table it was evident that the maximum soil pH was recorded in January and May (8.0) and minimum (7.0) in polluted sites and in unpolluted site (Kakatiya University campus) it was usually 7.0. Nannipieri *et al.* (1978), Kay and Lilly (1970) measured the stability and kinetic properties of accumulated soil enzymes and their correlation with that of pH. The variation of bulk density was in between 1.13 to 1.32 gm/cm^3 with the mean value of 1.24 gm/cm^3 and Carter (1990), Wallace and Wallace (1990) a1so drawn some conclusions on the characterization of the soils based on the property of bulk density and micro-porosity with that of some enzyme kinetics. The average value of pore space is 53.8 per cent in aluminium industry, and 50.4 per cent in soils at Kakatiya University campus. The fluctuations in water holding capacity at polluted aluminium industry site was 20.6–25.6 per cent whereas this variation was between 19.7–24.9 in the control soils. The mean value of conductivity was 1.19 µmho in the soils at aluminium industry. The minimum conductivity level was recorded in the month of June in the soils collected at control site. Lavado and Tabaoda (1987) established soil salinity and sodicity which have a detrimental effect on both chemical and physical aspects of soil quality. The wilting coefficient values were recorded in maximum in the month of June (8.93). This range of variation in control soils was 6.29 to 8.93. Meghraj *et al.* (1986) established and concluded that the highest wilting coefficient was in the polluted soils to unpolluted soils.

Table 32.1: Physical Characteristics of Polluted Soils of Aluminium Industry in Warangal During 1995–96

	1995						1996						Mean Value	Standard Error
	July	Aug.	Sept.	Oct.	Nov.	Dec.	Jan.	Feb.	March	April	May	June		
pH														
A:	7.5	7.0	7.0	7.5	7.5	7.0	8.0	7.5	7.0	7.5	8.0	7.0	7.4	±0.27
B:	7.0	7.0	7.0	7.0	7.0	7.0	7.0	7.0	7.0	7.0	7.0	7.0	7.0	±0.00
Bulk Density (gm/cm^3)														
A:	1.16	1.24	1.31	1.17	1.26	1.31	1.26	1.13	1.32	1.23	1.16	1.28	1.24	±0.64
B:	1.19	1.23	1.26	1.39	1.26	1.36	1.42	1.41	1.23	1.32	1.38	1.28	1.32	±0.52
Pore Space (%)														
A:	56.2	53.2	50.6	55.9	55.1	50.6	75.1	57.4	50.2	53.6	56.2	51.7	53.8	±0.76
B:	55.1	53.6	52.5	47.6	52.5	46.7	46.4	46.7	53.6	50.2	47.9	51.7	50.4	±0.95
Water Holding Capacity (%)														
A:	20.6	21.4	23.2	22.8	23.2	21.2	20.8	22.4	24.6	24.7	25.1	25.4	22.9	±0.52
B:	24.2	23.4	22.8	20.3	20.3	21.9	19.7	20.2	21.8	20.6	24.8	24.9	22.4	±1.73
Conductivity (µm Mho)														
A:	0.95	0.65	0.54	0.22	0.40	0.32	0.68	1.28	4.23	0.92	1.42	2.68	1.19	±0.44
B:	0.62	0.46	0.17	0.37	0.19	0.20	0.68	0.72	1.02	0.98	0.44	1.28	0.49	±0.33
Permanent Wilting Co-efficient														
A:	7.13	7.38	8.00	7.89	8.00	7.31	7.27	7.72	8.48	8.52	8.68	8.93	7.94	±0.34
B:	8.35	8.07	7.85	7.00	7.55	6.29	6.96	7.52	7.79	8.38	8.89	8.93	7.84	±0.28

A: Polluted; B: Unpolluted.

The chemical properties of the soils from 1995 July to 1996 June was presented in Table 32.2. The alkalinity of the soils varied among the different soil samples (polluted and unpolluted) during 1995–96. Generally, it was observed that, the maximum alkalinity was recorded in March. The range of variation was in between 0. 01 to 0. 62 mg/g and 0.01–0.20 mg/g in control soils. The chloride range of variation was in between 0.85 to 6.12 mg/g in polluted soil at aluminium industry and 0.92–2.13 mg/g in control soils. Slobe and De (1986) in their study on effects of plant use, agriculture and forestry compared chloride concentrations with that of degradation of environment and soil pollution. The calcium range in the soils varied from 1.46 to 10.5 mg/g in industrial amended soils effluent and were recorded 1.46–9.40 mg/g range in control soils. Ross (1973) studied the glycoside hydrolase activities and its significant relation with calcium. The magnesium varied in between 1.9 to 4.09 mg/g in aluminium industry soils. The calcium carbonate range was in between 24.0–60.5 per cent in polluted and 20.0–42.5 per cent in control soils. The aluminium content was high in September month, and the range was between 0.02 to 0.46 mg/g in aluminium industry effected soils and 0.01 to 0.11 mg/g in control soils. The range of variation in the content of silicate were high in the month of July in polluted soils with the range of 0.29–0.74 mg/g. The sodium content was high (22.4 mg/g) in polluted site near aluminium industry. The potassium content was high in the month of June with a range between 1.4 to 24.3 mg/g in polluted sites. The iron; content was high in the month of July 1.2 mg/g in polluted soils and 0.9 mg/g in September in control soils. The sulphates were high in the month of April and the range of variation was 20 to 186 mg/g in polluted soils and 17 to 95 mg/g in unpolluted soils. Foth (1990) in his extensive study on fundamentals of soil science, demonstrated sulphates as basic nutrients in changing the productivity levels and which were constantly and rapidly decreasing through industrial effluents. The maximum phosphorus content was recorded in the month of February and may months and the minimum amounts were recorded during July month in polluted soils. The range was between 9.0–28.0 mg/g in polluted and 9.0–14.0 mg/g in control soils. Sanchez and Uehera (1980) suggested the management practices to avoid soils with high phosphorus fixation. Raghava *et al.* (1980), Bhattachalya and Deshpandey (1993) successfully demonstrated the relations between polluted and unpolluted soils in various industrial belts on development and yield of crops studied by them. The highest nitrate was recorded in the month of February in polluted soils and March for unpolluted soils and the variation was in between 0.32 to 0.98 mg/g in polluted soils. Somasekhar *et al.* (1984), observed that the industrial waste water treatment increases the pH, organic carbon, nitrogen of the soil and decreased the water holdings capacity. No significant quantitative changes were seen in nitrites in between polluted and unpolluted, soils. Mc Clung *et al.* (1983) explained the nitrification inhibition and decrease nitrite content in soils amended with sewage sludge and certain industrial effluents. The ammonia contents was high in control compared to unpolluted soils. In polluted soils the range was in between 0.20 to 0.50 mg/g. Maximum ammonia was recorded in the month June in control soils. Parr (1972) and Brady (1995) reviewed the status of ammonia in chemical and biochemical considerations for minimising the efficiency of fertiliser nitrogen. The highest amounts of organic matter recorded in the soils flooded with effluents from aluminium industry, where the range was 10.44 to 89.9 per cent. The heavy metals such as mercury, cadmium, cobalt, chromium, lead were analysed and found to be very meagre in their concentrations.

The soil microorganisms which determines the form and arrangement of the soils were enumerated and presented in Table 32.3. The variation for fungi was 0.7 to 1.10 × 10^5 in the soils effected by aluminium effluents and the minimum and maximum number of colonies fluctuated in between 0.5 to 2.3 × 10^5 in control soils. Anders *et al.* (1986) in their studies established the relationship between pollution of the soils and soil microflora. The range of variation of bacterial density was in between 3.4 to 25.0 × 10^6 g among polluted soils whereas it was 2.8 to 12.3 × 10^6 g in control soils.

Table 32.2: Chemical Characteristics of Polluted Soils of Aluminium Industry in Warangal During 1995–96

	1995						*1996*						*Mean Value*	*Standard Error*
	July	*Aug.*	*Sept.*	*Oct.*	*Nov.*	*Dec.*	*Jan.*	*Feb.*	*March*	*April*	*May*	*June*		
Alkalinity														
A:	0.08	0.14	0.12	0.06	0.08	0.12	0.46	0.46	0.62	0.08	0.01	0.38	0.21	± 0.05
B:	0.06	0.10	0.08	0.04	0.06	0.14	0.08	0.20	0.16	0.06	0.01	0.20	0.09	± 0.01
Chlorides														
A:	1.98	1.91	1.42	2.27	1.42	2.55	3.97	2.55	0.85	1.27	1.49	6.12	2.32	± 0.41
B:	1.77	2.05	1.42	1.27	1.56	1.42	1.98	1.98	2.13	1.21	0.92	2.13	1.66	± 0.11
Calcium														
A:	1.60	2.00	1.92	1.46	1.76	4.32	10.5	4.80	9.45	7.40	5.31	7.47	2.32	± 0.41
B:	4.84	2.48	5.29	1.44	1.76	1.92	6.41	9.45	7.85	4.64	7.05	9.29	1.66	±0.17
Magnesium														
A:	0.29	0.87	0.97	0.19	0.87	0.68	1.16	0.87	0.97	1.20	1.77	4.09	1.16	±0.29
B:	3.89	1.26	3.99	4.71	0.48	1.55	1.36	5.83	0.49	1.12	1.36	0.48	2.21	±0.53
Calcium Carbonate														
A:	51.0	60.5	24.0	31.0	46.5	37.5	34.5	46.0	51.5	49.0	43.5	35.0	42.5	±2.98
B:	37.5	28.0	27.5	22.5	36.0	42.5	30.0	36.0	31.0	29.0	40.5	20.0	31.7	±2.01
Aluminium														
A:	0.06	0.17	0.46	0.42	0.17	1.16	0.02	0.22	0.02	0.29	0.11	0.09	0.18	±0.04
B:	0.05	0.04	0.04	0.11	0.10	0.06	0.11	0.01	0.01	0.07	0.11	0.08	0.07	±0.01
Silica														
A:	0.74	0.35	0.42	0.38	0.29	0.68	0.44	0.50	0.60	0.43	0.74	0.35	4.93	±0.45
B:	0.45	0.25	0.45	0.35	0.2i	0.37	0.71	0.43	0.54	0.74	0.72	0.56	4.81	±0.51
Sodium														
A:	6.4	7.9	10.6	10.6	6.8	4.3	5.3	10.1	12.2	16.3	7.8	22.4	1.92	±0.47
B:	0.9	2.3	1.2	0.6	0.9	0.6	4.3	0.9	2.3	1.5	2.0	0.6	1.50	±0.31
Potassium														
A:	0.28	0.30	0.32	0.22	0.35	0.28	0.32	0.31	0.68	0.29	0.14	2.03	0.46	±0.14
B:	0.39	0.38	0.44	0.42	0.43	0.44	0.48	0.52	1.08	0.69	0.74	0.54	0.54	±0.05
Iron														
A:	0.12	0.06	0,02	0,09	0.02	0,05	0.02	0.09	0.03	0,09	0.04	0.08	0.04	±0.01
B:	0.01	0.07	0.09	0.08	0.01	0.01	0.01	0.01	0.01	0.02	0.01	0.04	0.02	±0.01
Sulphates														
A:	4.4	1.6	4.7	3.2	6.5	2.6	5.8	4.1	2.0	18.6	15.0	17.0	7.95	±0.17
B:	7.5	9.5	6.5.	4.3	5.5	1.7	3.5	3.3	6.1	5.5	6.8	5.6	5.48	±0.06
Phosphates														
A:	9.0	27.0	54.0	36.0	46.0	68.0	68.0	96.0	280.0	190.0	41.0	234.0	168.0	±2.60
B:	9.0	22.0	78.0	25.0	48.0	58.0	58.0	140.0	128.0	86.0	46.0	20.0	72.0	±1.11

Contd...

Table 32.2–Contd...

	1995						1996						Mean Value	Standard Error
	July	*Aug.*	*Sept.*	*Oct.*	*Nov.*	*Dec.*	*Jan.*	*Feb.*	*March*	*April*	*May*	*June*		
Nitrates														
A:	0.52	058	0.55	0.36	0.42	0.48	0.62	0.98	0.32	0.62	0.52	0.49	0.54	±0.05
B:	0.34	0.28	0.41	0.45	0.56	0.52	0.74	0.72	0.96	0.48	0.62	0.46	0.54	±0.01
Nitrites														
A:	0.24	0.28	0.18	0.16	0.40	0.20	0.60	0.19	0.24	0.16	0.16	0.18	0.25	±0.04
B:	0.15	0.22	0.14	0.19	0.25	0.21	0.38	0.09	0.16	0.22	0.17	0.22	0.19	±0.02
Organic Matter (%)														
A:	70.4	51.7	89.9	36.2	49.7	24.8	10.4	49.6	38.2	48.7	31.0	36.6	44.8	±6.00
B:	52.8	47.6	54.8	53.8	24.8	27.9	53.8	58.9	71.4	41.7	36.6	25.9	45.8	±4.22

A: Polluted; B: Unpolluted.

Table 32.3: Biological Characteristics of Polluted Soils of Aluminium Industry in Warangal During 1995–96

	1995						1996						Mean Value	Standard Error
	July	*Aug.*	*Sept.*	*Oct.*	*Nov.*	*Dec.*	*Jan.*	*Feb.*	*March*	*April*	*May*	*June*		
Fungi (1 × 10^6/g)														
A:	1.8	0.7	1.1	1.2	1.8	1.8	1.3	0.9	1.1	0.9	2.1	1.7	1.36	±0.15
B:	1.1	0.7	0.5	2.3	1.2	1.5	0.6	0.9	0.7	0.6	1.1	1.3	1.04	±0.16
Bacteria (1 × 10^6/g)														
A:	10.4	12.4	11.2	3.4	6.9	8.8	7.9	6.3	5.9	8.2	21.2	23.0	10.46	±1:39
B:	42	6.1	4.6	2.8	3.4	6.2	6.9	4.1	3.4	12.3	4.6	9.4	5.66	±0.83
Actinomycetes (1 × 10^5/g)														
A:	2.1	6.3	0.9	1.3	0.9	0.7	0.8	0.6	1.3	1.6	1.9	2.4	1.72	±0.46
B:	0.9	1.4	0.6	0.9	1.1	0.7	1.6	1.3	1.1	0.9	1.3	1.7	1.12	±0.09

A: Polluted; B: Unpolluted.

Talashilkar (1989) working on recycling of urban waste and relation to agriculture reviewed the application of sewage effluent, industrial waste water and sludge compost on agricultural land. The minimum and maximum population of actinomycetes were 0.6–6.3 × 10^5 gill in polluted soils and their range was 0.6–1.7 × 10^5/g m in control soils.

The soil enzyme which determines the physiological status of the soils were estimated (Table 32.4). The cellulase activity (Relative enzyme activity) was recorded in between 136.4 to 476.2 in polluted soils and the minimum and maximum range was in between 140.9 to 307.7 for control. Preece (1992) and Baize (1993) suggested that cellulase is to contribute the breakdown of some hazardous, and toxic insecticide. The catalase enzyme showed its maximum accumulation during the month of December. The maximum and minimum values of enzymes activity were 0.7 to 5.9 units in the polluted soils and this variation was 0.8 to 5.7 units in unpolluted soils. Broadman (1985) measured five soil types for catalase activity incubated with and without adding organic matter and Bentomite. The proteolytic activity was maximum in the month of March with the range of 0.23 to 3.56 in polluted soils and 0.12 to 3.59 in unpolluted soils. Gray and williams (1971) in their studies on microbial, productivity in soil ascertained the affinity of pyrophosphate for extraction of proteases from three different soil varied differently during the study period. The amylase activity was in between 0.05 to 1.0 units in the soils treated with aluminium industry effluents. The range of activity in the control soils was in between 0.08 to 3.42 units.

Table 32.4: Enzyme Characteristics of Polluted Soils of Aluminium Industry in Warangal During 1995–96

	1995						*1996*						*Mean Value*	*Standard Error*
	July	*Aug.*	*Sept.*	*Oct.*	*Nov.*	*Dec.*	*Jan.*	*Feb.*	*March*	*April*	*May*	*June*		
Cellulase–Relative Enzyme Activity (REA)														
A:	311.5	242.1	306.8	191.6	190.5	187.9	476.2	142.9	136.4	220.8	294.9	306.8	250.7	±27.5
B:	248.8	284.1	307.7	294.1	232.6	229.9	142.9	140.9	141.8	250.0	278.6	306.8	238.16	±18.4
Catalase Units														
A:	3.1	2.2	2.5	1.4	1.9	5.9	0.8	0.7	0.8	2.4	2.7	2.1	2.04	±0.4.1
B:	1.8	2.3	1.8	2.2	2.1	5.9	0.9	0.8	1.8	1.9	1.4	0.8	1.95	±0.39
Protease (mg/g)														
A:	0.23	0.24	1.0	0.31	1.32	1.19	1.41	1.51	3.56	0.84	0.72	0.48	1.07	±0.26
B:	0.62	0.12	0.34	0.56	1.00	1.11	1.92	2.66	3.59	0.56	0.41	0.28	1.05	±0.25
α-amylase units														
A:	0.98	0.65	0.13	0.17	0.05	0.12	0.90	1.00	3.70	0.28	0.31	0.72	0.71	±0.29
B:	0.36	0.17	0.43	0.13	0.96	0.50	0.48	0.80	1.20	0.52	0.68	0.74	0.58	±0.09

A: Polluted; B: Unpolluted.

The Nanniperi *et al.* (1980) signified the activities of amylase with extracted C and N suggested that a major proportion of the humic compounds were removed from soils with the amylases. The urease activity ranged between 0.29 to 1.24 mg/g in polluted soils. Kaspar and Bland (1992), Maun Taj *et al.* (1992) studied the relative effect of sewage study on urease activity, which was highly sensitive to small quantities of trace ions and because its susbstrate urea is added to soil as a synthetic fertilizer and in animal excreta.

Acknowledgement

Authors are thankful to Prof V. Thirupathaiah, Head, Department of Botany, Kakatiya University for provided laboratory facilities.

References

Barman, S.S. and Lal, M.M. 1994. Accumulation of heavy metals (Zn, Cu, Cd and, Pb) in soil and cultivated vegetables and weeds grown in judiestrainally polluted fields. *J. Env. Biol.*, 15: 107–115.

Bhattacharya, B. and Bhattacharya, A.K. 1993. Ecological study of plant species diversity in Delhi region. *J. Ecobiol.*, 5: 207–211.

Brady, N.C. 1995. *The Nature and Properties of Soils*, 10th Edition Organisms of the Soil, Prentice Hall, India Pvt. Ltd., New Delhi, p. 253–277.

Chadha, J. and Pandey, S.N. 1990. *Industrial Pollution and Plants.* Ashish Publishers, New Delhi.

David, I., Gregory, S., Tylka, L. and Lewis, C. 1996. Effects of fertilizers on virulence of *Steinernene Carpocap. Applied Soil Ecology*, 3: 27–34.

Gupta, A. 1991. A characteristic soil actinomycetes of Agra region. *Acta Botanica Indica*, 19: 81–83.

Guru Prasada Rao, M. and Nanda Kumar, N.V. 1981. Analysis of irrigation reservoir contaminated by tannery effluents. *Ind. J. Env. Hlth.*, 23: 239–241.

Hattingh, A.M. 1990. Role of clay minerology in the choice of an empluric equation for predicting field capacity of soils. *Appl. Plant Sci.*, 4: 21–24.

Kaspar, T.C. and Bland, W.L. 1992. Soil temperature and root growth. *Soil Sci.*, 154: 290–299.

Kay, G. and Lilly, M.D. 1970. The chemical attachment of chimolypsin to water insomable polymers using 2-Amino-4, 6-Dichloro-5-Triazine. *Acta*, 198: 276–289.

Lal, K., H.K. Parwana and Verma, P. 1993. Effect of waste water irrigation on soil properties. *Indian Env. Prot.*, 13: 274–378.

Mauntay, S.D. and Mishra, R.P. 1992. Dehydrogenase and urease activities of earthworm caste and the surrounding soil. *J. Indian Bot. Soc.*, 71: 303–304.

Messing, I. and Jarwis, N.T. Agropotentiality of paper mills waste water. *Soil Poll.*, 7: 97–120.

Mishra, B. and Srivastava, L.L. 1990. Physico-chemical characteristics of humic substances of major soil associations of Bihar. *Plant Soil.* 122: 185–192.

Omar, S.A., Abdul Sattar, M.A., Khathul, A.M. and Alle, M.A. 1994. Growth and enzyme activities of fungi and bacteria in soil salinized with sodium chloride. *Soil Micro Biologa*, 39: 23–28.

Page, A.L. 1974. Fate and effects of trace elements in sewage sludge when applied to agricultural land: A Literature Review. *Env. Prot. Agency* (EPA) Cincinnati, Ohio, USA.

Patel, J.D. and Shakunthala Devi. 1986. Variations in chloroplasts of bafmesophyll cells of *Zuzyobium cuminil* and *Tamarindus indica* L. growing under air pollution stress of a fertilizer complex. *Indian J. Ecol.*, 13: 124–130.

Patil, H.S. 1985. Studies on the physico-chemical and biological characteristics of sewage stabilisation pond. *J. Env. Biol.*, 6: 93–102.

Rainer, G.J. 1966. The fumigation extraction method to estimate soil microbial biomass, calibration of KEC values. *Soil Biol. Biochem.*, 28: 25–31.

Rajeswari, S. 1993. Effect of pesticides on soil organisms. *Indian J. Env. Hlth.*, 35: 227–231.

Ray, M. and Benerjee, S. 1986. Cytological studies of the water contaminated with industrial effluents effects of Tanla Nallah water on *Allium salvum. Genetics*, 5: 475–483.

Ross, D.J., Burean, S., Hutt, L. and Roberts, H.S. 1973. Biochemical activities in a soil profile under hard beech forest I. Invertase and amylase activities and relationships with other properties. *N.Z. J. Sci.*, 16: 209–224.

Selvarangan, R. 1981. Prevention of pollution from tannery. *Leather Sci.*, 28: 192–199.

Signa, G.C., Isensec, A.R. and Sadsshi, A.D. 1993. Influence of rainfall intensity and Crop residense leaching of atrazing through intact no till soil cores. *Soil Sci.*, 166: 225–232.

Trivedi, R.K., Goel, P.K. and Trisal, C.L. 1987. *Practical Methods in Ecology and Environmental Science.* Environ Media Publications, Karad, India.

Yadav, K., Jha, K.K., Prasad, C.R. and Sinha, L. 1989. Kinetics of carbon mineralisation from poultry manural sewage sludge in two soils at field capacity and submergence moisture. *J. Indian Soc. Soil. Sci.*, 37: 240–243.

Chapter 33

Alteration in Oxygen Consumption in Freshwater Snail *Bellamya bengalensis* (Lamarck) During Pesticide Exposure

P.H. Rohankar & K.M. Kulkarni

Environmental Physiology Laboratory, Department of Zoology, Govt. Vidarbha Institute of Science and Humanities, Amravati

Abstract

The freshwater gastropod *Bellamya bengalensis* (Lamarck) from Amravati region were exposed to lethal and sublethal concentration of phosphamidon as acute and chronic treatment at normal temperature. There is decrease in oxygen consumption with increase in pesticide concentration with slight fluctuation in early hours.

Keywords: Bellamya bengalensis (Lamark), Phosphamidon, Oxygen consumption.

Introduction

Oxygen is the most primary necessity of all living organisms for survival. It provides energy for vital activities of organisms. Extensive and indiscriminate use of pesticides have resulted in their entrance into the aquatic ecosystem affecting nontarget organisms (Butler, 1966). These compounds might act as physiological stressors for non-target animals (Patil *et al.*, 1992). The respiratory potential of an animal is important physiological parameters to assess the toxic stress as it is a valuable indicator of energy expenditure and metabolism in general.

Several workers have studied the physiological processes of animals exposed to pollutants (Lomte and Jadhav, 1982; Nagbhushanam, 1985; Lokhande, 1992; Patil and Mane, 1994; Bakthavathsalam, 2004). Since the physiological processes of molluses are correlated primarily with environmental fluctuations and molluscs are more diverse in form and physiology than any other invertebrate phyla, the present work was designed to findout the effect of lethal and sublethal concentration of phosphamidon on respiratory metabolism of *B. bengalensis*.

Material and Methods

The snails were collected from Wadali tank near Amravati city, healthy and active snails were used in study after acclimatization. The snails were divided into two major groups as control and experimental. Test solutions were prepared for oxygen consumption experiment as LC_{50} for 96 hr and two sublethal concentrations as 1/2 of 96 hr LC_{50} and 1/3 of 96 hr LC_{50}. The snails of experimental group were exposed to these lethal and sublethal concentrations of phosphamidon and the Conc. of solution is kept constant throughout the experiment.

The respiratory measurements were made following the Winkler's method (Welsh and Smith, 1960). The measurement of oxygen consumption was done in both control and experimental groups by measuring the dissolved oxygen contents of water according to sodium oxide modification of Winkler's method. At the end of experiment, the snails were removed from the jar, deshelled, blotted dry and weighed. The oxygen consumption was calculated in ml/gm of snail/hr.

Results and Discussion

The changes in oxygen consumption are recorded in Table 33.1.

Table 33.1: Oxygen Consumption in *Bellamya bengalensis* Exposed to Acute Concentration of Phosphamidon at Different Time Intervals (mg/gm body wet wt. and percentage change)

Time Interval (Hours)	*Control*	*Concentration of Phosphamidon (0.135 ml/l)*
1	0.085 ± 0.0025	0.085 ± 0.0062 (+ 0.124)
2	0.085 ± 0.0033	0.084 ± 0.0062 (–0.176)
6	0.0867 ± 0.0028	0.087 ± 0.0064 (+ 0.346)
12	0.087 ± 0.0046	0.097** ± 0.0057 (+ 11.49)
24	0.088 ± 0.0041	0.089 ± 0.0041 (+ 1.136)
48	0.092 ± 0.0044	0.067 ± 0.004 (–27.17)
72	0.092 ± 0.0042	0.044 ± 0.0047 (–52.17)
96	0.091 ± 0.0043	0.015* ± 0.0049 (–83.51)

t test *$p < 0.05$, **$p < 0.01$, ***$p < 0.001$

(Figures in parenthesis indicate percent change value).

During acute phosphamidon exposure (0.135 mill) there is significant rise in oxygen consumption from first hr. to 12 hr stage and gradually declines from 24th hour upto end of experiment. The fall in rate of oxygen uptake is maximum at 96 hr stage.

The snails exposed to 1/2 and 1/3 of 96 hr of phosphamidon. The time exposure was taken from one to thirty days. At higher sublethal conc. (0.00675 mill) there is significant fall in rate of oxygen consumption except first two days where these is significant increase in oxygen content. From 3rd day the decrease in oxygen consumption continues upto the 30th day. Similar results are also observed in lower sublethal conc. Of phosphamidon but per cent changes are less significant. The decrease is about half of the higher sublethal conc. of same pesticide. The snails *B. bengalensis* exposed to lethal concentration of phosphamidon show pronounced decrease in oxygen consumption at 96 hr exposure.

Table 33.2: Oxygen Consumption in *B. bengalensis* Exposed to Sublethal Concentration of Phosphamidon at Different Time Intervals (mg/gm body wet wt. and percentage change)

Time Interval (hours)	*Control*	*Concentration of Phosphamidon*	
		0.045 ml/l	*0.0675 ml/l*
1	0.092 ± 0.0042	0.099 ± 0.0050	0.102*** ± 0.0041
		(+7.60)	(+10.86)
2	0.093 ± 0.0055	0.097 ± 0.0057	0.097 ± 0.0055
		(+4.30)	(+5.16)
3	0.096 ± 0.0044	0.089 ± 0.0041	0.85 ± 0.0053
		(−7.29)	(−11.45)
4	0.099 ± 0.0039	0.082*** ± 0.0066	0.077*** ± 0.0090
		(−17.17)	(−22.22)
7	0.101 ± 0.0034	0.079* ± 0.0046	0.067** ± 0.0040
		(−21.78)	(−33.66)
10	0.105 ± 0.0041	0.075** ± 0.0064	0.050 ± 0.0048
		(−28.57)	(−52.38)
15	0.111 ± 0.0026	0.071*** ± 0.0075	0.036*** ± 0.0090
		(−36.036)	(−67.56)
30	0.115 ± 0.0039	0.064** ± 0.0071	0.022 ± 0.0074
		(−44.36)	(−80.86)

t test *$p < 0.05$, **$p < 0.01$, ***$p < 0.001$

(Figures in parenthesis indicate per cent change value).

The oxygen utilization by an organism is commonly used as an index of metabolic activity. The decrease in Oxygen consumption in pesticide exposed snail is the net result of metabolic depression. The initial increase in oxygen consumption after exposure to sublethal conc. may therefore exhibit a new steady state of metabolism to compensate physiologically with the stress caused by pollutants (Nagbhushanam, 1985; Jayasurya *et al.*,1991; Rajamannar and Manohar, 1998). According to Ghosh (1987), Singh (1994), Kumta *et al.* (1998), apart of increased oxygen consumption is required to support enhanced physiological activities in metabolizing and eliminating the pollutants by exposed animals.

Skidmore (1970) suggested that pesticide damages the surface cells and blood capillaries of the gill filament by a deposition of thick mucous layer which interferes with oxygen intake and transfer to body tissues in fish which may cause ultimately respiratory distress. In freshwater snail *B. bengalensis* gill damage was observed after exposure to pesticide (Lokhande, 1992). Similar results are also stated by Gurusamy and Ramdas (2000) and Mathivanan in fishes. The present study is in agreement with these workers, indicating that the pollutants interfere with respiratory metabolism.

References

Bakthavathsalam (2004). Effect of carbofuran on oxygen consumption and weight loss of the house cricket *Gryllodes sigillatus* (Walker). *Env. & Ecol.*, 22(3): 508–522.

Butler, P.A. (1966). Pesticides in the marine environment. *J. Appl. Ecol.*, 3: 263.

Ghosh, T.K. (1987). Toxicogenic evaluation of ekalux, Nuvan and sequin on oxygen consumption, Succinate and lactate dehydrogenase activities in some tissues of *Oreochromis mossamibicus. Ad. Bios.*, 6(1): 45–56.

Gurusamy, K. and Ramdas, V. (2000). Impact of DDT on oxygen consumption and opercular activity of *Lepidocephalichthyes thermalis. J. Ecotoxicol. Environ. Moxit*, 10(4): 239–248.

Hanumante, N.M., Deshpande, V.B. and Nagbhushanam, R. (1980). Effect of DDT on oxygen consumption of normal and pharmacologically treated marine pulmonate *Onchidium verruculotum. Indian J. Exp. Biol.*, 18: 753–754.

Jayasuriya, S., Subramanian, M.A. and Varadraj, G. (1991). Effect of detergents on the oxygen consumption of catfish *Mystus vittatus. J. Ecobiol.*, 3(3): 217–220.

Kumta, A., Patil, M.M. and Rakesh Kumar (1998). Perturbation in oxygen consumption due to stress associated with sewage toxicity in Estuarine fish *Oreochromis mossambica. Env. & Ecol.*, 16(2): 409–412.

Lomte, V.S. and Jadhav, M.L. (1982). Effect of toxic compounds on oxygen consumption in bivalve *Corbicula regularis* com. *Physiol. Ecol.*, 7(1): 31–33.

Lokhande, P.J. (1992). Evaluation of toxicants and environmental factor combination stress on physiology of the snail *Bellamiya bengalensis* F. typica (Lamark). *Ph.D. Thesis*, Amravati University, Amravati.

Mathivanan, R. (2004). Effect of sublethal concentration of quinolphos on selected respiratory and biochemical parameters in the freshwater fish *Oreochromis mossambicus. J. Ecotoxicol. Environ. Monit.*, 14(1): 57–64.

Nagbhushanam, R., Bodkhe, M.K. and Sarojini, R. (1985). Effect of sevinol exposure on the respiration of freshwater crab *Barytelphusa cunicularis. Proc. Nat. Acad. Sci. (India)*, 55(B): 305–310.

Patil, S.S. and Mane, U.N. (1994). Oxygen uptake of the bivalve molluscs *Lamellidens marginalis* (Lamarck) from lentic environment due to mercury stress Abs. 6.3, Nat. Symp. "*Ecoenvironmental impact and Organism response*", Amravati, 15–17th Oct.

Rajamannar, K. and Manohar, L. (1998). Effect of pesticides on oxygen consumption of fish *Labeo rohita. J. Ecobiol.*, 10(3): 205–208.

Singh, Alaknanda (1994). Oxygen consumption and ventillation rate in *Channa punctatus* (Bloch) exposed to sublethal levels of mercuric chloride. *Env. & Ecol.*, 12(2): 252–255.

Skidmore, K.F. (1970). Respiration and osmoregulation in Rainbow trout with gills damaged by Zinc sulphate. *J. Exp. Biol.*, 52: 481–494.

Welsh, J.H. and Smith, R.I. (1961). *Laboratory Exercises in Invertebrate Physiology*. Burges publishing Co. Minneapolis, pp. 126.

Chapter 34

Estimating the Oxygen Requirement in a Mass Bathing Tank: A Case Study: Mahamaham 2004 (Southern Kumbhamela)

Ashutosh Das, V. Srihari*, S. Madhan Babu** and S. Ananthapadmanaban****

**Senior Lecturer, **Research Assistant, ***Dean,*
School of Civil Engineering, SASTRA Deemed University, Thanjavur – 613 402

ABSTRACT

Presently, the ongoing tradition of mass bathing is mostly being carried out in many of the temple tanks and rivers all over the country on any set of auspicious dates with base as sun or moon. But, with constant influx of pollutants to the water-bodies, the lack of the basic sanitary consciousness of galloping public have made the condition of water scary to dip into due to ever increasing environmental degradation. Monitoring studies have been carried out in case of many water bodies, starting from temple tanks to rivers, ponds, lakes and even in seas. Thus, the study of Mahamaham Tank bathing was aimed for a better sanitary and social objective and to know how of this paper study, for adoption at different conditions of near similar conditions can also be extended.

Introduction

The mass bathing has been one of the essential components in the cultural fabric of unity in diversity of our country. However, progressive deterioration of broad selfless consciousness of the general populace, quite contrary to the very spirit of the Indian culture (*vasudaiva kutumbakam;* universal brotherhood), and subsequent irresponsible activities have led to a new upsurge in pollutions and consequent disastrous scenario. When the echo of the all–pervading pollution (air, water, land) is heard in the religious ceremonies and festivities like mass bathing, we experience a dilemma which is never experienced before–faith Vs science. The very foundation of religious faith (that is self purification) is questioned!

Undoubtedly, the natural cleansing capacity of an ordinary water-body (what to say of a supposedly holy water-body) can take care of the so-called sporadic pollution due to mass–bathing. However, the major loss of aesthetic sense of people, incremental chemical pollutional loads (both from point and non-point sources), inaccessibility of nature's own cleansing agents (such as sunlight, wind, rainfall, aquatics, etc) and processes (sedimentation, dilution, etc.) have created a major concern in protecting water bodies in all the religious bathing places of our country.

Kumbakonam is located at latitude 10.98 North and longitude 79.4 East. It lies about 313 kilometres away from City of Chennai on the south, about 90 kilometres from Tiruchirapalli on the east and about 40 Kilometres from Thanjavur on the North East is one of the special grade Municipal towns of Tamil Nadu and the second bigger town in Thanjavur District. Popularly known as TEMPLE city, Kumbakonam is picturesquely located amidst two rivers "River Cauvery" on the north and "River Arasalar" on the south. One of the oldest towns in Tamil Nadu and is famous for its Mahamaham festival, celebrated once in 12 years in the Mahamaham tank located in the heart of the city, Kumbakonam has been the Capital of Chola Kings.

The famous tank of Mahamaham Kumbakonam covers an area of 20 acres and is in a shape of a pot. The tank is surrounded by 16 small mandapams and one "NAVA KANNIKA TEMPLE" in the eastern side. It is believed that this tank consist of 19 holy teertham at the centre. Since the Mahamaham festival is carried out once in 12 years, the tank remains neglected for eleven years nine months (leaving three months for renovation). Although there are few festivals of very small scale observed at different time of the year, maintenance remain poor all through. The holy wells of the pond remains completely isolated from the water table due to high sedimentation and silting, to such an extent that even during Mahamaham festival, the connection between the tank water and ground water could not be restored. During the Mahamaham festival period an active renovation is carried out, which comprises of complete drainage of water (if any), cleansing of the sludge, thorough super-chlorination, deposition of fine clay-bed, and refilling with treated municipal water pumped from Kollidam infiltration gallery, located at a distance of about five km.

Experimental Procedure

Chemical Modeling of Dissolved Oxygen

For study of the oxygen behaviour, a numerical-analytic model was framed, as follows. Since dissolved oxygen plays a more significant role, amongst the physico-chemical parameters, an attempt was made to evaluate its mass balance at the outlet, which is the known to have maximum organic loading.

Chlorine Dosage at the Inlet

One of the major contributor of qissolved oxygen is chlorine, provided by the bleaching powder added, probably because of the disinfecting capability of the chlorine (and thereby reduction of BOD) and its intrinsic effect of redox potential of water (and corresponding oxygen availability). Since chlorine dosage was provided batch-wise, the effective dosages were calculated for intermediate time (hourly basis) from the data obtained for batch-wise dosage from public health authorities (Figure 34.1).

Estimation of Typical Retention Time in the Tank

Inlet water is allowed to enter the tank through cascading over the steps, whereas the outlet was pumped out. Since the water level was maintained mostly constant, the inlet and outlet discharge rates were assumed to be same. Assuming no internal diversion of the water molecules, except at the boundaries, the minimum and maximum distances the water molecule (assumed to be sames as that of chlorinated water molecule) need to travel B C from the inlet to outlet, are given by (assuming an idealized trapezoidal tank ABCD, with Central Mandapam EFGH):

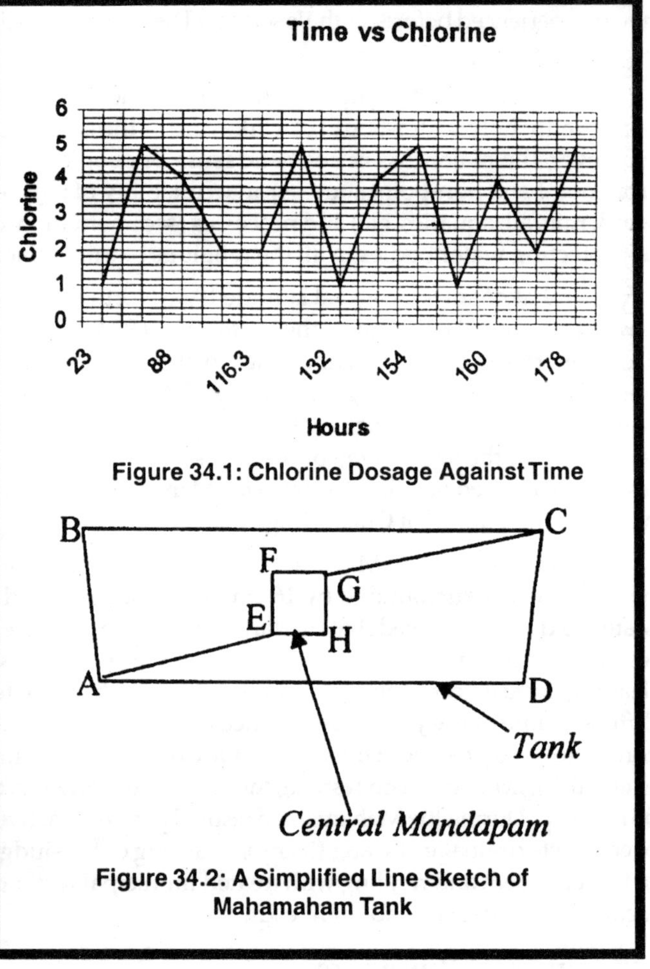

Figure 34.1: Chlorine Dosage Against Time

Figure 34.2: A Simplified Line Sketch of Mahamaham Tank

The *average distance* was calculated taking the arithmetic mean of the maximum and minimum distances, which was found to be 272.79 m.

The hourly retention time of the chlorinated water molecule in tank were estimated from the ratio of the mean distance traveled (*i.e.*, 272.79 m) and the hour-wise velocities obtained above. The average of the hourly detention time was found to be 55 minutes, with values ranging from 0.15 to 7.23 hours. Since the distribution of the detention time showed a high standard deviation of 0.84, and water-flow is mostly turbulent (due to mass bathing), a detention time of 1 hour is assumed in the modeling (which also had an additional advantage that it happens to coincide with the sampling interval).

Estimation of Dissolved Oxygen Contributed by Chlorine at the Outlet

Laboratory experiments were conducted with different dosages of chlorine (1 ppm to 6 ppm, which is the range of chlorine concentration estimated in the inlet water) in containers of equal capacities kept under atmospheric temperature (mean laboratory temperature was same as the mean maximum temperature at Mahamaham, *i.e.* 32 °C) using the typical Kollidam Infiltration gallery water (the water which was used to feed the Mahamaham tank), under constant stirring condition (analogous to stirring due to mass bathing). The residual chlorine after 1 hour duration (corresponding to the average detention time of the tank) was plotted against the initial chlorine dosage. The plot showed the following linear relationship between these two factors with a high correlation (> 0.88).

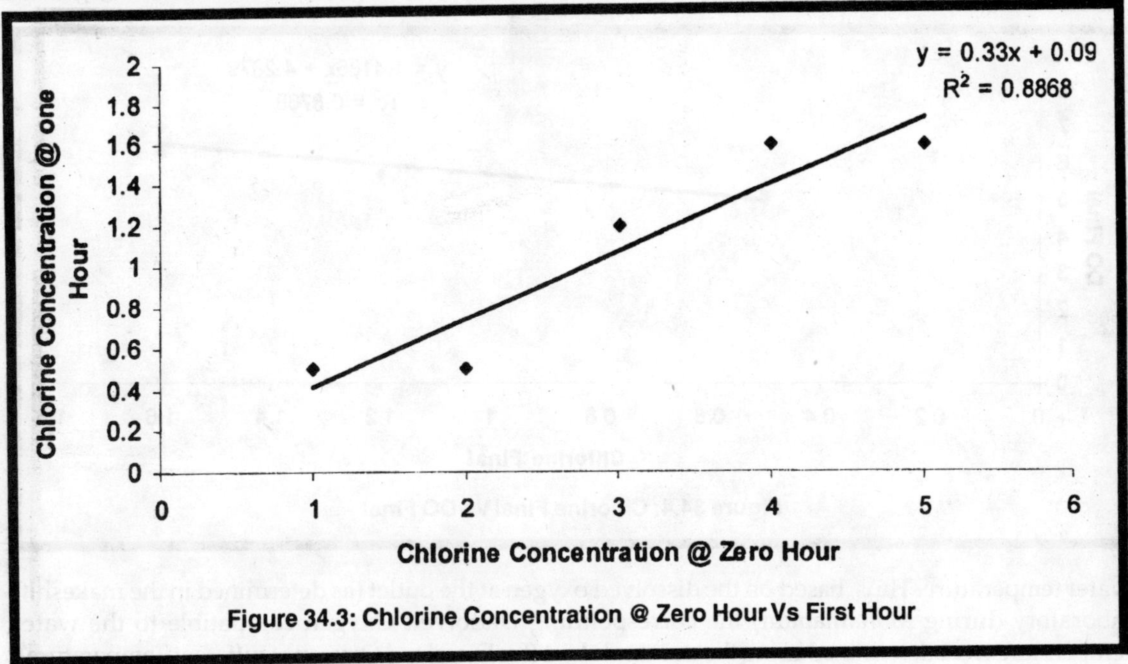

Figure 34.3: Chlorine Concentration @ Zero Hour Vs First Hour

Chlorine Outlet Concentration [@ (t + 1)hr] (Cl_{out})

= 0.33 × Chlorine Inlet Concentration [@ t hr](Cl_{in}) + 0.09

Studies conducted by Central Board for the Prevention and Control of water Pollution, New Delhi at Kumbha Mela (Jan–Feb, 1983; ADSORBS/8/1983–84) for control bathing by three persons in 70 litre of water contributed 10mg/litre of chlorine, which is contained in various human secretions, such as sweat, saliva, mothers milk, semen, blood (Vic. *et al.*, 1996; Magri and Pase, 1966; Montague, 1993). Based on this result, corresponding to the volume of water Mahamaham tank (*i.e.*, 90,00,000 litre), the total calculated chlorine contributed due to population would be $(0.037 \times 10^{-5}) \times$ Number of Population.

Thus, the total Chlorine at the Outlet (assuming chlorination and bathing as the main contributors) = [0.33*(Chlorine inlet) + {0.037*(10^5)*(population)} +0.09]

To determine the corresponding dissolved oxygen contributed due to chlorine; experiments were conducted in the laboratory to determine the effect of chlorine dosage on dissolved oxygen content of the water. Using the same water as the previous experiment (*i.e.*, typical Kollidam Infiltration gallery water) different dosages of chlorine were added and stirred and the corresponding DO was determined. The plot between the chlorine concentration and the increase in DO was again found to show a distinct linear relationship (with a high correlation coefficient of 0.87), given as follows:

Dissolved Oxygen Contributed by Chlorine

= 0.57 × Chlorine Dosage + 4.29

= 0.57 × [0.33*(Chlorine inlet) +{0.037*(10^{-5})*(Population)} + 0.09] + 4.29

Estimation of Dissolved Oxygen Contributed by Atmosphere

Since dissolved oxygen in the water tries to maintain equilibrium with the atmosphere, the total oxygen contributed by atmosphere depends both on available dissolved oxygen in water and the

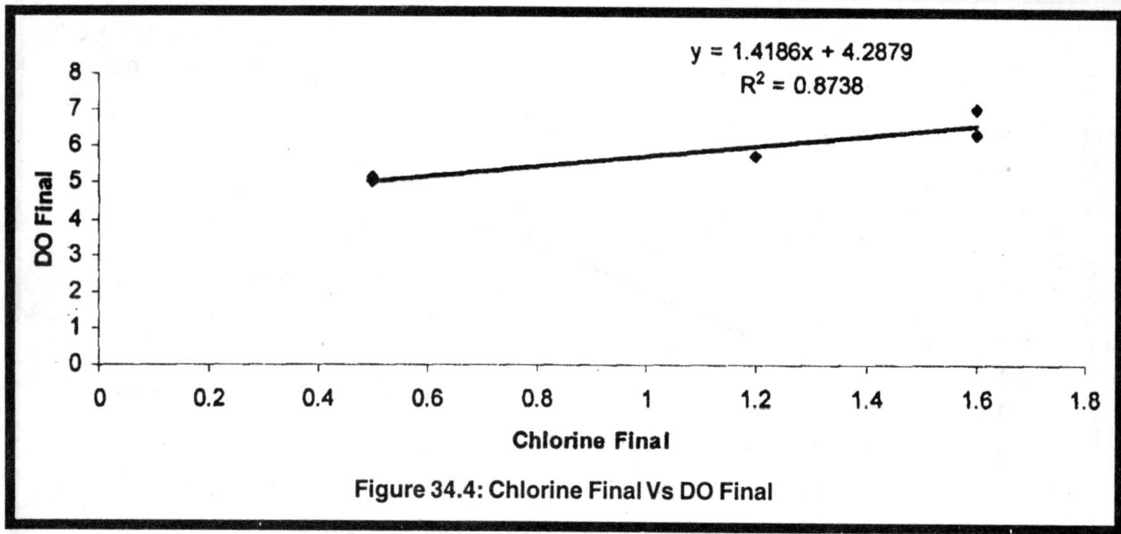

Figure 34.4: Chlorine Final Vs DO Final

water temperature. Thus, based on the dissolved oxygen at the outlet (as determined in the makeshift laboratory during Mahamaham) the corresponding Dissolved Oxygen compatible to the water temperature were determined using the standard chart (for Dissolved Oxygen at different Temperature). However, the total contribution of dissolved oxygen from atmosphere also depends upon the saturation percentage of dissolved oxygen in water, which was again dependent upon the temperature and the actual dissolved oxygen present in water. The percentage of saturation (hourly basis) was determined using the following Nomogram (Figure 34.5).

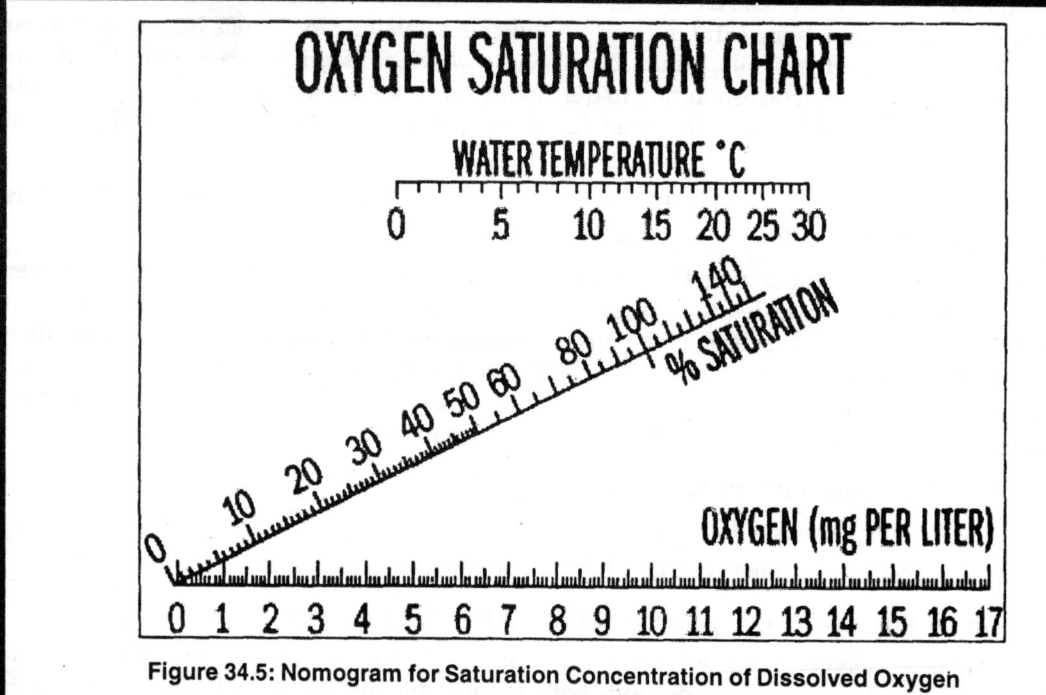

Figure 34.5: Nomogram for Saturation Concentration of Dissolved Oxygen

Based on the corresponding percentage of saturation (with respect to water temperature and actual DO), the dissolved oxygen contributed by atmosphere (due to mechanical stirring during mass bathing) was determined with due correction to the percentage of saturation, as obtained above (from the nomogram).

Thus, assuming the total dissolved oxygen was contributed mainly by the chlorine concentration (both due to chlorination and loading from the individuals undergoing bathing) and atmospheric aeration, with least contribution by wind. The theoretical dissolved oxygen contributed to the tank was determined on hourly basis.

When the actual dissolved oxygen was plotted against the estimated dissolved oxygen (Figure 34.6), they show a strikingly linear relationship (with a correlation coefficient more than 0.93), which is as follows:

Actual Dissolved Oxygen

= 0.6984 × Estimated Dissolved Oxygen – 4.38

It is also interesting to note that, at higher (>~ 4 mg/l) concentration of actual dissolved oxygen, the major contributor to theoretical dissolved oxygen was found to be the atmosphere and at low concentration, the major contributor is chlorine. This means, at low dissolved oxygen content in the water, which is actually the time for alarm, the main cause for the low contribution is air, therefore specific forced aeration would be highly recommended, than higher chlorination (to already Chlorine-rich water).

The reduction of dissolved oxygen (that is, the difference between the theoretical and actual dissolved oxygen) was plotted against time (Figure 34.7), which indicated a general trend which showed a hike to a peak from 5th to 6th March and then drastic reduction from 6th March.

Since the main cause of reduction of dissolved oxygen is organic loading; this may be, because of higher influx of population, between 5th and 6th March leading to high consumption of dissolved

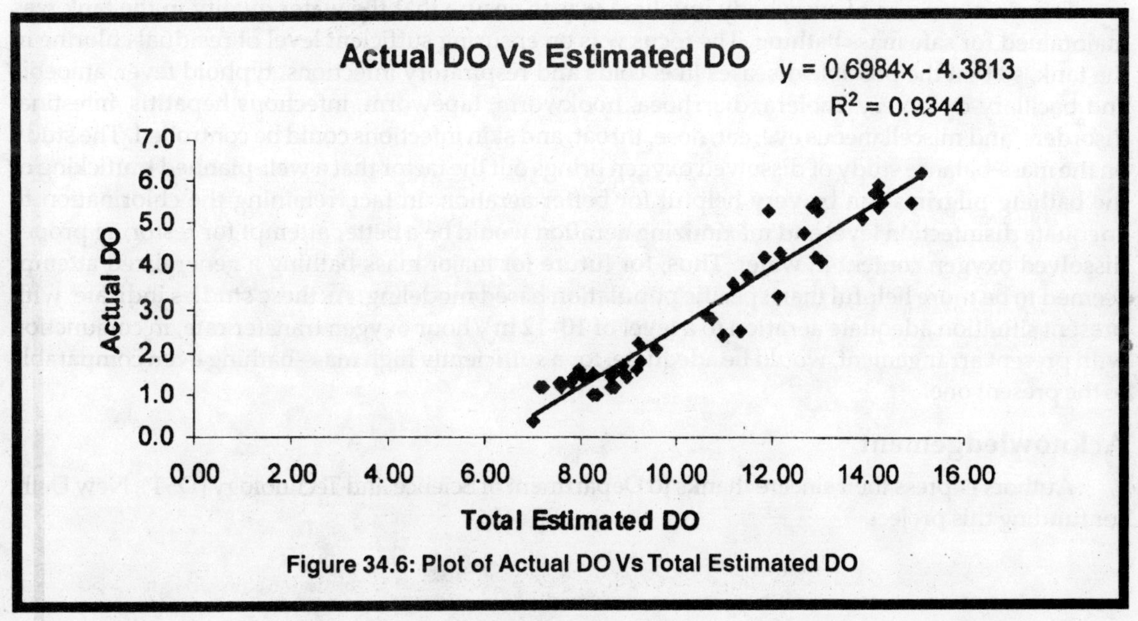

Figure 34.6: Plot of Actual DO Vs Total Estimated DO

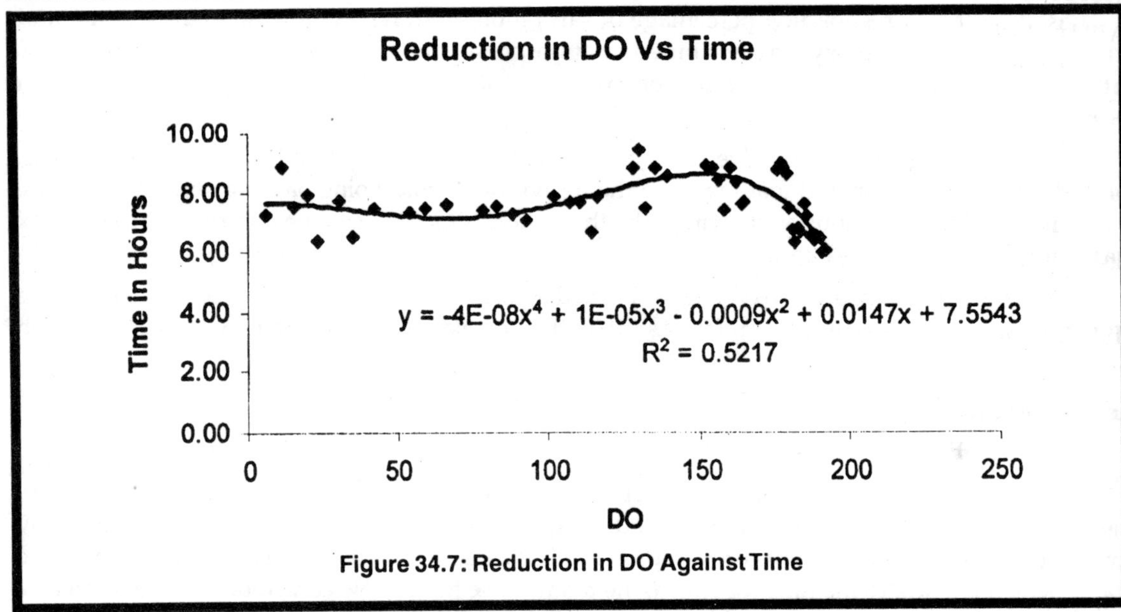

Figure 34.7: Reduction in DO Against Time

oxygen and subsequent reduction on 6th March could be because of strict trafficking of the population (allowing only one-third of the total tank area used for bathing) causing to lesser organic loading rate in spite of higher population and similar degree of chlorination as the day before.

Thus, effective trafficking seems to be an effective method for preservation of dissolved oxygen, especially for very high population flux.

Conclusion

The main concern of everybody involved was to ensure that the water quality in the tank was maintained for safe mass bathing. The focus was on ensuring sufficient level of residual chlorine in the tank, so that the possible diseases like: colds and respiratory infections, typhoid fever, amoebic and bacillary dysentery, cholera, diarrhoea, hookworm, tapeworm, infectious hepatitis, intestinal disorders, and miscellaneous eye, ear, nose, throat, and skin infections could be controlled. The study on the mass-balance study of dissolved oxygen brings out the factor that a well-planned trafficking of the bathing pilgrims can be very helpful for better aeration. In fact, retaining the chlorination to adequate disinfection level and maximizing aeration would be a better attempt for restoring proper dissolved oxygen content in water. Thus, for future for major mass-bathing a generalized attempt seemed to be more helpful than specific population based modeling. As these studies indicate, with present situation adequate aeration to a level of 10–12 m^3/hour oxygen transfer rate, in conjunction with present arrangement, would be adequate, for a sufficiently high mass-bathing even comparable to the present one.

Acknowledgement

Authors express their sincere thanks to Department of Science and Technology (DST), New Delhi for funding this project.

References

APHA, AWWA, WPCF. (1992). *Standard Methods for the Examination of Water and Wastewater*, 18th edn. American Public Health Association, Washington D.C.

Choudhuri, N. and K.R. Ranganathan (1980). Classification and zoning: A tool for water quality management. *Proc. Symp. Hydrology in Water Resources Development*, New Delhi, 1: 80.

Environment Canada. (1972). *Guidelines for Water Quality Objectives and Standards*, Technical Bulletin No. 67 Inland water branch, Department of Environment Canada.

ISI. (1973). *Quality Tolerances for water for Swimming Pools*. IS: 3328-1965 Bureau of Indian Standards, New Delhi.

ISI. (1977). *Tolerance Limits for Inland Surface Waters Subject to Pollution* (First revision) IS: 2296-1974 Bureau of Indian Standards, New Delhi.

Kudavayil, Subramanian (1990). Thanjavur Nayakkar Varalaur History of Thanjavur Nayakkas, Saraswathi Mahal Library, Thanjavur, p. 181–183.

Morris, R. (1991). EC bathing water viorological standard. *Water Science and Technology*, 24(2): 49–52.

NTAC (1968). Water Quality Criteria Report of the National Technical Advisory Committee, Federal Water Pollution Control Administration, Washington. D.C.

Raman, V. (1980). Bacteriological Quality of Pushkarini Waters: A study conducted by National Environmental Engineering Research Institute, Nagpur (Personal Communication).

Stevenson, A.H. (1952). Studies of "Bathing Water Quality and Health" presented, before the 2nd session of the Engineering section.

USEPA (1976). Quality Criteria for water U.S. Environmental Protection Agency Washington DC.

Water Pollution from Mass Bathing case studies in Ganga. Central Board for the, Prevention and Control of Water Pollution, New Delhi. ADSORBS/8/1983–84.

WHO. (1998). Detailed Guidelines on Self Recreational Water Environment.

WHO. (2000). *Guidelines on Monitoring and Bathing Water*.

Chapter 35

A Report on Clinical Importance of Hypertension, Diabetes Mellitus and Heart Rate Associated with Acute Myocardial Infarction

Rakesh K. Pandey, Arun K. Pandey**, G.G. Potey***, Arvind K. Pandey*

**Department of Environmental Science, A.P.S. University, Rewa, M.P.,*
***R.K.D.F. Homoeopathic Medical College, Bhopal, M.P.,*
****Head of Department Biochemistry, R.D.G. Medical College, Ujjain, M.P.*

ABSTRACT

Blood Pressure was associated with a two-fold increase in the risk of coronary Heart disease in males and six folds in females. Approximately two fold-increased incidence of hypertension in diabetics may be an important contributory factor.

Diabetes mellitus increases the susceptibility to all clinical manifestation of coronary artery disease and its manifestation in greater among men than women.

That person has in resting heart rate of 90 or more per minute has more incidences or coronay heart disease and mortality.

Keywords: *Acute Myocardial Infarction (AMI), Cardiac Heart Disease (CHD), Diabetes Mellitus (DM).*

Introduction

Hypertension is found in about 50 per cent of men and in about 75 per cent or women with Acute Myocardial Infarction. Conversely there is a high incidence of coronary heart disease in-patient with hypertension. Prevalence of hypertension in adult to be 24 per cent and ranging from 4 per cent in these aged 18 to 29 years to 65 per cent in those aged more than 80 years.

Association of diabetes mellitus with coronary heart disease is a well-recognized fact. The relative risk for cordial death was 1.9 in diabetic men and 3.3 in diabetic women compared with non-diabetic men and women after adjustment for other cordial risk factor. There is at least a two-fold increase in incidence of myocardial infarction in diabetics compared with non-diabetics. Tachycardia in resting state leads to increase oxygen requirement be the myocardium thus had more incidence of coronary heart disease.

Materials and Methods

The clinical materials for the present study consists of normal healthy subjects and diseased subjects, suffering from acute myocardial infarction, admitted in S.S. Medical College and Gandhi Memorial Hospital, Rewa, M.P.

Hypertension

It was taken after 5 minute of relaxed recumbent position. Patients were classified according to diastolic blood pressure: normal (60–90), mild (91–110), Moderate (111–120), Severe (121–140), Gross (More than 140).

Diabetic Status

Fasting and post-prandial blood sugar examined 2 categories may according to diabetic state:

Over Diabetic

Past history of diabetes, signs and symptoms present, 2 hours post prandial blood sugar more than 120 mg per cent.

Latent Diabetic

No symptoms, fasting blood sugar more than 100 mg per cent and post prandial blood sugar more than 120 mg per cent.

Heart Rate

Normal = 60–90/minute

Tachycardia = More than 90/minute

Results and Discussion

Table 35.1, 35.2 and 35.3 showed the statistical significance of hypertension, diabetes mellitus and heart rate respectively along with Acute Myocardial Infarction. Hypertension was present in 56 (35 per cent) cases in the present series of 160 cases. It seems to be highly statistical significant risk factor. 43 (26.87 per cent) of total case in present studies where diabetic, while 117 (73.12 per cent) were non-diabetic. Therefore it seems to be a moderately significant risk factor.

Table 35.1: Showing Distribution of Diastolic Blood Pressure in AMI

Sl.No.	*Distribution of Diastolic B.P.*	*No. of Cases*	*Percentage %*	*"P" Values*
1.	Less than 90 mm of Hg	104	65	
2.	91–110 mm of Hg	47	29.37	> 0.05
3.	111–120 mm of Hg	7	4.375	> 0.05
4.	121–140 mm of Hg	2	1.25	> 0.05
5.	Above 140 mm of Hg	–	–	–
	Total	**160**	**100**	

Table 35.2: Showing Blood Sugar Levels in AMI

Sl.No.	Blood Sugar in mg. %	No. of Cases	Percentage	"P" Values
1.	80–120	117	73.13%	> 0.05
2.	More than 120	43	26.87	
	Total	**160**	**100**	

Table 35.3: Showing Heart Rate in AMI

Sl.No.	Heart Rate/Minute	No. of Cases	Percentage	"P" Values
1.	Normal 60–90	111	67.37	< 0.001
2.	Tachycardia More than 90	49	30.63	
	Total	**160**	**100**	

Tachycardia was present in 49 (30.63 per cent) cases in a total of 160 cases. It seems to be a significant risk factor.

References

Bradley, R.F. and Bryfogle, J.W. (1956). Diabetics and ischemic heart disease. *Amer. J. Med.*, 20: 207.

Deming, O.B. and Daly, M.M. (1958). Effect of hypertension in artherosclerosis. *J. Exper. Med.*, 107: 581.

Framingham Heart Study: Habits and coronary heart disease. Public health service Publication, 1515, U.S. Govt. Printing Office, 1996.

Kannel, W.B., Dawber, T.B., Kagan, A., Nicholas, Revotaski, Joseph, Stokes. (1961). Factors of risk in development of coronary heart disease: The Framingham study. *Annul of Internal Medicine*, 55: 33.

Koop Acute MI causes and risks 1998–2001.

Macmohan, S., Pelo, R., Culler, J. *et al.* (1990). Blood, stroke and coronary heart disease. *Lancet*, 335: 765.

Pandey, A.K. (2003). Acute myocardial infarction associated with lipid profile and its management. *Ph.D. Thesis*, A.P.S. University, Rewa, M.P., India.

Reaven, G.M. (1998). Role of insulin resistance in human disease. *Diabeter*, 37: 1595.

Stamler, J., Berkson D.M., Lindberg, H.A. Risk factors, their role in aetiology and pathogenesis of atherosclerolic disease. *Inhissler.*

Chapter 36

Quenching of Diphenylamine by Benzoic Acid and Carbon Tetrachloride in Chloroform

*S. Bakkialalakshmi**, *B. Shanthi***, *D. Chandrakala** *and R. Santhi**

**Department of Physics, Annamalai University, Annamalai Nagar – 608 002*

***Centralized Instrumentation and Services Laboratory (CISL), Annamalai University, Annamalai Nagar – 608 002*

ABSTRACT

An attempt has been made to obtain the fluorescence quenching spectra of Diphenylamine using benzoic acid and carbon tetrachloride as quenchers in chloroform. The concentration of fluorescer solution was kept constant (0.01 m) in each experiment, by adding fixed adiquot of stock solutions prepared in the solvent. Quenchers (benzoic acid and carbon tetrachloride) in the respective solvent was added to the above solution in small volumes (0.1 ml to 0.5 ml) and the fluorescent intensities were measured. The stern-volmer constants for all quenching processes have been plotted. Regression analysis for stern-volmer plot has been carried out. Regression coefficient (r) and the stern-volmer constant (K_{sv}) were calculated. Stoke's shift, ionization potential, electron affinity and the solvent parameter (z) have also been calculated.

Keywords: *Fluorescence quenching, Benzoic acid, Carbon-tetrachloride, Diphenylamine, Stern-volmer polt, Stoke's shift, Ionization potential, Electron affinity and The solvent parameter.*

Introduction

The reduction in the intensity of fluorescence (Ewing, 1998) is called as quenching. The quenching processes may take place either in the ground state or in the excited state.

The quenching can also be affected by solvents in the absence of many other added quenchers.

The fluorescence of a number of dyes is quenched by ions in the order, that,

$$I > CNS > Br > C_2O_4^{2} > SO_4^{2} > NO_3 > F$$

This is correlated with the increasing ionization potential, which shows that the fluorescence quenching efficiency of these ions is related to the case of charge transfer from these ions.

An experimental study on the fluorescence quenching in solutions of chlorophyll 'a' was made using the theory of the quenching of fluorescence by means of the quencher tensoactivity experimentally established by Emanoil Lucatu (Emanoil and Marilena, 1970).

Various authors (Beems *et al.*, 1967; Mataga *et al.*, 1966; Murra and Koizumi, 1966; Kameta and Koizumi, 1967) have shown that even in the absence of ground state complex formation, quenching of fluorescence with concurrent emission from a complex in the excited state may be observed. This has been attributed to charge transfer in the excited state where changes in ionization potential and electron affinity render possible the formation of charge transfer state in cases where no such complex formation is possible in the ground state. Fluorescence quenching of some aromatic amines by p-chloranil in cyclohexane and methanol was studied in our earlier paper (Bakkialakshmi *et al.*, 2002).

Experimental Materials and Methods

AR grade sample (benzoic acid and carbon tetrachloride) were distilled and used. Melting point is 121°C and 76–77°C respectively. The aromatic amine (Diphenylamine) and the solvent chloroform were purified by standard methods. The fluorescence spectra of fluoresphors were measured with JASCO model FP-550 spectrofluorometer, operating with 150W xenon lamp as light source. The absorption spectra of diphenylamine were measured using a JASCO-UVIDEC-650 spectrophotometer.

Results and Discussion

The fluorescence quenching spectrum of diphenylamine in chloroform without the quenchers (Benzoic acid and carbon tetrachloride) and with various concentrations of the quenchers (Benzoic acid and carbon tetrachloride) have been shown in Figures 36.1 and 36.2.

$[I_0/I]$ values for different concentrations of the quenchers (benzoic acid and carbon tetrachloride) were calculated. The calculated values are given in Table 36.1. The quenching ratios I_0/I were plotted against quencher concentration [Q] and these plots are shown in Figure 36.3.

Table 36.1: Fluorescent Intensity Ratios of Dipheylamine of Different Concentration in Chloroform

Solvent	*(Q) in 10^{-5} M/L*	*(I_0/I)*	
		(1) Benzoic Acid	*(2) Carbon Tetrachloride*
Chloroform	2	1.11	1.30
	4	1.20	1.50
	6	1.31	1.61
	8	1.48	1.89
	10	1.65	2.15

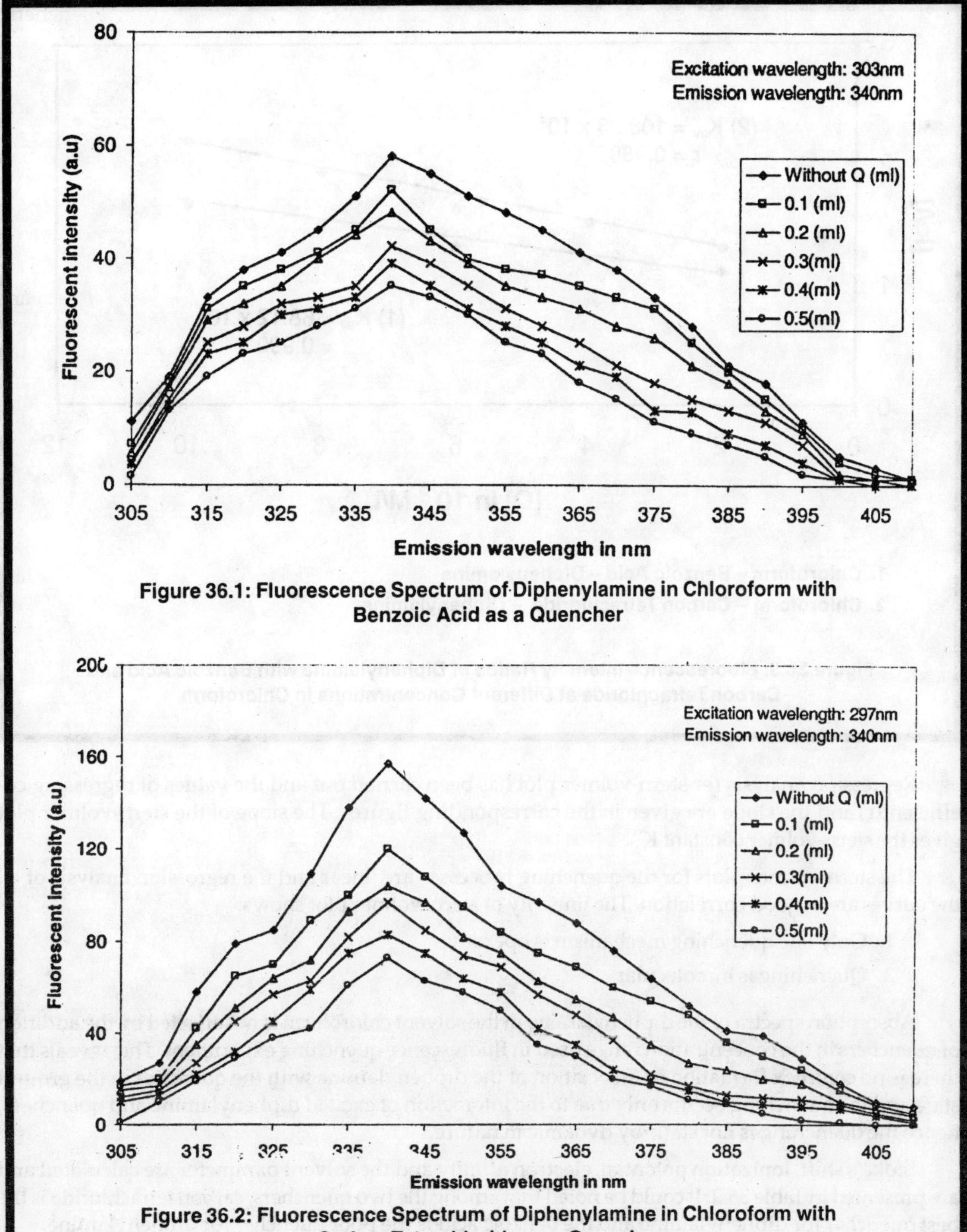

Figure 36.1: Fluorescence Spectrum of Diphenylamine in Chloroform with Benzoic Acid as a Quencher

Figure 36.2: Fluorescence Spectrum of Diphenylamine in Chloroform with Carbon Tetrachloride as a Quencher

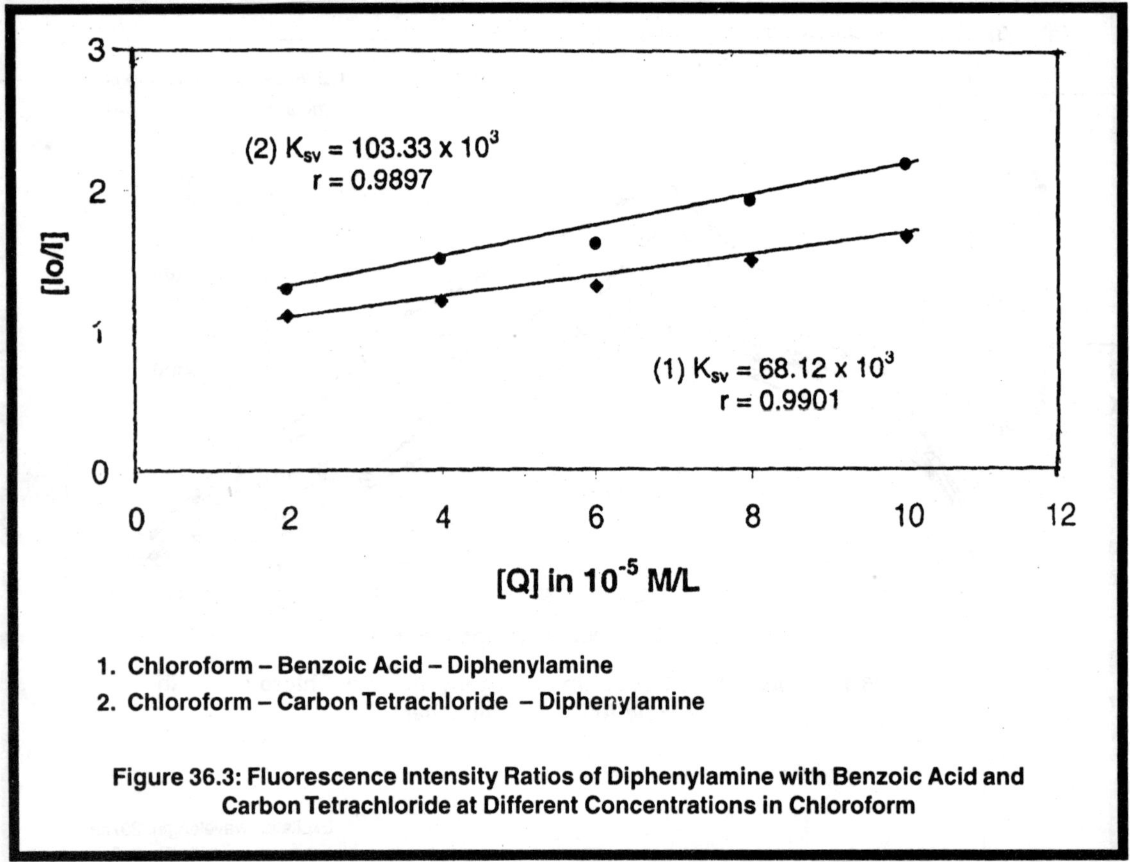

Figure 36.3: Fluorescence Intensity Ratios of Diphenylamine with Benzoic Acid and Carbon Tetrachloride at Different Concentrations in Chloroform

Regression analysis for stern-volmer plot has been carried out and the values of regressing co-efficient(r) and the slope are given in the corresponding figures. The slope of the stern-volmer plot gives the stern-volmer constant K_{sv}.

The stern-volmer plots for the quenching processes are linear and the regression analysis of all the curves are in good correlation. The linearity of stern-volmer plot shows.

1. Only one quenching mechanism is operative
2. Quenching is bimolecular

Absorption spectra of the diphenylamine in the solvent chloroform is not affected by the addition of quenchers in the concentration range used in fluorescence quenching experiment. This reveals that there is no complex formation or association of the diphenylamine with the quencher in the ground state and the quenching occurs only due to the interaction of excited diphenylamine and quenchers, hence the quenching is not static by dynamic in nature.

Stoke's shift, ionization potential, electron affinity and the solvent parameter are calculated and are presented in Table 36.2. It could be noted that among the two quenchers, carbon tetrachloride is the best quencher for diphenylamine and the benzoic acid is the poor quencher for diphenylamine.

Table 36.2: Energy, Ionization Potential, Electron Affinity, Stoke's Shift and Solvent Parameter Values of the Complexes in Chloroform

Solvent	Fluro-phore	Quen-chers	λ_{CT} (nm)	hV_{CT} (ev)	λ_D (ev)	E_A (ev)	λ_{abs} (nm)	λ_{flu} (nm)	Stoke's Shift Δv (cm^{-1})	Solvent Parameters (kca/mole) (z)
Chloroform	Diphenyl amine	Benzoic acid	589	2.1093	7.8866	2.0478	303	340	3592	48.5399
		Carbon tetrachloride	605	2.0534	7.8167	2.1051	297	340	4259	47.2561

Conclusion

Fluorescence quenching studies of diphenylamine by benzoic and carbon tetrachloride in chloroform have been successfully carried out. The stern-volmer constants for the quenching processes have been determined. The stern-volmer plots for the quenching proceses are linear and the regression analysis of all the curves gives a very good correlation. We can conclude that among the two quenchers, carbon tetrachloride is one of the best quenchers and the benzoic acid is the one of the poor quenchers for diphenylamine.

References

Galen, W. Ewing (1998). *Instrumental Methods of Chemical Analysis*, 5th edition. McGraw–Hill International Editions.

Emanoil, Lucatu and Marilena vasilache. (1970). *Journal of Luminescence*, 3: 132–136.

Beems, H., Kribbe, H. and Weller. (1967). *J. Chem. Phys.*, 47: 1183.

Mataga, N., Okada, T. and Yafuomoto, N. (1966). *Bull. Chem. Soc.*, Japan, 39: 2562.

Murra, T. and Koizumi, M. (1966). *Bull Chem. Soc.*, Japan, 39: 2588.

Kameta, K and Koizurni, M. (1967). *Bull Chem. Soc.*, Japan, 40: 2254.

Bakkialakshmi, S., Shanthi, B. and Deepa, S. (2002). *Bull. Pure & Appl. Sci.*, 21D(1): 5–11.

Chapter 37

Nutrient Status of Kanale Tank, Sagara Taluk with Reference to Diversity of Aquatic Macrophytes

R. Purushothama, J. Narayana, B.R. Kiran and K. Harish Kumar

Department of Environmental Science, Kuvempu University, Jnana Sahyadri, Shankaraghatta – 577 451, Karnataka

ABSTRACT

This article analyses species diversity of aquatic macrophytes in Kanale tank, Sagara taluk, Karnataka, during January 2004 to December 2004. A total of 19 species of macrophytes belonging to 2 free floating, 5 rooted floating, 6 submerged, 1 emerged and 5 semi aquatic categories were recorded. The Nutrients *viz.*, Ca^{2+}, Mg^{2+}, Na^{+}, K^{+}, and PO_4^{3-} that influences the distribution and diversity of aquatic macrophytes is discussed in details. Further, integrated approach of control of pollution load and ecozoing of tank with an emphasis on sustainable utilization of aquatic macrophytic resources is enlightened in brief.

Keywords: *Nutrient, Aquatic macrophytes, Kanale Tank.*

Introduction

Aquatic macrophytes play a key role in biogeochemical cycles and food webs of living things in the lentic water bodies. On the other hand aquatic plants are primary producers in aquatic ecosystem, and further the absence of aquatic flora may also leads to absence of fauna in oceans. Distribution of macrophytic plants and abundance also indicate the water quality. In natural ecosystem, macrophytes witnessed to remove both toxic and non-toxic elements in the sediment and water (Narayana and Somashekar, 1997). Despite, trophic status is mainly influenced the variety of communities and indicator species occur at the source. Moreover, metabolic activities of macrophytic communities accelerate the physico-chemical condition of the stream.

Submerged aquatic macrophytes are one of the most important components of the littoral zones of lakes. Field studies conducted by (Carpenter and Lodge, 1986) Balls *et al.* (1989) and Sondergaard *et al.* (1996) evidenced that submerged macrophytes have an important role in restoration of shallow eutrophic lakes. Therefore, from an engineering point of view, evaluation of life cycles of macrophytes is a key issue before using them in restoration of lakes (Carpenter and Adams, 1977; Asaeda and Bon, 1997).

Study Area

Studies on nutrient status of Kanale tank, Sagara taluk were conducted during the period from Jan 2004 to Dec 2004. It is a perennial water body as it receives the water by rainfall. It lies between 14°12′ latitude and 74°84′ longitude, catchment area of the tank is 1.30 sq km, water spread area is 22.1 Hectare. Hence, the present study was under taken to investigate the nutrient status of the perennial tank water and diversity of aquatic macrophytes. The results obtained are discussed in light of available literature. This paper deals with species diversity of morphometrically and limnological features of a perennial tank Kanale (Table 37.1) from conservation and management perspective.

Table 37.1: Morphometric and Limnological Features of Kanale Tank

Morphometric Parameters	*Kanale Tank*
Area (sq. km)	1.30
Catchment area (km^2)	1.30
Maximum depth (mt)	4.0
Mean depth (mt)	2.0
Water spread area (Hectare)	22.1
Limnological Parameters	*Range*
Parameters	
pH	6.5–8.3
Nitrate	0.11–0.26
Phosphate	0.001–0.21
Ca^{2+}	2.52–5.89
Mg^{2+}	0.87–4.0
Na^+	1.6–4.9
K^+	1.1–4.2

Methods

The study was carried during January 2004 to December 2004. Surface water samples were collected on a basis in accordance to methods adopted by APHA (1998). For sampling purpose 2-liter capacity of plastic carbuoys were used. The different physico-chemical parameters such as Ca^{2+}, Mg^{2+}, Na^+, K+, NO_3^+ and PO_4^{3-}. Content was analyzed using standard suggested method (APHA, 1998). However, aquatic macrophytes were collected and identified by using standard books and manuals (Fasset, 1975; Cook, 1974; Haslam, 1978; Britto, 1991).

Results and Discussion

The data on nutrient status of Kanale tank water has been given in Table 37.3. The tank water exhibits wide range of calcium 2.52–5.89 mg/L, concentration of magnesium ranged between 0.87–4.0 mg/L, and phosphate between 0.001–0.21 mg/L. Similarly Nitrate concentration ranged between 0.11–0.26 mg/L. Despite other factors like Sodium and potassium exhibits 1.6–4.9 mg/L 1.1–4.2 mg/L.

In this study, macrophytic groups involves submerged, floating, emerged and semi aquatic plants of which one species of floating, five species of rooted but floating, six species of submerged, one species of emerged and five species of semi aquatic plants are encountered in Table 37.2.

Table 37.2: Macrophytic Diversity in Kanale Tank

Species Name	*Jan*	*Feb*	*Mar*	*Apr*	*May*	*Jun*	*July*	*Aug*	*Sept*	*Oct*	*Nov*	*Dec*
Free Floating												
Azona pinnata R.Br.	–	–	–	–	–	–	–	–	+	–	–	–
Salvinia natans L.	+	+	+	+	+	+	–	–	+	+	+	+
Rooted Floating												
Nelumbo nucifera Gaertn	+	+	+	+	+	+	+	+	+	+	+	+
Nymphaea nouchali Burm	+	+	+	+	+	+	+	+	+	+	+	+
Nymphea stellata	+	+	+	+	+	+	+	+	+	+	+	+
Nymphoides indicum (L) O. Kuntze	+	+	+	+	+	+	+	+	+	+	+	+
Trapa bispinosa (L) Roxb	+	+	+	+	–	–	–	+	+	+	+	+
Submerged												
Ceratophyllum demersum L.	+	+	+	+	+	+	+	+	+	+	+	+
Chara sps. L.	+	+	+	+	–	–	–	+	+	+	+	+
Valisnaria sps.	+	+	–	+	–	–	+	+	+	+	+	+
Hydrilla verticillata (L.F.) Royle	+	+	+	+	+	–	–	–	–	+	+	+
Nitella sps.	+	+	+	–	–	–	–	+	+	+	+	+
Utricularia vulagaris L.	+	+	+	+	+	–	–	+	+	+	+	+
Emergent												
Limnophyton obtusifolium (L.) Miq.	+	+	+	+	+	–	–	–	–	–	–	–
Semi Aquatic												
Centella asiatica Urban	+	–	–	–	–	+	+	+	+	+	+	+
Eleocharis dulcis (Ell.)	+	+	+	+	+	–	–	–	–	–	+	+
Eriocaulan odaratum	+	–	–	–	–	–	–	+	+	+	–	+
Ludwigia sps. (L.) Ell.	+	–	–	–	–	–	–	+	+	+	+	+
Limnophila heterophylla (L.)	+	+	+	+	–	–	–	–	–	–	–	+

Note: +: Present; –: Absent.

The number of investigator studied on lentic ecosystem and concluded that physico-chemical characters influence the growth, species distribution, indicator groups and pollution tolerant species (Narayana and Somashekar, 2002) the increasing concentration of phosphorus and nitrogen have important effects on the primary production. Waglenska *et al.*, 1987 observed a close relationship between phytoplankton abundance and diversity of submerged macrophytes.

Table 37.3: Nutrient Status of Kanale Tank

Parameters	*Jan*	*Feb*	*Mar*	*Apr*	*May*	*Jun*	*July*	*Aug*	*Sept*	*Oct*	*Nov*	*Dec*	*Range*
pH	7.3	8.0	8.0	7.3	8.2	6.6	7.4	7.1	7.4	6.5	8.3	6.9	6.5–8.3
PO_4	0.2	0.21	0.16	0.12	0.14	0.04	0.002	0.1	0.2	0.001	0.001	0.001	0.001–0.21
NO_3	0.12	0.13	0.11	0.14	0.18	0.26	0.19	0.21	0.11	0.15	0.21	0.18	0.11–0.26
Ca^{2+}	2.52	4.2	5.05	3.36	5.89	2.77	4.2	4.2	4.2	5.05	4.2	2.52	2.52–5.89
Mg^{2+}	1.1	4.0	3.43	0.87	2.26	1.39	1.68	0.3	0.36	1.36	2.31	2.2	0.87–4.0
Na^{+}	4.3	4.9	3.6	3.1	3.6	4.1	4.0	4.1	3.6	1.9	2.1	1.6	1.6–4.9
K^{+}	3.8	4.2	1.1	2.9	3.2	3.1	3.2	3.4	3.2	1.5	1.4	1.1	1.1–4.2

Note: All the parameters are in Mg/L.

The sources of pollutant through agricultural runoff were identified. Due to increased dissolved nutrients organic matter has resulted. Further increase of sedimentation promoted the growth of macrophytic vegetation. So that reduces the volume of water and decreases the depth of water body. If this process continuous water in this tank turned to eutrophic. This will support unwanted weeds. Therefore, impact of biodiversity can be witnessed. Finally beauty of the lake may disappear.

Higher species diversity and abundance of aquatic macrophytes in agricultural land use area having shallow depth and fine sediment as compared to forested land use area having more depth and coarse sediment in opposite shoreline indicate distinct role of environmental factors in affecting the growth and distribution of aquatic macrophytes (Schmid, 1965: Nicholes, 1992). The high species diversity of aquatic macrophyte beds in area featured by laundry spots, inlets or municipal drains and sewage disposal correspond to impact of cultural eutrophication from adjacent urban area (SAIC, 1992). This sort of variation is not distinct in case of shallow and small tank Kanale. The urban drains, though having low nutrient input, deliver about 7, 5 times nitrogen and 9 times phosphorus of the agricultural inputs in lake Phewa (SAIC, 1992). In case of Kanale tank, rapid succession is characterized by formation of floating islands and increase in cover of dense macrophyte beds (Rai *et al.*, 1995). Relevant conservation and management measured to maintain biodiversity and health of the present water body representative tropical aquatic ecosystem of Sagara (Karnataka): include an integrated approach of control of pollution load and ecozoning Kanale, whereas construction of checkdams and catchment ponds at tank site combined with an emphasis on sustainable utilization of aquatic macrophyte resource in future.

References

APHA. 1998. *Standard Method for the Examination of Water and Waste Water*, 18th Ed. New York.

Aseada, T. and Bon, T.V. 1997. Modeling the effects of macrophytes on algal blooming in eutrophic shallow lakes. *Ecol. Modeling*, 104: 261–287.

Balls, H.B., Moss, B. and Irvine. 1989. The loss of submerged plants with eutrophication I. Experimental design, water chemistry, aquatic plant and Phytoplankton biomass in experiments carried out in ponds in the Norfolk Broad land. *Freshwater Biol.*, 22: 71–87.

Britto, S.J. 1991. *Macrophytes Collection Procedure*. St. Joseph College, Tiruchiraapally, p. 1–33.

Carpenter, S.R. and Lodge, D.M. 1986. Effects of Submerged macrophytes on ecosystem process. *Aqua. Bot.*, 26: 341–370.

Cook, C.D.K. 1974. *Water Plants of the World: A Manual for the Identification of the Genera of Fresh Water Macrophytes*. W. Junk. B.V. Publishers, The Hague, pp. 561.

Fasset. 1975. *A Manual of Aquatic Vascular Plants*. The University of Wisconsin press, Madison, pp. 405.

Haslam, S.M. 1978. *River Plants: The Macrophytic Vegetation and their Identification and Management*. Cambridge University Press, Cambridge.

Narayana, J. and R.K. Somashekar. 1997. Heavy metal composition in the sediment and plants of the river cauvery. *J. Env. & Poll.*, 4(4): 325–328.

Narayana, J. and R.K. Somashekar. 2003 Macrophytes diversity in relation to water quality: Investigation on river Cauvery. In: *Ecology and Conservation of Lakes, Reservoirs and Rivers*. ABD Publishers, Jaipur.

Schmid, W. 1965. Distribution of aquatic vegetation as measured by line intercept with SCUBA. *Ecology*, 46(6): 816–822.

SAIC (Science Applications International Corporation) 1992. Environmental protection study Phewa Lake. Pokhara Project, Nepal. Final report submitted to Asian development Bank. Philippines by Science Application International Corporation USA, pp. 202.

Rai, A.K., Shreshtha, B.C., Joshi, P.L., Gurung, T.B. and Nakanashi, M. 1995. Bathymetric maps of lakes Phewa, Begnas and Rupa in Phokhara valley, Nepal. Memories of the faculty of Science. Kyoto University (Series of Biology) 16: 149–154.

Sondergaard, M. Brunn, L., Lauridsen, T., Jeppesen, E., Madsen, T.V. 1996. The impact of grazing waterfowl on submerged macrophytes in situ experiments in a shallow eutrophic lake. *Aqua. Bot.*, 53: 73–84.

Waglenska, T., Dylinska, L.B., Kartin, T.E. and Spoiska, I. 1987. *Ecological Pollution*, p. 173–208.

Chapter 38
Antimicrobial Activity of *Syzygium aromaticum* (Clove) and *Millingtonia hortensis* (Indian Cork Tree) Against Human Pathogens

A. Sujatha, B. Sandhya

P.G. Department and Research Centre, Lady Doak College, Madurai – 2, Tamil Nadu, India

ABSTRACT

The antimicrobial activity of *Syzygium aromaticum* and *Millingtonia hortensis* have been tested against human pathogens such as *Pseudomonas* sp., *Staphylococcus aureus, Escherichia coli, Klebsiella pneumoniae* and *Candida albicans*. Solvents used for extraction are 95 per cent ethanol, Petroleum ether and Acetone. Dried flower bud powder of clove and dried leaf powder of cork tree was used for extraction. Out of dried flower bud extract of S. *aromaticum* and leaf extract of *M. hortensis, S. aromaticum* is found to be more potent and out of the three solvents used, 95 per cent ethanol is found to be most effective.

Keywords: Antimicrobial activity, Disc diffusion method, Medicinal plants.

Introduction

Microbes are closely associated with the health and welfare of human beings. Some are beneficial and some are detrimental. As preventive and curative measures, plants and their products are used in the treatment of infections for many centuries before. WHO estimated that 80 per cent of people worldwide rely on plant based medicines for their primary healthcare (Famsworth *et al.*, 1985) and

India happens to be the largest user of traditional medical cure, using 7000 plant species. There are nearly 300,000 species of plants that are chemically distinct in the world.

Currently, the world trade in plant medicine runs in to billions of dollars. According to the WHO estimates, the present demand for medicinal plants are about 14 billion Dollars per year. Projected demand by the year 2050 is estimated to be around 5 trillion US dollars! (Trivedi and Sharma, 2004).

India is one of the major exporters of crude drugs, mainly to developed countries like U.S.A, Germany, U.K., Japan, etc. As plants are commented as potent biochemists, man is able to obtain from them a wondrous assortment of industrial chemicals. Less than 5 per cent of all plant species have been analyzed as potential medicines. As the World's human population continues to explode, there will be wild spread of new microbial disasters. So, the present investigation is focussed at seeking out plant species, which are most likely candidates to combat predicted diseases of future.

Synthetic antibiotics have been effectively used for the control of diseases. However, due to indiscriminate use of these drugs, various pathogenic organisms have developed resistance to many of the currently available antibiotics (WHO scientific working group, 1983). Hence the research focus is turned on to plant products as alternatives, as they seemed to cause fewer side effects, more effective, and readily available. Antibiotic principles are distributed widely among angiospermic plants. A variety of compounds are accumulated in plant parts accounting for their constitutive antimicrobial activities (Callow, 1983).

Antimicrobial compounds are also produced in plants in response to stimuli from microbial infections (Dixon, 1983). In the present study, two plants were tested for antimicrobial activity against five human pathogens. This would be helpful in search of novel plant antimicrobial compounds.

Materials and Methods

Preparation of Extracts

Selected plant parts were dried in shade and ground and these powders were used for extraction. Extracts were prepared using three solvents namely, 95 per cent Ethyl alcohol, Acetone and petroleum ether. Extraction was done at room temperature by simple extraction method as follows. (Deshpande *et al.*, 2004) Dried powder (10 g) was mixed with 100 ml solvent and kept on shaker for 24 hrs. Then it was allowed to stand for five hours, the supernatant was filtered and the solvent was evaporated at 60°C. After total evaporation of the solvent, the residue was dissolved in dimethyl formamide and used for antimicrobial activity testing. Different concentrations of (5 per cent, 10 per cent, 15 per cent, 20 per cent) of the extracts were used for testing the antimicrobial activity.

Test Micro-organisms

Both gram positive and gram-negative microorganisms were used for the test. The gram-positive organisms include *Staphylococcus aureus, Candida albicans, Gram-negative* organisms include *Pseudomonas* sp., *Escherichia coli* and *Klebsiella pneumoniae.* 0.5 ml of all bacterial cultures were inoculated in 5ml of Nutrient Broth and yeast (*Candida albicans*) in 5ml of SDA broth and were incubated for 5hrs at 37°C.

The microorganisms were procured from Vijaya clinical lab, Madurai, Tamil Nadu.

Detection of Antimicrobial Activity

During this work, Nutrient agar and Sabouraud's Dextrose Agar (SDA) were used for testing antimicrobial activity, using Disc diffusion method (Kirby-Bauer method, 1966). In this method each sterile paper disc of what man's filter paper of 5 mm diameter was dipped in the extract separately. It was then placed on the surface of the solid culture media, which was inoculated with 100 µl of growing pure culture of each test microorganism in separate plates by spread plate method. Duplicates were maintained. The plates were incubated at 37°C for 24 hrs, and then screened for zone of inhibition. Negative controls were maintained for all the solvents used. As a positive control, Standard antibiotic–Streptomycin (Hi-Media, 10 mcg) was used.

Results and Discussion

In the present investigation, antimicrobial activity of 2 plant extracts against 5 microbial species was recorded. Considerable antimicrobial activity was detected in acetone and 95 per cent ethanol extract of *Syzygium* sp. On the other hand, significant activity was detected in acetone extract of *Millingtonia* sp. Petroleum ether extracts of both the plants showed no significant antimicrobial activity. Effect of plant extracts on different organisms as depicted in Tables 38.1 and 38.2.

Table 38.1: Antimicrobial Activity of *Syzygium aromaticum*

Plant Used		*Syzygium aromaticum (Clove) (Dried Flower Bud)* Zone of Inhibition (in mm) Using Disc Diffusion Method											
Solvents Used		*95% Ethanol* 1				*Acetone* 2				*Pet. Ether* 3			
Test Organisms	Standard Streptomycin 10 mcg	5%	10%	15%	20%	5%	10%	15%	20%	5%	10%	15%	20%
Pseudomonas sp.	–	10	**14**	**16**	**17**	7	9	9	**15**	–	–	–	–
S. aureus	*28*	–	–	–	8	8	10	10	**12**	–	–	–	7
E. coli	*31*	**12**	**16**	**17**	**21**	6	6	8	8	–	–	–	–
K. pneumoniae	*26*	10	**12**	**16**	**23**	6	9	10	**13**	–	–	–	8
C. albicans	*31*	–	–	–	8	–	10	**16**	**18**	–	–	–	–

Note: Zone of Inhibition ≥ 12 mm, indicated in Bold; –: Indicates NO activity.

Table 38.2: Antimicrobial Activity of *Millingtonia hortensis*

Plant Used		*Millingtonia hortensis (Indian Cork Tree) (Leaves)* Zone of Inhibition (in mm) Using Disc Diffusion Method											
Solvents Used		*95% Ethanol* 1				*Acetone* 2				*Pet. Ether* 3			
Test Organisms	Standard Streptomycin 10 mcg	5%	10%	15%	20%	5%	10%	15%	20%	5%	10%	15%	20%
Pseudomonas sp.	–	–	–	–	–	–	–	–	–	–	–	–	–
S. aureus	*28*	–	–	–	–	6	8	9	10	–	–	–	–
E. coli	*31*	–	–	6	9	–	–	9	11	–	–	7	9
K. pneumoniae	*26*	–	–	–	–	6	6	7	9	–	–	–	–
C. albicans	*31*	–	–	–	–	–	–	–	–	–	–	–	–

Pseudomonas sp.

All concentrations (5 per cent, 10 per cent, 15 per cent and 20 per cent) of 95 per cent ethanol, acetone extracts of S. *aromaticum* were found to be potent as an inhibitory agent against *Pseudomonas* cultures. The growth was considerably affected with the formation of inhibitory zone ranging from 7-15 mm at different concentrations. In case of petroleum ether extract, no inhibitory activity was seen.

In case of leaf extract of *M. hortensis* no inhibitory activity was recorded against *Pseudomonas* sp. at different concentrations of 95 per cent ethanol, acetone, petroleum ether extracts.

Staphylococcus aureus

Acetone extract of *S. aromaticum* proved its inhibitory activity against *S. aureus* with a zone ranging from 8–12 mm while 95 per cent ethanol and petroleum ether extracts responded with inhibition zone of 8 mm and 7 mm respectively.

In case of *M. hortensis* no activity was reported with 95 per cent ethanol and petroleum ether extracts, while acetone extract showed moderate activity with an inhibition zone of 6 mm–10 mm

Escherichia coli

S. aromatium 95 per cent ethanol extract showed potent antimicrobial activity with an inhibition zone of 12 mm–21 mm. On the contrary, acetone extract showed moderate response giving a zone of 6 mm–8 mm and no zone was formed using petroleum ether extract.

The growth of *E. coli* organism was inhibited in 15 per cent and 20 per cent concentrations of all extracts of *M. hortensis.*

Klebsiella pneumomae

All concentrations of *S. aromaticum* ethanol and acetone extract proved activity against *K. pneumoniae. S. aromaticum* is found to be potent providing a zone of maximum inhibition 23 mm in case of ethanolic extract and a zone of 13 mm in case of acetone extract. Petroleum ether extract of *S. aromaticum* started responding only at 20 per cent concentration with a zone of 8 mm.

In case of acetone extracts of *M. hortensis,* zone of 6–9 mm was noted. No zone of inhibition was recorded against *K. pneumoniae* using ethanol and petroleum ether extracts.

Candida albicans

S. aromaticum dried flower bud extracted with petroleum ether showed no inhibitory effect against the yeast *C. albicans.*

It is found that using acetone as an solvent for extraction, a maximum zone of 18 mm was obtained at 20 per cent concentrations and a zone of 8 mm using 95 per cent ethanol for extraction. All the extracts at all the concentrations of *M. hortensis* showed no zone of inhibition at all.

In summary, *Syzygium aromaticum* (clove) is found to be having more antimicrobial activity than *Millingtonia hortensis* (Indian cork tree) and it's activity is most significant against *K. pneumoniae, E. coli* and *Pseudomonas* sp.

References

Alankara Rao, G.S. and Rajender Prasad Y. (1981). Antimicrobial property of *Acorus calamus* Linn. *in vitro* studies. *Indian Perfum.,* 25: 18–24.

Bauer, A.W., Kirby, W.M.N., Sherrier, J.C. and Yurek, M. (1966). Antibiotic susceptibility testing by standardized single disc method. *Am. J. Clin. Pathol.*, 45: 493.

Deshpande, A.R., Mohd Musaddiq, Bhadange, D.G. (2004). Studies on antibacterial activity of some plant extracts. *J. Micro. World*, 6: 45–49.

Dixon, R.A., Dey, P.M. and Lamb (1983). *Phytoalexins: Enzymology and Molecular Aspects Advances in Enzymology*. 55: 1–69.

Hufford, C.D. and Jia, Y. (1993). Antimicrobial compounds from *Petalostemum Purpureum*. *J. Nat. Prod.*, 56: 1878–18.89.

Ibrahim, D. and Osman, H. (1995). Antimicrobial activity of *Cassia alata* from Malaysia. *J. Ethnopharmcol.*, 45: 151–156.

Meyer, J.J. and Afolayan, A.J. (1995). Antibacterial activity of *Helichrysum aureonitens*. *J. Ethnopharmcol.*, 47: 109–111.

Miller, J.H. (1992). *A Short Course in Bacterial Genetics*. CSHL Press.

Trivedi, P.C., Sharma, N. (2004). *Ethnomedicinal Plants*. Pointer Publishers, India.

Uma Devi, P. and Gnana Soundari, A. (1995). Radio protective effect of leaf extract of Indian medicinal plants *Ocimum sanctum*. *Indian J. Exp. Biol.*, 33: 205–208.

WHO Scientific Working Group (1983.) Antimicrobial resistance. *Bulletin of the World Health Organization*, 61(3): 383–394.

Chapter 39

Production and Biochemical Analysis of Bacteriocin (Nisin) from Immobilized and Non-immobilized *Lactococcus lactis* Subsp. *Lactis*

D. Jegadeeshkumar, Suman Gulati**, K. Revathi**, K. Anbarasu*, P. Vijayalakshmi** and R. Kungumapriya***

*Department of Microbiology, P.G.P. College of Arts and Science, Namakkal – 637 206
**Post Graduate and Research Department of Zoology, Ethiraj College for Women, Chennai

ABSTRACT

Conservation and preservation of food are the prerequisites for food security and have remained a priority to mankind since time immemorial. Food borne diseases are a major problem today. In a country like India there is a strong need to conserve food. There are various methods of food production such as thermal sterilization, frozen dehydration and chemical preservatives. In recent times, biopreservation is the alternative to physical and chemical preservatives. Bacteriocin is a bacterial protein, which are bactericidal to some strains. Many bacteriocins are inhibitory to a wide variety of gram-positive bacteria including many food borne pathogens. In the present study, bacteriocin production is attempted from immobilized and non-immobilized *Lactococcus lactis* subsp. *lactis*. The antimicrobial activity of the bacteriocin produced and its biochemical nature has also been studied.

Keywords: *Bacteriocin, Nisin, Lactococcus lactis.*

Introduction

A number of microorganisms are involved for food spoilage such as *Pseudomonas aeruginosa*, *Bacillus cereus*, *Escherichia coli*, *Salmonella typhi*, *Clostridium botulinum*, etc. The hot and humid climate of our country is quite favourable for the growth of numerous insects and microorganisms that causes spoilage of food every year. It is estimated that 20–40 per cent of our agricultural production is lost due to spoilage by insects, pests and microbes. Thermal sterilization, frozen dehydration and chemical preservation are some the methods of preservation of food products to extend their shelf life. Fumigants like methyl bromide, ethyletle dibromide etc. are also used. Drying, salting and fermentation were some of the traditional methods of prevention of food materials from spoilage. However, canning and freezing were relatively of recent development.

In recent times biopreservation is a good alternative for the physical and chemical preservation techniques of food. Bacteriocin is a bacterial protein that has an inhibitory effect on several species of food borne pathogens. In the present study *Lactococcus lactis* subsp. *lactis* has been employed for bacteriocin production. The antimicrobial effect of lactic acid has beat appreciated by man for many years and has been enabled him to extend the shelf life of many foods. A number of microflora such as *Lactococcus lactis, Lactobacillus plantum, Pediococcus acidilactici, Streptococcus cremoris, Enterococcus faecium, Diplococcus pneumonia, Clostridum butyricum, Clostridium botulinum, Clostridium perfringens, Staphylococcus epidermidis, Proteus mirabilis* etc. produce bacteriocins and such as Nisin, Pedoicin, Lactostapcin, Enterocin, Burylicin, Bioticin, Perfringocins, Staphylococcin and Proticin that has been used as a biopreservative.

Among the various kinds of bacteriocins, Nisin produced by *lactococcus lactis* is a worldwide accepted and licensed one. Nisin is most soluble and stable at a pH of 2. It is a type 1 antibiotic composed of 34 amino acids and a molecular mass of 3354 Da. The Nisin precursor is synthesized easily in the active growth phase and production rate is maximal towards the end of the exponential growth phase. Jacob and his co-workers first coined the term Bacteriocin in 1953. In the present study bacteriocil production by the immobilized and non-immobilized cultures of *Lactococcus lactis* was brought about and the bacteriocin quality was determined by thin layer chromatography. The bactericidal activity of the bacteriocin produced was also studied against a wide variety of gram-positive pathogens.

Materials and Methods

Lactococcus lactis subsp *lactis* used in the present study for the production of biopreservative bacteriocin has been obtained from Institute of Microbial Technology (IMTECH), Chandigarh. The selective media used for the growth of *Lactococcus lactis* is brain heart infusion agar, incubated at 37°C for 24 hrs. For the preparation of immobilized culture 6 per cent sodium alginate and 4 per cent calcium chloride solutions were autoclaved separately. The bacterial culture obtained from the brain heart infusion agar was mixed with 6 per cent sodium alginate solution in the ratio of 25 : 40. The mixture was introduced as drops into the 4 per cent calcium chloride solution with the help of a 21G sterile syringe. Sodium alginate reacts with calcium chloride to produce beads in which bacterial cells are entrapped. The beads were washed with sterile distilled water, and the size, shape, wet weight, and the number of beads produced per ml were calculated and tabulated. The washed beads were transformed to the Modified Dairy Based Medium (MDBM) and incubated at 37°C for 34 hrs at a pH of 6.8. In the same manner the normal non-immobilized strain was also introduced the MDBM for a comparative study.

After the incubation period for the separation of bacteriocin the bacterial cells were killed by heating to 75°C for 15 minutes and the pH of the medium was adjusted to 2.5. The treated cultures were centrifugated at 3000 rpm for 15 minutes and the active cell free supernatant was subjected to salt precipitation with ammonium sulphate followed by centrifugation at 6000 rpm for 15 minutes. The protein precipitate recovered was suspended in 5 ml of 50 mM sodium phosphate buffer at a pH of 7 and was subjected to salting out procedure for the bactetriocin isolation. This was followed up by dialysis for purification. Lowry *et al.*, method, estimated the amount of bacteriocin protein produced by the immobilized and non-immobilized cultures.

The bacteriocin was subjected to bactericidal activity again and food borne pathogens by agar diffusion method. The following organisms, *Staphylococcus aureus, Escherichia coli, Salmonella typhi* and *Bacillus cereous* were isolated from cheese and meat and identified by standard biochemical tests such as Indole, MR, VP, Citrate, Urease, TSI, Oxidase and Catalase. The zone of inhibition was measured to detect the antimicrobial activity of bacteriocin. Bacteriocin was also subjected to biochemical analysis by Paper chromatography. The Rf values of the various spots appeared in the chromatogram was detected and compared with the standard.

Results

The outbreaks of food borne diseases have led to considerable illness. Nisin is one of the widely accepted and used food preservative and its antimicrobial activity against a wide variety of food pathogens has been proved repeatedly. In the present study two different methods were employed for bacteriocin production such as the non-immobilized and immobilized cultures. The culture of *Lactococcus lactis* grown on MDBM contained 8.5×10^7 cfu/ml. Some of these were immobilized using 6 per cent sodium alginate were the bacteria were trapped in beads and the rest were cultured as the non-immobilized ones, the bacteria containing beads of sodium alginate were about 2.5 ± 0.5 mm in size, weighing about 14 ± 1.2 mg and containing about 2.83×10^6 cells per bead. Results showed that the beads were intact for 90 hrs. Table 39.1 reveals the bacteriocin content of the two different methods of production adopted in this study. The immobilized culture produced 255 µg of bacteriocin as compared to 155 µg of bacteriocin protein produced by the non-immobilized culture.

Table 39.1: The Bacteriocin Protein Content of the Cultures Used in Present Study

Sl.No.	*Name of Sample*	*O.D. Value*	*Protein (Nisin) Concentration (µg.ml)*
1.	Non-immobilized	0.23	155 µg
2.	Immobilized	0.36	255 µg

Table 39.2 gives the results of the biochemical characterization of the food borne pathogens employed in the present study. The antimicrobial activity of the bacteriocin has been proved by the agar diffusion test which shows that the bacteriocin is highly sensitive against *Staphylococcus aureus, Bacillus cereus, Salmonella typhi,* and has intermediary effect against *E. coli* (Table 39.3).

Table 39.2: The Biochemical Characterization of Food Borne Pathogens

Sl.No.	*Organisms*	*Biochemical Character*								
		G	*C*	*O*	*I*	*MR*	*VP*	*C*	*U*	*TSI*
1.	*S. aureus*	+	–	–	+	+	+	–	–	A/A
2.	*S. typhi*	–	+	–	–	+	–	+	–	A/A, G, H_2S
3.	*B. cereus*	+	+	–	–	–	–	+	+	Ak/A
4.	*E. coli*	–	+	–	+	+	–	–	–	A/Ak

Note: G: Gram stain; C: Catalase; O: Oxidase; I: Indole; MR: Methyl Red; VP: Vogous proskauer; C: Citrate; U: Urease; TSI: Triple sugar iron test.

Table 28.3: The Antibacterial Activity of Bacteriocin Against Various Food Pathogens

Sl.No.	Bacteria	Zone of Inhibition (mm)		Result
		Immobilized Culture	Non-immobilized Culture	
1.	*S. aureus*	22	21	Sensitive
2.	*S. typhi*	15	13	Sensitive
3.	*B. cereus*	21	20	Sensitive
4.	*E. coli*	12	10	Intermediate

The biochemical analysis by paper chromatography showed that the bacteriocin developed from both the cultures contained a component at an Rf value of 0.6 and a component at an Rf of 0.4 against butanol (4): acetic acid (1): water (5) mobile phaste. Based on the Rf values it was concluded that this component with an Rf of 0.6 was tryptophan and the component with an Rf value of 0.4 was concluded to be leucine.

Discussion

Food preservation is a challenge task for human beings to fulfil the consumer demands. The outbreak of food borne diseases has led to considerable illness and even death (Albrecht, 1986). It is estimated that about 28.81 million cases of food borne illness are reported worldwide every year and approximately 50 per cent of these cases are associated with meat and meat production (Archer and Kvenbery, 1982). India is the second largest producer of food and about 22–40 per cent of our agriculture production is lost due to spoilage by insects, pests and microbes. The microbial contaminants not only spoil the food but also cause human illness. A number of traditional methods of food preservation such as salting, drying and fermentation have been adopted but this makes the food less acceptable to consumers.

Corlett (1989) reported that minimally processed, vacuum packed, refrigerated meat products have become increasingly popular to fulfil consumer demands for natural foods as many of these products contain no preservatives. Vadenbergh, 1993 reported the use of nisin in meats. The effect of nisin-sodium chloride interaction on the outgrowth of *Bacillus* and *licheniformis* spres has also been proved (Bell and Karan, 1985). Gram-positive bacteria and their spores are sensitive to nisin, whereas yeast, moulds and Gram-negative bacteria are generally resistant (Mattick and Hisch, 1947). The present study is an approach to preserve food by biopreservation methods against Gram-positive and Gram-negative pathogans.

Bacteriocin is a proteineous compound displaying bactericidal activity against closely related species. It is a ribosomally synthesized antimicrobial polypeptide produced by Gram-negative bacteria. The present study reveals that both the non-immobilized and immobilized produce bacteriocin protein, the immobilized culture being more efficient producing 255 µg/ml of bacteriocin as compared to 155 µg/ml by the non-immobilized culture.

The anti-microbial test of bacteriocin revealed that the bacteriocin developed from *Lactococcus lactis* the ability to kill both Gram-positive and Gram-negative pathogens. However the bactericidal activity varied for different pathogens. The bacteriocin was found to be highly active against *Staphhylococcus aureus, Bacilius cereus, Salmonella typhi,* and *E. coli* in hierarchy. However, the zone of measurement showed that the bacteriocin produced by the immobilized culture had greater antimicrobial effect as compared to the bacteriocin produced by the non-immobilized culture.

The biochemical analysis showed that both the immobilized and non-immobilized culture produced bacteriocin containing typtophan and leucine. Ivanora *et al*,. 2000 reported that the bacterical activitiy of bacteriocess varies between species and the activity is more in closely related species than the different genera. Further the same author reported that the antimicrobial activity also depends on the source of organism collection and the media used for assessment. In the recent years large amount of research has been carried out on the natural antimicrobials for food application. Bacteriocins comprise of just one group of compounds being studied. Bacteriocin has the advantage that it has been given the GRAS (Generally Recognized as Safe) status with the regulatory agencies. Other than Nisin, many other purified bacteriocins have not been licensed and thus applied studies on the other bacteriocins are lacking.

Convincing evidence of the inhibition of pathogens and spoilage bacteria is required to stimulate commercial interest in bacteriocin as agents for biopreservation. Unfortunately, except for a few bacteriocins most of the bacteriocins have a narrow antibacterial spectrum and are not active against Gram-negative bacteria. Use of nisin with a chelating agent expands the antimicrobial spectrum of nisin to include a wide variety of Gram-negative bacteria.

References

Albrecht, J. (1986). Business and technology issues in USA science and technology. *Food Technol.*, 40: 122–127.

Archer, D.I. and Kvenbery, J.E. (1982). Incidence of food borne diarrhoeal diseases in the United States. *J. Food Production*, 48: 887–894.

Bell, R.G. and Karan, M. (1985). The effect of nisin-sodium chloride interaction on the outgrowth of *Bacillus licheniformis*. *J. Appl. Bacteriol.*, 59: 127–132.

Corlett, D.A. Jr. (1989). Refrigerated food and use of hazard analysis and critical control point principles. *Food Technol.*, 43(2): 91–94.

Mattick, A.T.R. and Hirisch, A. (1947). Further observation on the inhibitory substance (Nisin) from *Streptococci*. *Lancet*, 11: 5–12.

Vadenberg, P.A. (1993). Lactic acid bacteria: Their metabolic products and interference with microbial growth. *FEMS Microbiology Reviews*, 12: 221–238.

Chapter 40
Species Distribution of the Family Verbenaceae in the Valley Districts of Manipur

Y. Nanda Devi and P.K. Singh

Ethnobotany and Physiology Laboratory, Department of Life Sciences, Manipur University, Canchipur, Imphal –795 003, Manipur

ABSTRACT

A survey of the available species distribution of the family Verbenaceae was conducted during the period of 2001 to 2003 from the valley districts (Imphal East, Imphal West, Bishnupur and Thoubal) of Manipur state. During the survey, a total of 23 species were collected. The species were categorized into–Commonly available (CA)–13; Wild (W)–14; Cultivated (C)–17; Ornamental (O)–12; Timber (T)–2; Edible (E)–4; and Medicinal (M)-16. *Phyla nodiflora* (Linn.) Greene, is restricted only in Bishnupur district. Phytosociological parameters were incorporated for all the species of the family Verbenaceae and selected associated plants. For this study 1 sq m quadrat was taken and IVI was calculated. The highest IVI value among the five Verbenaceae species was recorded in *Clerodendron fragrans* (Vant.) R.Br. (74.14) and minimum IVI in *Clerodendron glandulosum* Colebr ex Wall. (25.76). Amon the non-Verbenaceae species maximum IVI is recorded in case of *Bidens* species (86.82) and minimum in *Achyranthus aspera* Linn (5.98).

Keywords: Verbenaceae, Distribution, IVI, Phytosociological parameters.

Introduction

The family Verbenaceae is predominantly distributed in tropical and sub-tropical regions, though a few species spread over the temperate regions and a few species in the cooler parts of the world. This family is a large family composed of about 77 genera and 3,020 species or more (Saxena and Saxena,

2001). In India, about 23 genera and 128 species were recorded (Hooker, 1973). Twelve genera and 66 species had been recorded from Assam (Kanjilal, 1982). In Manipur, the former taxonomists of Manipur like-Deb, 1961; Singh, 1987; Singh, 1990 and Sinha, 1996, recorded 16 genera and 39 species. The area of Manipur state is 22,327 km² whereas; Manipur valley has a total area of 2,067 km². The valley area is carved out by Imphal River and its tributaries and is surrounded on all sides by the Manipur hills, which forms central segment of the eastern tertiary folding in India, located between 24°14'N latitudes and 93°42'E –94°11E longitudes. Manipur valley is a repository of rich biological diversity. The valley is supposed to be form by the depositions of stream that was blocked by some convulsion of earth movement. The valley might be found as a result of a lake being filled up by the river form sediments. The present Loktak Lake is said to be the remnant of the original lake in the past. Loktak Lake occupies South East corner of the valley districts. In the southern part of the valley districts, some lakes such as Pumlen pat, Kharung pat, Ikop pat and Waithou pat etc. were occupied. Langol, Heingang, Nongmaijing ching, Langthabal, Waithou and a series of islands, *viz.*, Sendra, Ithing, Thanga and Karang were also included in this district. The alluvial soil type which have clayey loam texture and grey to pale brown coloured were found in an area of 1600 km². The organic soil type occurred along the Loktak and other lakes and marshes.

Materials and Methods

The investigation was performed in the valley districts of Manipur (Imphal East, Imphal West, Bishnupur and Thoubal) during the period of 2001 to 2003. Phytosociological parameters of five Verbenaceae species (*Callicarpa macrophylla* Vahl., *Clerodendron fragrens* (Vant.) R.Br., *Clerodendron glandulosum* Colebr ex Wall, *Stachytarpheta jamaicensis* Vahl. and *Vitex trifolia*, Linn.) with its associated plants were studied in Manipur University Campus using quadrat method having the size of 1 sq. m. Quantitative analysis of vegetation for density (D), abundance (A) and frequency (F) was calculated following the standard methods of Curties and Me Intosh (1950), and Oosting (1958). Their relative values were also calculated and summed to get Importance Value Index (IVI) out of 300 by adopting the method of Curtis (1959) and Misra (1968).

Results and Discussion

During the survey, a total of 23 Verbenaceae species were collected. The species were categorized into: Commonly available (C)–13; Wild (W)–14; Cultivated (C)–17; Ornamental (O)–12; Timber (T)–2; Edible (E)–4 and Medicinal (M)–16 (Table 40.1). The Density (D), Abundance (A), Frequency (F) with their relative values, Abundance/Frequency (A/F) and Importance Value Index (IVI) out of 300 for five different Verbenaceae species with its associated plant species were calculated (Table 40.2A–40.2E).

In Table 40.2A, the maximum density and abundance with their relative values were recorded in *Alternanthera philoxeroides* Griseb and minimum in *Lantana camara* Linn. The maximum IVI value was also found in *Alternanthera philoxeroides* Griseb (61.31) and was followed by *Stachytarpheta jamaicensis* Vahl. (50.59). The minimum IVI value was also recorded in *Lantana camara* Linn. (5.28). *Alternanthera philoxeroides* Griseb (61.31), *Stachytarpheta jamaicensis* Vahl. (50.59), *Mikania cordata* (Burm.) B.L. Robinson (46.77), *Rungia parviflora* (30.24), *Callicarpa macrophylla* Vahl. (28.23) etc. were recorded as dominant species.

In Table 40.2B, highest values of density, abundance with their relative values and IVI were obtained in *Clerodendron fragrans* (Vant.) R.Br. The lowest values were recorded in *Achyranthus aspera* Linn. *Clerodendron fragrans* (Vant) R.Br. (74.14), *Solanum* species (40.6), *Gynura cusimba* (D. Don) Moore (38. 37), *Ipomoea* species (37.22) were found to be dominant species.

Table 40.1: Distribution and Uses of Some Commonly Available Verbenaceae Species in the Valley Districts of Manipur

Sl.No.	Name of the Species	Uses	Parts Used	Application
1.	*Callicarpa macrophylla* Vahl.	CA/W/M	Leaves	Rheumatic joints
2.	*Clerodendron fragrans* (Vant.) R.Br.	CA/W/C/E/M	Leaves	Injuries
3.	*C. glandulosum* Colebr ex Wall.	CA/W/C/O/E/M	Leaves	For maintaining B.P.
4.	*C. inerme* Gaertn.	CA/C/O		
5.	*C. nutans* Wall.	C/O/M	Leaves	Applied to infections of the skin
6.	*C. serratum* Spreng.	CA/W/C/E/M	Leaves and inflorescence	Cold and cough
7.	*C. siphonanthus* R.Br.	CA/W/C/O/M	Leaves, stem and root	Asthma and cough
8.	*C. speciosissimum* Vant	C/O		
9.	*C. splendens*	C/O		
10.	*C. squamatum* Vahl	C/O/M	Leaves	Dizziness problem
11.	*C. thomsoniae* Balf	C/O		
12.	*Duranta repens* Linn	CA/W/C/O/M	Fruit	Lethal to mosquito larvae
13.	*Gmelina arborea* Roxb.	CA/W/C/T/M	Leaves	Snake bite and Scorpion sting
14.	*Holmskioldia sanguinea* Retz. (Linn.) Greene	CA/W/C/O/M	Leaves and Inflorescence	Headache and dizziness due to high B.P.
15.	*Lantana camara* Linn.	CA/W/M	Leaves	Injuries for stopping bleeding
16.	*L. montevidensis* (K. Spreng) Briq.	C/O		
17.	*L. new-gold* (K. Spreng) Briq.	C/O		
18.	*Phyla nodiflora* (Linn.) Greene	W/M	Leaves and tender stalks	Indestion for children
19.	*Premno herboceoe* Roxb.	W		
20.	*Stachytarpheta jamaicensis* Vahl.	CA/W/M	Leaves and Bark	Diarrhoea and dysentery
21.	*Tectona grandis* Linn.	CA/C/T/E/M	Young leaves	Normal blood circulation
22.	*Verbena officinalis* Linn.	W/M	Leaves and young shoots	Gastric troubles and worms
23.	*Vitex trifolia* Linn.	CA/W/C/M	Leaves and inflorescence	Tuberculosis and anticancer activity

CA: Commonly available; W: Wild; C: Cultivated; O: Ornamental; T: Timber; E: Edible; M: Medicinal.

In Table 40.2C, the maximum and minimum values of density and abundance with their relative values were found in *Bidens* species and *Xanthium strumerium* Linn. respectively. The maximum IVI value was also recorded in *Bidens* species (86.82) however, the minimum IVI value was found in *Mikania cordata* (Burm.) B.L. Robinson (7.74). The dominant species were recorded in species like *Bidens* species (86.82), *Hydrocotyl javanica* Thunb. (58.31), *Oxalis corniculata* Linn. (46.66), *Fragaria* species (29.15), *Clerodendron glandulosum* Colebr ex Wall (25.76) etc.

Table 40.2A: Phytosociological Studies of *Callicarpa macrophylla* Vahl. with its Associated Plants

Sl.No.	Name of the Species	Density (1 sq.m.)	RD (%)	Abundance (1 sq.m.)	RA (%)	Frequency (%)	RF (%)	IVI Out of 300	A/F Ratio
1.	*Callicarpa macrophylla* Vahl	3.6	8.04	4.5	7.19	80	13	28.23	0.057
2.	*Stachytarpheta jamaicensis* Vahl.	9	20.09	9	14.37	100	16.13	50.59	0.09
3.	*Mimosa pudica* Linn.	2.6	5.81	6.5	10.38	40	6.46	22.65	0.17
4.	*Rungia parviflora*	4.2	9.38	7	11.18	60	9.68	30.24	0.117
5.	*Mikania cordata* (Burm.) B.L. Robinson	8	17.86	8	12.78	100	16.13	46.77	0.08
6.	*Alternanthera philoxeroides* Griseb	11.8	26.34	11.8	18.84	100	16.13	61.31	0.118
7.	*Lantana camara* Linn.	0.2	0.45	1	1.6	20	3.23	5.28	0.05
8.	*Xanthium strumerium* Linn.	1.4	3.13	2.34	3.8	60	9.68	16.61	0.039
9.	*Oxalis corniculata* Linn.	3	6.7	7.5	11.98	40	6.46	25.14	0.188
10.	*Bidens* species	1	2.24	5	7.99	20	3.23	13.46	0.25
	Total	**44.8**		**62.69**		**620**			

Table 40.2B: Phytosociological Studies of *Clerodendron fragrans* (Vant.) R.Br. with its Associated Plants

Sl.No.	Name of the Species	Density (1 sq.m.)	RD (%)	Abundance (1 sq.m.)	RA (%)	Frequency (%)	RF (%)	IVI Out of 300	A/F Ratio
1.	*Clerodendron fragrans* (Vant.) R.Br.				35	100	20.84	74.14	0.14
2.	*Xanthium strumerium* Linn.	0.4	1	2	2.62	20	4.17	7.79	0.1
3.	*Lantana camara* Linn.	2	10	3.34	4.38	80	16.67	31.05	0.042
4.	*Solanum* species	6	15	10	13.1	60	12.5	40.6	0.167
5.	*Gynura cusimba* (D.Don) Moore	5.2	13	13	17.03	40	8.34	38.37	0.325
6.	*Acnyranthus aspera* Linn.	0.2	0.5	1	1.31	20	4.17	5.98	0.05
7.	*Ageratum conizoides* Linn.	4.2	10.5	7	9.17	60	12.5	32.17	0.117
8.	*Drymeria cordata* Willd.	2.4	6	12	15.72	20	4.17	25.89	0.6
9.	*Hibiscus rosa-sinensis* Linn.	0.6	1.5	1.5	1.97	40	8.34	11.81	0.037
10.	*Ipomoea* species	5	12.5	12.5	16.38	40	8.34	37.22	0.313
	Total	**40**		**76.34**		**480**			

In Table 40.2D, the maximum density and relative density were recorded in *Stachytarpheta jamaicensis* Vahl. The highest and lowest IVI values were recorded in *Stachytarpheta jamaicensis* Vahl. (62.34) and *Knoxia roxburghii* (Spreng). M.A. Rau (6.67) respectively. The minimum IVI values were found in *Knoxia roxburghii* (Spreng) M.A. Rau (6.67), *Jasminum* species (9), *Eupatorium odoratum* Linn. (11.33), *Mimosa pudica* Linn. (18.42).

In Table 40.2E, the maximum density and abundance with their relative values and IVI were recorded in *Ageratum conyzoids* Linn. (IVI–66.7). The maximum values were found in *Mimosa pudica* Linn. (IVI–6.68). On the basis of maximum IVI values, *Ageratum conizoides* Linn. (66.7), *Drymeria cordata*

Willd. (43.46), *Rungia parviflora* (41.58), *Vitex trifolia* Linn. (35.18) etc. were found to be dominant species.

Table 40.2C: Phytosociological Studies of *Clerodendron glandulosum* Colebr ex Wall. with its Associated Plants

Sl.No.	Name of the Species	Density (1 sq.m.)	RD (%)	Abundance (1 sq.m.)	RA (%)	Frequency (%)	RF (%)	IVI Out of 300	A/F Ratio
1.	*Cleodendron glandulosum* Colebr ex Wall.	10.6	6.5	10.6	6.1	100	13.16	25.76	0.106
2.	*Stachytarpheta jamaicensis* Vahl.	4.6	2.82	5.75	3.31	80	10.53	16.16	0.072
3.	*Bidens* species	62	38	62	35.66	100	13.16	86.82	0.62
4	*Oxalis corniculata* Linn.	28.2	17.28	28.2	16.22	100	13.16	46.66	0.282
5.	*Fragaria* species	14	8.58	17.5	10.14	80	10.53	29.15	0.219
6.	*Hydrocotyl javanica* Thunb	38	23.29	38	21.86	100	13.16	58.31	0.38
7.	*Xanthium strumerium* Linn.	0.8	0.5	1.34	0.78	60	7.9	9.18	0.023
8.	*Plantago erosa* Wall.	2.4	1.48	4	2.31	60	7.9	11.69	0.067
9.	*Mikania cordata* (Burm.) B.L. Robinson	1.2	0.74	3	1.73	40	5.27	7.74	0.075
10.	*Rungia parviflora*	14	0.86	3.5	2.02	40	5.27	8.15	0.0875
	Total	**163.2**		**173.89**		**760**			

Table 40.2D: Phytosociological Studies of *Stachytarpheta jamaicensis* Vahl. with its Associated Plants

Sl.No.	Name of the Species	Density (1 sq.m.)	RD (%)	Abundance (1 sq.m.)	RA (%)	Frequency (%)	RF (%)	IVI Out of 300	A/F Ratio
1.	*Stachytarpheta jamaicensis* Vahl.	19.2	29.73	19.2	17.9	100	14.17	62.34	0.192
2.	*Bidens* species	13.2	20.44	13.2	12.3	100	14.17	47.45	0.132
3.	*Rungia parviflora*	5.4	0.36	5.4	5.04	100	14.17	20.11	0.054
4.	*Mimosa pudica* Linn.	5.4	0.36	6.75	6.29	80	11.77	18.42	0.085
5.	*Eupatorium odoratum* Linn.	1.4	2.17	3.5	3.27	40	5.89	11.33	0.0875
6.	*Knoxia roxburghii* (Spreng) M.A. Rau	0.6	0.93	3	2.8	20	2.94	6.67	0.15
7.	*Desmodium triflorum* (Linn.) DC	9.2	14.25	23	21.44	40	5.89	41.58	0.575
8.	*Jasminum* species	0.8	1.24	2	1.87	40	5.89	9	0.05
9.	*Crotalaria* species	4	6.2	6.67	3.73	60	8.83	18.76	0.112
10.	*Xanthium strumerium* Linn.	5.4	0.36	5.4	5.04	100	14.71	20.11	0.054
	Total	**64.6**		**107.32**		**680**			

From the above results, it is confirmed that the five Verbenaceae species are found to be dominant species. Since the values of Abundance/Frequency ratio (A/F) of five different Verbenaceae species were recorded more than 0.05, these five species were found to be distributed aggregately (Curtis, 1959).

Table 40.2E: Phytosociological Studies of *Vitex trifolia* Linn. with its Associated Plants

Sl.No.	*Name of the Species*	*Density (1 sq.m.)*	*RD (%)*	*Abundance (1 sq.m.)*	*RA (%)*	*Frequency (%)*	*RF (%)*	*IVI Out of 300*	*A/F Ratio*
1.	*Vitex trifolia* Linn.	5.4	11.21	5.4	7.3	100	16.67	35.18	0.054
2.	*Ageratum conizoides* Linn.	14.6	30.3	14.6	19.73	100	16.67	66.7	0.146
3.	*Rungia parviflora*	6.4	13.28	16	21.63	40	6.67	41.58	0.4
4.	*Centella javanica* Thunb (Linn.) Urban	4.4	9.13	.5.5	7.44	80	13.34	29.91	0.,069
5.	*Drymeria cordata* Willd.	8	16.6	10	13.52	80	13.34	43.46	0.125
6.	*Gynura cusimba* (D. Don) Moore	2.4	4.98	6	8.11	40	6.67	19.76	0.15
7.	*Stephania rotunda* Hook.f. and Thoms.	3.2	6.64	4	5.41	80	13.34	25.39	0.05
8.	*Xanthium strumerium* Linn.	0.8	1.66	4	5.41	20	3.34	10.41	0.2
9.	*Mimosa pudica* Linn.	0.4	0.83	2	2.71	20	3.34	6.88	0.1
10.	*Knoxia roxburghii* (Spreng) M.A. Rau	2.6	5.4	6.5	8.79	40	6.67	20.86	0.1625
	Total	**48.2**		**74**		**600**			

In the present study, 23 Verbenaceae species were recorded from valley districts of Manipur (TabLe 40.1). Phytosociological studies of five selected Verbenaceae species were performed and found to be associated with some non-Verbenaceae family species (Tables 40.2A–40.2E). They are: *Achyranthus aspera* Linn. *Ageratum conizoides* Lion, *Alternanthera philoxeroides* Griseb, *Bidens species, Centella javanica Thunb* (Linn.) Urban, *Crotalaria* species, *Desmodium trijlorum* (Linn.) DC., *Drymeria cordata* Willd., *Eupatorium odoratum* Linn., *Fragaria* species, *Gynura cusimba* (D. Don) Moore, *Hibiscus rosa-sinensis* Linn., *Hydrocotyl javanica Thunb, Ipomoea* species, *Knoxia roxburghii* (Spreng) M.A. Rau, *Jasminum* species, *Mikania cordata* (Burm.) B.L. Robinson, *Mimosa pudica* LiNn., *Oxalis corniculata* Linn., *Plantago erosa* Wall, *Rungia parvijlora, Solanum* species, *Stephania rotunda* Hook. f. and Thoms., *Xanthium strumerium* Linn. A particular genus has its affinity to grow mutually in a particular area. In this study, seasonal variation of the associated species is not discussed. However, associated species vary from one site to another (Tables 40.2A–40.2E). The data of variation of associated plants was also supported by various workers like Trivedi (1983), Dix (1959) and Rath and Misra (1980).

The fluctuation in the number and density of species is primarily related to environmental factors, annual and perennial nature of the species (Singh and Yadava 1974). The same case was also noticed in IVI values of five different sites (Tables 40.2A–40.2E). Joshi *et al.* (1990) recorded IVI values ranging from 5–188 in a herbaceous layer in a tropical deciduous forest in Orissa where 24 species had been recorded. In the present investigation, the IVI values ranging from 5.28–86.82. However, least values were recorded in Dager's (1987) report in protected and grazed ecosystem at Ujjain, the IVI values ranges from 1.9–40.9 and 1.5–51.9. The value of IVI can be fluctuated because of the Alpine herbaceous layer ranging from 3.8 to 185.2 (Sundriyal and Joshi, 1990).

From the above conditions, it can be concluded that in herbaceous layer, the IVI values ranges from 5 to above. It is also noticed in the present investigation. So, the phytosociological study of the species of Verbenaceae family will be able to highlight the continuation, sustainability of rich bioresource by studying their distribution, uses, utility etc.

Acknowledgements

The authors are thankful to the Head, Department of Life Sciences, Manipur University, Canchipur for laboratory facilities. Thanks goes to BSI Hawrah and BSI Shillong for identification of species and to my colleagues, S. Padmabati, A. Ksheroda and S. Sangeeta for their valuable helps during the tenure of the programme.

References

Curtis, J.T. 1959. *The Vegetation of Wisconsin University*. Wisconsin Press, Madison.

Curtis, J.T. and Mc Intosh, R.P. 1950. The interrelation of certain analytic and synthetic phytosociological characters. *Ecology*, 31: 434–455.

Dagar, J.C. 1987. Species composition and plant biomass of an ungrazed and a grazed grassland at Ujjiain, India. *Trop. Ecol.*, 28(2): 208–215.

Deb, D.B. 1961. Dicotyledonous plants of Manipur Territory. *Bull. Bot. Sur. India*, Vol. 3: 314–315.

Devi, A.R. 1997. Comparative study of net primary production and nutrient status in the grasslands of Canchipur. *Ph.D. Thesis*. Manipur University, Canchipur.

Hooker, J.D.S. 1973. *The Flora of British India*. 4: 560–604.

Joshi, S.K., D.P. Pati and Balura N. 1990. Primary production of herbaceous layer in a tropical deciduous forest in Orissa, India. *Trop. Ecol.*, 31(2): 73–83.

Kanjilal, U.N. 1982. *Flora of Assam*. 3: 458–496.

Lawrence, G.H.M. 1978. *Taxonomy of Vascular Plants*. Oxford and IBH Publishing Co., New Delhi.

Misra, R. 1968. *Ecology Workbook*. Oxford and IBH Publ. Co., New Delhi.

Oosting, H.J. 1958. *The Studies of Plant Communities*. W.H. Freeman and Co., San Francisco, U.S.A.

Phillips, E.A. 1958. *Methods of Vegetation Study*. Henry Holt and Co. Inc., New York, USA.

Rath, S.P. and Misra, B.N. 1980. Effect of grazing on the floristic composition and life form of species in the grassland of Berhampur. *Indian J. For.*, 3(4): 336–339.

Saxena, N.B. and Saxena, S. 2001. *Plant Taxonomy*. Pragati Prakashan, Meerut.

Singh, N.P. Chauhan, A.S. and Mondal, M.S. 2002. *Flora of Manipur*. BSI, Braboume Road, Kolkata, 1: 8.

Sinha, S.C. 1996. *Medicinal Plants of Manipur*. MASS and MCIC Publication, Imphal.

Sundriyal, R.C. and Joshi, A.P. 1990. Effect of grazing on standing crop, productivity and efficiency of energy capture in an alpine grassland ecosystem at Tungnath (Garhwal Himalaya), India. *Trop. Ecol.*, 31(2): 84–97.

Trivedi, B.K. 1983. Architecture and above ground production of a grassland community at Jhansi (India). *Int. J. Ecol. Env. Sci.*, 9(3): 111–122.

Vedaja, S. 1998. *Manipur: Geography and Regional Development*. Rajesh Publications, New Delhi.

Chapter 41

Role of Zinc in Experimental Myocardial Infraction Influence of Abana: An Ayurvedic Formulation

C. Sheela Sasikumar and C.S. Shymala Devi***

**P.G. Department of Biochemistry, D.G. Vaishnav College, Chennai – 600 106*
***Head (Retd), Department of Biochemistry, University of Madras, Guindy Campus, Chennai – 600 025*

ABSTRACT

The present study is conducted to elucidate the role of zinc in isoproterenol induced myocardial Infarction. This work also reveals the cardioprotective effect of abana an Ayurvedic formulation in the pathogenesis induced by isoproterenol administration of abana maintained the levels of serum zinc and activities of marker enzymes to near normal.

Keywords: *Abana, Zinc, Myocardial infarction, Isoproterenol, Marker enzymes.*

Introduction

Zinc a membrane stabilizing agent is an essential component to several enzyme systems and a cation mainly bound to proteins. The availability of zinc controls tissue concentration and activity of zinc metallo-enzymes and also the rate of synthesis of nucleic acid and proteins (Singh, 1983).

Abana is a cardiotonic formulation containing extracts of several important medicine plants used in Indian system of medicine (Yeganarayan *et al.*, 1997). It possess anti-thrombotic, anti-hyper cholesterolaemic, antioxidant like activity (Sheela, 2000) properties (Tiwari, 1990).

Isoproterenol is a β-adrenergic agonist which has been reported to cause oxidative stress in the myocardium resulting in infarct like necrosis of heart muscle (Wexler, 1976). Isoproterenol induced myocardial infarction has been used as model for the evaluation of cardioprotective agents.

The present study was undertaken to screen the role of zinc in experimental myocardial infarction and to understand the protective effect of ayurvedic formulation abana.

Materials and Methods

Male albino rats of wistar strain, weighing 100–150 g were used for the study. They were fed with commercial pelleted rat chow and given water ad libitum. The rats were divided into 4 groups of 6 animals each. Group 1 served as control. Group 2 rats were administered isoproterenol (20 mg/100 g body wt, subcutaneously twice at an interval of 24 hrs) in saline. Group 3 rats received abana (75 mg/100 g body wt) orally for a period of 60 days. Group 4 rats were orally administered with abana at the above mentioned dosage for 60 days and isoproterenol (20 mg/100 g) at the end of experimental period. After the experimental period the animals were sacrificed by cervical decapitation. Blood was collected and serum separated was used for various biochemical estimations. Assay of creatine kinase (CK) by Okinaka *et al.* (1961) method and assay of aspartate aminotransferase (AST), alanine aminotransferase (ALT) and lactate dehydrogenase (LDH) by King (1965) method. Heart was dissected out immediately, washed in ice-cold saline and homogenised in 0.1 M Tris HCl buffer pH 7.4 and homogenate is used for the assay of all the above enzymes. Zinc level in serum and heart homogenate was determined by Atomic Absorption Spectroscopy, RSIC IIT, Chennai. Total protein was estimated by Lowry *et al.* (1951) method.

Statistical analysis was performed using students 't' test.

Results and Discussion

Isoproterenol administration resulted in significant decrease in serum zinc level which might have been due to necrosis produced in myocardium by isoproterenol (Table 41.1). A sharp decrease in serum zinc level has been shown to occur after myocardial infarction in both clinical and experimental conditions (Low, 1976). Significant correlation exist between the fall of serum zinc levels and enzymes like creatine kinase, lactate dehydrogenase and transaminases.

Table 41.1: Levels of Zinc in Serum and Heart of Control and Experimental Animals

Zinc	*Group*		*Description*	
	Group 1 (Control)	*Group 2 (Isoproterenol)*	*Group 3 (Abana)*	*Group 4 (Abana + Isoproterenol)*
Serum (mg/Litre)	1.28±0.06	0.78±0.02^{a***}	1.32±0.06^{aNS}	1.20±0.04^{b***}
Heart (μg/g wet tissue)	15.6±1.3	18.8±1.7^{a***}	14.4±1.3^{aNS}	14.6±1.2^{b***}

Values are expressed as mean ± SD for 6 animals in each group.

Statistically Significant Variations

As compared to Group 1: a

As compared to Group 2: b

p < 0.05; **: p < 0.01; ***: p < 0.001; NS: Non Significant.

Diagnostic markers of myocardial infraction are CK, LDH, AST and ALT. Isoproterenol administration caused a significant decrease in the activity of these enzymes in heart with concomitant increase in serum (Tables 41.2 and 41.3). An increase in the activities of marker enzymes could be due to leakage of enzymes from infarcted heart and the amount of enzymes appeared in serum is in proportion to the number of cells (Sheela, 2000). The observed decrease in serum zinc concentration might be due to the mobilization of zinc to the area of tissue injury to participate in myocardial reparative process and also to take part in the synthesis of zinc metallo–enzymes. Zinc had been reported to be an important element in wound healing and it is a factor essential in the biosynthesis and integrity of connective tissues (Lindman *et al.*, 1973).

Table 41.2: Activities of Aspartate Transaminase, Alanine Transaminase, Creatine Kinase and Lactate Dehyrogenase in Serum of Control and Experimental Animals

Parameters (IU/Litre)	*Group*		*Description*	
	Group 1 (Control)	*Group 2 (Isoproterenol)*	*Group 3 (Abana)*	*Group 4 (Abana + Isoproterenol)*
Aspartate transminase	26.9±2.2	46.2±3.43^{a***}	27.8±2.3^{aNS}	30.8±2.6^{b***}
Alanine transaminase	12.5±1.0	21.6±1.6^{a***}	12.1±0.8^{aNS}	14.1±1.2^{b***}
Creatine Kinase	287.8±9.3	389.5±11.73^{a***}	282.4±8.9^{aNS}	299.3±8.4^{b***}
Lactate dehydrogenase	75.6±5.4	123.6±7.6^{a***}	72.5±5.2^{aNS}	82.5±6.4^{b***}

Values are expressed as mean ± SD for 6 animals in each group.

Statistically Significant Variations
As compared to Group 1: a
As compared to Group 2: b

p < 0.05; **: p < 0.01; ***: p < 0.001; NS: Non Significant.

Table 41.3: Activities of Aspartate Transaminase, Alanine Transaminase, Creatine Kinase and Lactate Dehyrogenase in Heart of Control and Experimental Animals

Parameters (IU/Litre)	*Group*		*Description*	
	Group 1 (Control)	*Group 2 (Isoproterenol)*	*Group 3 (Abana)*	*Group 4 (Abana + Isoproterenol)*
Aspartate transminase (nmoles of pyruvate liberated/min/mg protein)	43.8±1.4	28.8±2.0^{a***}	42.1±1.2^{aNS}	40.5±1.9^{b***}
Alanine transaminase (nmoles of pyruvate liberated/min/mg protein)	18.2±1.5	12.1±0.83^{a***}	17.4±1.3^{aNS}	16.4±0.6^{b***}
Lactate dehydrogenase (nmoles of pyruvate liberated/min/mg protein)	114.2±7.2	82.9±5.3^{a***}	119.7±7.4^{aNS}	109.8±6.9^{b***}
Creatine Kinase (µmoles of phosphorous liberated/min/mg protein)	10.3±0.5	6.3±0.08^{a***}	9.8±0.48^{aNS}	9.2±0.5^{b***}

Values are expressed as mean ± SD for 6 animals in each group.

Statistically Significant Variations
As compared to Group 1: a
As compared to Group 2: b

p < 0.05; **: p < 0.01; ***: p < 0.001; NS: Non Significant.

Abana pretreated rats maintained the activities of diagnostic marker enzymes and zinc concentration in serum and heart to near normal which could be due to presence of various phytochemicals and minerals present in the formulation. *Terminalia arjuna, Withania somnifera Glycrrhiza glabra, Nardostachys jatamansi* present in the formulation are widely used cardio protective agents in ancient system of medicine (Khanna, 1991).

With all these findings it would be concluded that zinc play an important role in reparative process and ayurvedic formulation abana can afford significant protection to myocardium induced by isoprpterenol.

References

Khanna, A.K., Chander, R. and Kapoor, N.K. 1991. Hypolipidaemic activity of Abana in rats. *Fitoterapia*, 62(3): 271–275.

King, J. 1965. *Practical Clinical Enzymology*. D van Nostrand Co. Ltd., London, pp. 121.

Lindemna, R.D., Yunice, A.A., Bazter, D.J., Milter, L.R. and Nordgiust, J. 1973. Myocardial zinc metabolism in experimental myocardial infraction. *J. Lab. Clin. Med.*, 79: 452–457.

Low, W.I. and Ikram, H. 1976. Plasma zinc in acute myocardial infarction: Diagnostic and prognostic implications. *Br. Health J.*, 38: 1339–1342.

Lowry, O.H., Rosebrough, N.J., Farr, A.L. and Randall, R.J. 1951. Estimation of proteins by folin-phenol reagent. *J. Biol. Chem.*, 193: 265–275.

Okinaka, S., Kumagai, H., Ebashi, E., Sugaita, M. and Momoi, Y. 1961. Serum creatine phospho kinase activity in progressive muscular dystrophy and neuromuscular disease. *Arch Neurol.*, 4: 520–526.

Sheela Sasikumar, C. and Shyamala Devi, C.S. 2000 Effect of abana an ayurvedic formulation on lipid peroxidation in experimental myocardial infarction in rats. 38: 827–830.

Sheela Sasikumar, C. and Shyamala Devi, C.S. 2000. Protective effect of Abana a polyherbal formulation on isoproterenol induced myocardial infarction in rats. *Indian J. Pharmacol.*, 32: 198–201.

Singh, R. 1983. Serum zinc in myocardial infarction: Diagnostic and Prognostic significance. *Angiology*, 34: 215–219

Tiwari, A.K., Agarwal, A. Shukula, S.S. and Dubey, G.P. 1990. Favourable effects of Abana on lipoprotein profiles of patients with hypertension and Angina pectoris. *Alternative Med.*, 3: 139–142.

Wexler, B.C. and Greenberg, B.P. Protective effects of clofibrate on isoproterenol induced myocardial infarction in arteriosclerotic and non arteriosclerotic rats. *Atherosclerosis*, 29: 373–375.

Yeganarayan, R., Sangle, S.A., Sirsikar, S.S. and Mithra, D.K. 1997. Regression of cardiac hypertrophy in hypertensive patients-comparison of Abana with propanol. *Phytother. Res.*, 11(3): 257–259.

Chapter 42

Certain Herbal Ethno-medicines Used by the Karbi Tribe of Assam, India

*Sanjib Kumar Gogoi**, *Robindra Teron** and *Pratap J. Handique***

*Department of Botany, Diphu Government College, Diphu, Karbi Anglong Dist., Assam, India

**Department of Biotechnology, Gauhati University, Guwahati – 781 014, India

ABSTRACT

This article includes 30 indigenous and wild medicinal plants used by the Karbi tribe inhabited in the Karbi Anglong district of Assam (India). The plants/plant parts are used as medicinal preparation to cure a number of ailments. It is based on a through survey conducted during 2004–2005 for spot collection, identification and ethno-medico-botanical enumeration of indigenous plants utilized by the Karbi people. The plants have been collected with information on habit, habitat, place of collection, rarity status along with the local names and nature of their utilization. The collected plant species have been identified with latest taxonomic classification with the help of standard herbaria and have been preserved as herbarium specimens.

Keywords: *Karbi ethnomedicine, Karbi Anglong, Herbal medicine.*

Introduction

The North East India, particularly the hilly regions provide home for many species of wild plants with potent medicinal properties and higher economic value. Although a good number of plant species have been identified and reported from these regions, a good number of plant species found in

---

** *Corresponding Author.*

unexplored areas are yet to be reported. Karbi Anglong district is one such area which is not explored up to the fuller extent for medicinal plant wealth though a number of workers earlier reported several ethno-medicinal plants from this area (Borthakur, 1976; Chopra *et al.*, 1956; Handique *et al.*, 1987, Sarker *et al.*, 1989). Karbi Anglong is the major tribal dominated district of Assam. Karbi is a semi-nomadic tribe living in the Karbi Anglong district of Assam with their own age-old material culture. They have a clear understanding of the forest and various forest resources. They use a number of wild plant species to meet their various day-to-day requirement such as food, fiber and medicines. However, verbal communication inherited from one generation to the other is the only medium communication of their rich traditional knowledge.

Karbi Anglong district. It is a rain shadow area situated in 24.54°N to 26.41° and 92.45°E to 93.53°E. It occupies an area of about 10,3320 sq km. The altitude varies from 600 m in the north to 900 m in the south and in the valley 75 to 150 m. The average annual rainfall is 1041.40 mm. The maximum and minimum temperature ranges between 25.8°C and 17.2°C. Ridges and hilly terrain cover the major part of the district. Soil is red loamy and laterite. The hill soil is rich in nitrogen and organic matter but is mostly acidic. Alluvial soil is common in the valleys. The vegetation of the district is tropical dry deciduous forest with an admixture of semi-evergreen type.

During the field works plant species have been collected along with the detailed information on medicinal uses, local name, and distribution and rarity status. Available literature indicates that many of these plant species are not used at least for the same purpose elsewhere in India. (Saikia and Dey, 1954; Chopra *et al.*, 1956; Jain 1967, 1981).

In this article a total 30 plant species are enumerated which are being used for medicinal purposes by the people of Karbi tribe inhabited in the Karbi Anglong district of Assam, India). It is based on a thorough survey conducted during 2004–2005 for spot collection, identification and ethno–medico-botanical enumeration of indigenous plants utilized by the Karbi people.

Methodology

During the field trips, help was taken from the local medicine man to identify the plants of their use. The local name and uses of each species were collected on the spot. Habit, habitat and distribution status of each species was recorded based on the field observation. All the collected plant species were preserved as herbarium specimen. Identification and taxonomic classification of these plants have been made with the help of Flora of British India (Hooker, 1872–1879) and Flora of Assam (Kanjilal *et al.*, 1934–1940) and by consulting Kanjilal Herbarium at the Botanical Survey of India, Eastern Circle, Shillong. The name of the plant species are arranged alphabetically giving the name of families within parenthesis following the system of classification as of Bentham and Hooker with splitting families accepted internationally.

Results and Discussion

A total of 30 plant species have been collected from various places of Karbi Anglong District. These plants are being used mostly as medicine/medicinal preparation in the treatment of jaundice, dysentery, pneumonia, common-cough and wound healing. The name of the plants, their uses and rarity status are given in the Table 42.1. All the plant species reported here are naturally occurring in the Karbi Anglong district. Mentioned should be made that the Karbi people, particularly who live in the remote hill areas are solely dependent on herbal medicine. It was observed that the people never tried for cultivation of the wild medicinal plants in their homestead. Social belief is that the plant collected from wild is only effective as medicine.

Table 42.1: List of Medicinal Plants Used by the Karbis

Sl.No.	Name of the Plant	Local Name in Karbi	Traditional Uses	Distribution and Rarity Status
1.	*Alpinia gelanga* Willd. (Zingiberaceae)	Mahabir	Raw rhizome is given to cure fever, cough and flu	Found in Singhason range only
2.	*Alstonia scholaris* R.Br. (Apocyanaceae)	Thengmu	Latex used to cure mold	Common
3.	*Aristolochia saccata* Wall (Aristolochiaceae)	Gastric arekang	Soft stem is eaten to relieve gastritis problems	Common in hilly regions especially virgin forest
4.	*Cissus repanda* Vahl. (Vitaceae)	Samphat	Roots used for hair washing	Common in hilly regions
5.	*Cleyera japonica* (Ternstreoimiaceae)	Therangri	Roots are used as fish poison	Hamren sub-division
6.	*Coffea bengalensis* (Rubiaceae)	Mirherai	Fruits and leaves are used for curing dysentery in infants	Common
7.	*Crassocephalum crepidioides* (Benth). S. Moore (Asteraceae)	Delon	Leaves and roots heal wounds effectively	Common in Singhason forest
8.	*Croton caudatus* Geiseler. (Euphorbiaceae)	Kung Kung	Sap of stem used for curing tongue sore	Common
9.	*Cyanotis vaga* (Lour). J.A. and J.W. Schult. (Commelinaceae)	Birlap	Aerial portion is cooked and eaten to cure stomach upset	Common
10.	*Derriea cuneifolia* Benth. (Papilionaceae)	Hisu	Root paste used in river water for poisoning of fish	Rare, found on hill slopes
11.	*Elastostema platyphylla* (Urticaceae)	Hanthai	Leaves used as hair conditioner and vegetable	Common in forest
12.	*Garcinia anomala* Plach. (Clusiaceae)	Praspri	Large yellow fruits are eaten to cure stomach upset	Rare
13.	*Grewia multiflora* Mast. (Tiliaceae)	Singnam longdak	Bark used as hair conditioner	Rare, found in Hamren sub-division
14.	*Gynocadia odorata* R.Br. (Flacourtiaceae)	Thebongkok	Powder of peri-carp kills or cause temporary unconsciousness to bees	Common in hilly region
15.	*Hedyotis scandens* Roub. (Rubiaceae)	Be akeng Kung	Decoction of leave when given to pregnant women removes complexity during childbirth	Rare and found in Singhason range
16.	*Hibiscus sabdariffa* Linn. (Malvaceae)	Hanserong	Leaves and flowers are used as antidote for poison	Common throughout the area
17.	*Impatiens parviflora* DC. (Balsaminaceae)	Tengnap	Paste of shoots cure wounds effectively	Common in hill area
18.	*Kaempferia galanga* Linn. (Zingiberaceae)	Bithiphaknur	Rhizome used for curing dog and pig bite	Rare, Found only in Singhason hill
19.	*Laportea crenulata* Gaudich. (Urticaceae)	Bapkangsam	Roots used for curing worms in dogs	Common throughout the area
20.	*Lygodium japonieum* (Thunb) Sw. (Schzeaceae)	Banchek	Paste of shoots cure wounds effectively	Common throughout the area

Contd...

Table 42.1–Contd...

Sl.No.	Name of the Plant	Local Name in Karbi	Traditional Uses	Distribution and Rarity Status
21.	*Marlea bigonniifolius* Roxb. (Alangiaceae)	Himipi a-ingke-an	Whole plant is eaten to cure pneumonia	Rare, Found in Singhason forest
22.	*Mirabilis jalapa* Linn. (Nytaginaceae)	Hunmili amir	Paste of flowers used for removing moulds	Cultivated
23.	*Orozylum indicum* Vent. (Bignoniaceae)	Nopak ban	Bark used to cure jaundice	Common throughout the area
24.	*Plectranthus incanus* Link. (Lamiaceae)	Rongpang reho	Leaves consumed to cure dysentery	Common in hilly regions
25.	*Premna latifolia* Roxb. (Verbenaceae)	Ankelok arongk	Leaves used for killing parasites for poultry	Distribution restricted
26.	*Smilax ocreata* A.DC. (Smilacaceae)	Phelangtung	Roots used for curing urinary problems	Common in forest
27.	*Spilanthes clava* DC. (Astecaceae)	Eso Keso abap	Inflorescence relieves toothache	Common throughout the area
28.	*Thunbergia grandiflora* Roxb. (Acanthaceae)	Nong-Nong	Antidote for snake bites and also used for curing eye sore	Common throughout the area
29.	*Vernonia cineria* Less. (Asteraceae)	Bang aram abap	Used as antidote for snake bike	Found in Singhason forest
30.	*Wendlandia coriacea* DC. (Rubuaceae)	Genterlong	Seeds are eaten to cure stomach upset	

Acknowledgement

We sincerely thank the Principal, Diphu Government College, Diphu and the Head, Department of Biotechnology, Gauhati University for providing necessary facilities. Thanks are due to all the village medicine-man and local inhabitants for providing information on medicinal uses of the collected plant species.

References

Borthakur, S. 1981. Less known medicinal uses of plants among the tribes of Karbi Anglong (Mikir Hills), Assam. *Bull. Bot. Surv. India*, 18(1–4): 166-171.

Chopra, R.N., Nayar, S.L. and Chopra, I.C. 1956. *Glossary of Indian Medicinal Plants* (Reprint 1986). CSIR, New Delhi.

Handique, P.J., Medhi, K.K., Goswami, P.K., Goswami, L.C., and Chowdhury, S. 1987. A preliminary study on the utilization of indigenous plants of Karbi Anglong district of Assam. *J. Assam Sci.*, 29(2): 8–15.

Hooker, J.D. 1872–1897. *Flora of British India*, Vols. I–VII, London.

Jain, S.K. 1967. Ethnobiology: Its scope and study. *Ind. Mus. Bull.*, 2(1): 39–43.

Jain, S.K. 1981. *Glimpses of Indian Ethnobotany*, New Delhi.

Kanjilal, U.N., Kanjilal, P.C., Das, A., De, R.N. and Bore, N.L. (1934–40). *Flora of Assam*, 5 Vols. Government Press, Shillong.

Saikia, M.L. and Dey, R.N. 1954. *Preliminary List of Medicinal Plants and Herbs Found in Assam*. Assam Govt Publ.

Sarker, S., Handique, P.J., Goswami, P.K., Ahmed, A., Goswami, L.C., and Chowdhury, S. 1987. A study on the utilization of indigenous plants of Karbi Anglong district of Assam: II–Food and vegetables, medicinal plants. *J. Assam Sci.*, 31(2): 43–54.

Chapter 43

Effect of Multifloral Honey on Blood Glucose Profile of Rabbits after Induced Lipidosis

Shashikala, D. Belsare and B. Jayanti

Biochemistry Laboratory, Department of Bioscience, Barkatullah University Bhopal, India

ABSTRACT

The saturated fat fed rabbits develop hypoglycemia and glucose intolerance. The animals tolerated blood glucose load after dietary administration of honey and did not develop hypoglycemia. It is suggested that honey contains glucose tolerance factor (GTF) that restores hypoglycemia to normal blood sugar level in saturated fat fed animals.

Keywords: *Saturated fat diet, Hypoglycemia, Honey, Glucose tolerance factor.*

Introduction

The rabbits given hydrogenated saturated fat in diet develop lipidosis (Belsare, 1980) as well as glucose intolerance during fat mobilization by ACTH (Belsare, 1981). The elevated level of plasma free fatty acids (FFA), which occur during it, may be secondary to the increased utilization of lipid, a metabolic condition that diminishes peripheral oxidation of glucose (Randle *et al.*, 1963). Treatment with honey prevents hepatocellular damage (Lutomski, 1987) which is due to bioelements (Dobrowolski, 1987) present in honey. More recent studies indicate that sugars present in honey do not enter cells by osmosis, but require' active carrier' to move them across the membrane barrier which means less of a 'rush' of sugar to the body with honey (Ladas and Rapitis, 1999). In the present investigation attempts have been made to study the effect of honey on blood glucose profile during dietary-induced lipidosis (fatty liver) in rabbits.

Materials and Methods

Albino rabbits of about 1 to 1.5 kg/BW were used in the present experiment. Eighteen animals were divided in 3 groups of 6 each and were used for blood glucose profile. Six animals of group I were maintained on standard diet and six of group 2 was given saturated fat diet (casein-vitamin free 20 per cent, starch 10 per cent sucrose 28 per cent, hydrogenated saturated fat 10 per cent, Brewer's yeast dried 10 per cent, 10 units of vitamin A and I unit of vitamin D per g of diet). Six animals were given saturated fat diet with honey (1.5g/kg). The animals were kept in cages individually with free access to water *ad lib.*

The peripheral blood was drawn from marginal ear vein for glucose estimation according to Nelson (1944) and Somogui's (1945) methods.

At the end of experiment the animals were fasted overnight and a loading dose of glucose, (0.5 g/kg) was administered by intravenous route as reported earlier (Belsare, 1981). The honey used in these experiment was collected from natural beehives of Rock-bee, *Apis dorsata* from the Pachmarhi forest. The saturated fat was purchased from Hindustan Lever Co., India. Administration of quantity of honey in diet of animals (per kg BW) was calculated from dietary requirement for human (100 g/60 kg /day). Significance of data was determined by Student's t-test.

Observations

One week after treatment the blood glucose level in control animals was 98 mg/dl and showed marginal increase six weeks after the treatment, but in saturated fat fed animals it showed significant decrease in its level as compared to the initial control one (Table 43.1). The dietary treatment with honey with saturated fat diet for six weeks resulted in restoration of the blood glucose level to its normal one.

Table 43.1: Blood Glucose Level (mg/dl) in Rabbits

Group	Duration in Weeks					
	1	2	3	4	5	6
Controls	97.8±1.4	98.2±2.3	96.9±2.1	98.0±3.0	95.1±1.5	97.8±2.5
Saturated fat fed animals	97.5±3.4	96.2±2.8	*88.0±4.2	*78.0±4.5	*68.0±3.8	*60.7±3.5
Honey + Saturated fat fed animals	97.7±3.1	97.8±2.9	97.0±3.1	97.2±6.2	97.1±4.8	97.8±6.2

*: Mean ± S.E.; $P < 0.001$.

After loading oral dose of glucose, the blood glucose level was significantly increased in control as well as in honey treated animals within one hour, but it reached the initial level three hours after treatment in controls (Figure 43.1), However, in the saturated fat–fed animals, it was also increased within one hour, but reached initial level five hours after giving a loading dose of glucose. In honey and saturated fat fed diet, the animals tolerated loading dose of glucose like those, which were maintained on normal diet.

Discussion

There is tendency to hypoglycemia in saturated fat-fed animals which is due to decrease in oxaloacetate and thus inhibits gluconeogenesis (Horst *et al.*, 1961). The honey treatment restores this hypoglycemia to normal condition, because honey enters the cells directly with the help of 'active

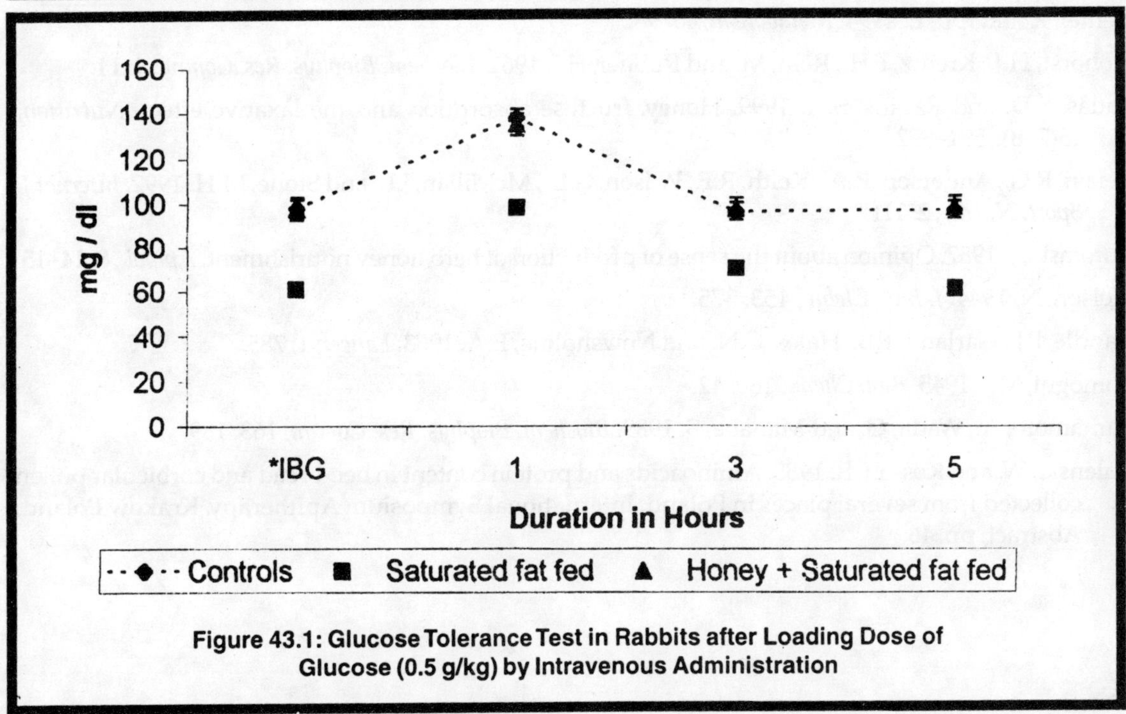

Figure 43.1: Glucose Tolerance Test in Rabbits after Loading Dose of Glucose (0.5 g/kg) by Intravenous Administration

factor'. This factor was postulated as glucose tolerence factor of which trivalent chromium was identified as its active ingredient (Yamamoto *et al.*, 1989). Lefavi *et al.* (1999) reviewed the literature and concluded that chromium plays a necessary role in the normal physiological system by acting as a "cofactor' for insulin and amplify its action as well as the efficiency of insulin' s effects on glucose, amino acid and fatty acid flux into cell with subsequent glycogen, protein and triglyceride synthesis which is dependent on the maintenance of adequate chromium store. Although the exact structure of GTF and its exact location is still unknown (Anderson, 1987), a nicotinic acid-chromium is essential in the GTF complex (Lefavi *et al.*, 1999). Honey contains nicotinic acid, glycine, glutamic acid and cysteine (Zelenski and Kossan, 1985), which are essential for the synthesis of GTF complex. There are several reports that administration of honey in animals and human restores hypoglycemia to normal state (Genter and Ipp, 1994).

References

Anderson, R.A.1987. In: *Trace Elements and Human Nutrition,* (Ed.) W. Mertz. New York, Academic Press, pp. 225.

Belsare, S.D. 1980. Chronic effects of some drugs on histopathology and lipid content of liver in rabbits. Z. Tierphysiol. Tierernahrg. u. Futtermittelkde. 45: 245–250.

Belsare, S.D. 1981. Studies on glucose tolerance during fat mobilization: Influence of heparin and salicylate. *J. Expt. Biol.*, 19: 88–89.

Dobrowolski, J.W. 1987. Bee made products, environmental protection, prophylaxis of diseases caused by deficiency of elements. *Apipol.* 8: 4–11.

Genter, P. and Ipp, E. 1994. *Metabolism*, 43: 98.

Hohorst, H.J., Krentz, F.H., Rein, M. and Rubner, H.J. 1961. *Biochem. Biophys. Res. Comm.*, 4: 14.

Ladas, S.D. and Rapitis, S.A. 1999. Honey, fructose absorption and the laxative effect. *Nutrition*, 15(7–8): 591–592.

Lefavi, R.G., Anderson, R.A., Keith, R.F., Wilson, G.D., McMillan, J.L. and Stone, M.H. 1992. *Internat J. Sport. Nutrit.*, 2: 111

Lutomski, J. 1987. Opinion about the sense of production of herb honey nourishment. *Apipol.*, 8: 14–15.

Nelson, N. 1944. *J. Biol. Chem.*, 153: 375.

Randle, P.J., Garland, P.B., Hales. C.N. and Newsholme, E.A. 1963. *Lancet*, 1: 785.

Somogui, M.J. 1945. *Biol. Chem.*, 160: 42.

Yamamoto, A., Wada, O. and Manabe, S. 1989. *Biochem. Biophys. Res. Comm.*, 163: 189.

Zalenski, W. and Kossan, R. 1985. Aminoacids and protein content in bee-bread and corbicular pollen collected from several places in Poland. International Symposium Apitherapy. Krakow Poland; Abstract, pp. 46.

Chapter 44
Role of Soft Computing and Neural Network in Medical Analysis

J. Justin Anand, M.S. Durairajan*, J. Justin Suresh** and P. Dhasarathan****

**Department of Computer Science, Sri Kaliswari College, Sivakasi – 626 130*
***Department of Computer Science, Adaikalamatha College, Vallam, Thanjavur*
****Microbial Biotech Unit, Department of Biotechnology, Sri Kaliswari College, Sivakasi – 626 130*

ABSTRACT

In the digital age, the computer applications are penetrated into all sectors of the knowledge based society, especially in the field of medicine. Many problems in real life domains are always accompanied by imprecision or uncertainly factors. Sometimes the information is far from complete, but a precision decision is required. In such situation, human can make a correct decision, while computer requires complete calculation to make a mathematical model of the problems. The solutions to problems with their properties are a family of algorithm named "Soft Computing". In general, Medicine has always benefited from the Forefront Technology. Applying the Soft Computing Techniques to the Medical Images has boosted the medicine to the extraordinary levels of achievements. Neural Networks, a model of the brain artificially connects many nonlinear neuron model and processes information in a parallel-distributed manner. The Neural Network has many characteristics such as nonlinear mapping, parallel processing learning, and self–organization. It is applied to pattern recognition, control and so on. The Neural Network that consists of three layers (input, output layers and one hidden layer) is able to express any functions while using enough hidden units. The present study analyzed medical image using soft computing techniques and artificial neural network.

Keywords: Neural network, Soft computing, Algorithm and Medical image.

Introduction

Medical Imaging in the broadest sense encompasses the key diagnostic techniques on which modem medical treatment depends (Gerig *et al.*, 1991). The information carried in these images is crucial to treatment, Cardiology, Neurology, Surgery, Obstetrics, Orthopedics, and Pulmonary medicine. The overall objective of the ICMIT is to develop and implement medical imaging technology, which will lead to improved diagnosis and health care delivery as well as reduction in costs as well (Hillman *et al.*, 1995). Medical image analysis is a rapidly expanding area that brings together experts from a number of disciplines. Image processing and analysis is required to process, interpret and understand the data, requiring familiarity with computer vision techniques.

Soft Computing is an emerging approach to computing which parallels the remarkable ability of the human mind to reason and learn in environment of uncertainty and imprecision. Soft Computing differs from conventional (hart) computing in that, unlike hard computing, it is tolerant of imprecision, uncertainty and partial truth. In effect, the role model for soft computing is the human mind (Namasivayam and Hall, 1996). An image analyzer comprises an image acquisition device, a means for converting the image to digital form, and software/hardware to process the data in order to extract the desired information from it. Neural Networks have immense ability to learn complex and non-linear relationships including noisy or less precise information. Neural Networks are well suited for solving problems in biomedical engineering and in analysis of biomedical signals.

Materials and Methods

Quantification of Diabetic Retinopathy Using Neural Networks and Sensitivity Analysis

The design of neural network classifiers for the identification of diabetic retinopathy is discussed. Red-free digitized fondle images are tiled, and a neural network is trained to distinguish exudates from drusen. By quantifying the degree of retinopathy, the approach can be used to screen diabetic patients for referral. A novel form of hierarchical feature selection using sensitivity analysis is presented. The resulting neural network is compact, and achieves 91 per cent sensitivity and specificity on a test set (Kikinis *et al.*, 1992).

Artificial Neural Networks was to understand and shape the functional characteristics and computational properties of the brain when it performs cognitive processes such as sensorial perception, concept categorization, concept association and learning. However, today a great deal of effort is focused on the development of neural networks for applications such as pattern recognition and classification, data compression and optimization. There are several applications are used for expand in Medical Images (Kischell *et al.*, 1995).

Results and Discussion

Soft Computing is an emerging approach to computing which parallels the remarkable ability of the human mind to reason and learn in environment of uncertainty and imprecision. Image Analysis with Genetic Programming applied to Brain Images, the algorithm gives impressive results in comparison to the ones obtained on the same images by a simple Neural Network (Velthuizen *et al.*, 1993).

The detection of specific features in medical images is often a key support for diagnosis. Taking advantage of large bases of images where features of interest have been localized by clinicians, a modular system has been developed to spot similar features on new images. Images are scanned through a window, the size of which being previously fitted to the feature of interest (Leeds and Jackson, 1994). The recognition process involves a coding phase followed by a classification phase. These phases rely on unsupervised and supervised learning respectively for their implementation.

This application presents a computer-based decision support system for automated interpretation of diagnostic heart images, which is made available via the Internet. The system is based on image processing. techniques, artificial neural networks, and large and well-validated medical databases (Taxt and Lundervold, 1994). The performance of the neural networks detecting infarct and ischemia in different parts of the heart, measured as areas under the receiver operating characteristic curves, was in the range 0.76–0.92. These results indicate a high potential for the tool as a clinical decision support system. The system is currently evaluated by a group of pilot users in different European countries.

Neural Network has been successfully implemented in many applications including medicine. Neural Network, which simulates the function of human biological neuron, has potential of ease implementation in many applications domain. The main consideration of Neural Network implementation is the input data (Taxt and Lundervold, 1994a). Once the network is train, the knowledge could be applied to all cases including the new cases in the domain. Studies have shown that NN predictive capability is a useful capability in medical application. Such capability could be used to predict patient condition based on the history cases. The prediction could help doctor to plan for a better medication and provide the patient with early diagnosis.

Acknowledgement

The authors are thankful to Thiru A.P. Selvarajan, Secretary, the Principal and the Management of Sri Kaliswari College for providing facilities to carry out the work.

References

Gerig, J. Martin and R. Kikinis. 1991. Automating segmentation of dual-echo MR head data. *The 12th International Conference of Information Processing in Medical Imaging* (IPMI).

Hillman, C. Chang, H. Ying, *et al.* 1995. Automatic system for brain MRI: Analysis using a novel combination of fuzzy rule based and automatic clustering techniques. *Medical Imaging, Image Processing*, p. 16–25.

Kikinis, M. Shenton and G. Gerig. 1992. Routine quantitative analysis of brain and cerebrospinal fluid spaces with MR imaging. *JMRI*, 2: 619–629.

Kischell, N. Kehtarnavaz, G. Hillman, H. Levin, Lilly, M. and T. Kent. 1995. Classification of brain compartments and head injury lesions by neural networks applied to MRI. *Neuroradiology*, 37: 535–541.

Leeds, N. and E. Jackson. 1994. Current imaging techniques for the evaluation of brain neoplasms. *Curr. Sci.*, 6: 254–261.

Namasivayam and L. Hall. 1996. Integrating fuzzy rules into the fast, robust segmentation of magnetic resonance images. *New Frontiers in Fuzzy Logic and Soft Computing Biennidl Conference of the North American Fuzzy Information Processing Society, NAFIPS*, p. 23–271.

Taxt, T. and A. Lundervold. 1994. Multispectral analysis of the brain in magnetic resonance imaging. *IEEE Workshop on Biomedical Image Analysis*, Los Alamitos, CA, USA, p. 33–42.

Taxt, T. and A. Lundervold. 1994a. Multispectral analysis of the brain using magnetic resonance imaging. *IEEE TMI*, 13: 470–481.

Velthuizen, R., Hall, L. and L. Clarke. 1993. Unsupervised fuzzy segmentation of 3D magnetic resonance brain images. *Proceedings of the IS and TSPIE 1993 International Symposium on Electronic Images: Science and Technology*, 1905: 627–635.

Chapter 45
Symploca: A New Report

*P.R. Sushama, Mary Esther Cynthia Johnson and P. Surekha Rani**

Associate Professor in Botany, Bydrobiology Laboratory, Botany Department, Osmania University College for Women, Koti, Hyderabad – 500 195, A.P., India

ABSTRACT

***Symploca* is a blue green alga. Trichome single in a thin sheath, filaments at first prostrate, later mostly forming erect bundles, sheath form or later gelatinizing, end cell not capitate. Cells longer than broad. Filaments mostly in erect bundles.**

Keywrods:* *Simploca, Hormogone, Damp wall.

Introduction

This alga was observed growing as blue green patches on the wall of residential house, whenever there is a white wash on the wall this alga is not vanishing but growing again and again on the wall. It shows its withstanding in adverse conditions. Withstanding in adverse conditions is a common phenomenon in cyanophyceae members it might be because of their vegetative mode of reproduction, an hormogone. Earlier *Symploca* was reported on submerged logs of tree, salt lakes in Kolkata, on wet rocks and damp tree trunks. Symploca was found growing on submerged parts of a plant from Borivli near Mumbai. It was also reported from the place Shembaganur and Skuja (1949) from Rangoon. Desikachary (1958) described this alga in his book Cyanophyta. Hormogone a fragment of trichome can divide and redivide and develop into a trichome.

Materials and Methods

The algal sample collected by scrapping the surface of the wall in to a polythene bag and examined in the laboratory.

* *Corresponding Author.*

Results and Discussion

Symploca belongs to Oscillatoriceae family. Their filaments bent, forming a thin prostrate or creeping thallus, seldom erect, in bundles or forming a gelatinous *Phormidium* like thallus blue green or blackish, sheath thin, firm or more or less gelatinous, up to 2 μ thick, coloured violet by chlor–zinc–iodide, trichomes not constricted at the cross walls, cross-walls not granulated, cells 5-8 μ long, end cell mostly broadly rounded, with slightly thickened outer membrane.

Trichomes simple in a narrow thin sheath. Sheath firm or later gelatinizes. Trichomes straight, sometimes slightly attenuated, end cell not capitate, some times with a thickened outer membrane. Four species are included by Kutzing. Cells longer than broad, shows typical cyanophycean cell structure *i.e.*, prokaryotic nature. The protoplast is clearly divisible into two parts, peripheral pigmented region, 'chromatopasm', and a central colourless region, the 'centroplasm'. In the chromatoplasm are found large refractive granules, cyanophycin granules, and these are of the nature of reserve food materials. Chlorophyll is known to occur in the blue green algae. It has been identified to be chlorophyll a, chlorophyll b is absent. C-phycocyanin and C-phycoerythrin present and is mainly responsible for the blue colour of the alga. Nuclear material is present in the colourless centroplasm, the central body lacks a nuclear membrane and the nucleoli. Mode of reproduction is vegetative propagation by hormogone formation. Hormogones are fragments of trichome, formed by the death of one or more intercalary cells or by the formation of special biconcave separation discs, on detaching from trichome they develop into anew trichome of *Symploca* by repeated divisions. In *Symploca* hormogones move slowly. Honnogone formation is the main feature seen in Oscillatoriaceae members as sexual reproduction is absent.

Acknowledgement

The authors thank the Head and staff Department of Botany, Osmania University, College for Women, Koti, Hyderabad for the laboratory facilities and help rendered.

References

Desikachari, T.V. 1958. *Cyanophyta*. Indian Council of Agricultural Research, New Delhi, p. 1–685.

Skuja, H. 1949. Zur Susswasseralgen flora Burmas. Nova Acta Reg. Soc. Sci. Upsal. Ser., 4,14(5): 1–188.

Chapter 46

Effect of Levels of Sodium Alginate on Chemical Quality of Dahi

P.G. Chaudhari, D.K. Kamble**, A.D. Kale*** and B.K. Pawar*****

**Post Graduate Student, **Assistant Professor of Animal Science and Dairy Science,*
****Ex. Deputy Director of Research (Animal Science and Dairy Science),*
*****Assistant Professor of Animal Science and Dairy Science*
Department of Animal Science and Dairy Science, Mahatma Phule Krishi Vidyapeeth,
Rahuri – 413 722, Distt. Ahmednagar, M.S.

ABSTRACT

The present investigation entitled, "Effect of levels of sodium alginate on chemical quality of dahi" was undertaken in the laboratory of the Department of Animal Science and Dairy Science, Post Graduate Institute, M.P.K.V., Rahuri during the year 2001–2002. The composite sample of homogenized cow and buffalo milk, were collected from private producers. In the present investigation attempt were made to incorporate stabilizer in milk for the production of quality dahi, so as to gain desired consistency to the product. Three stabilizers *viz.*, sodium alginate, gelatin and pectin were examined. From these, sodium alginate was selected since it completely stopped syneresis of dahi. Three levels of sodium alginate viz; control (S_1), 0.1 per cent (S_2) and 0.2 per cent (S_3) were included in the study. Dahi was prepared using the standard method under laboratory condition. The dahi samples from above treatment were analysed for titratable acidity, pH., fat content and total solid content as per IS: 1479, (Part–I), 1960, IS: 1479 (Part–II), 1961 by Elico pH meter, IS: 1224 (Part–I), 1977 and IS: 1419 (Part–II) 1961 by gravimetric method, respectively. The effect of levels of sodium alginate on development of acidity, pH and total solid content was significant while non-significant on fat content.

Keywords: *Sodium alginate, Dahi, Chemical quality.*

Introduction

The production of fermented milk products is rapidly increasing in all major developing countries of the world. They are known throughout the world for their taste, nutritive values and therapeutic properties. Dahi is one of the important Indian fermented milk product. The importance of dahi in a daily diet of Indian people has been well recognized from the vedic times. Its curative effect against gastro-intestinal disorder has been utilized by many civilization (Sinha and Sinha, 2000).

In India, 7.1 per cent of total annual milk production is utilized for dahi making. (Shreshta and Gupta, 2001). Dahi has been recommended for curing dyspepsia and dysentery also has anticholesterolemic, anticarcinogenic value. It is consumed directly as a end product or utilized as a intermediate product in the preparation of butter, ghee, lassi, shrikhand etc.

Until now preparation of dahi is mostly conceived to individual household and sweet shops without paying much attention on its quality. In order to have a production of uniform quality this has also led to search for cheap stabilizers that provide the necessary characteristics (binding, gelling and thickening) to the preserved product.

Several stabilizers like gelatin, sodium alginate, hydrocolloids, gums, starches and other derivatives have been tried for preparing fermented milk products to improve their body and texture characteristics and reduce wheying off (Gupta and Prasad, 2000).

In the present investigation the attempts were made to incorporate stabilizer in milk for the production of quality dahi, so as to gain desired consistency of product.

Materials and Methods

The present investigation entitled, "Effect of levels of sodium alginate on chemical quality of dahi" was carried out in laboratory of Department of Animal Science and Dairy Science, Post Graduate Institute, Mahatma Phule Krishi Vidyapeeth, Rahuri, during the year 2001–2002.

The composite samples of cow and buffalo milk were procured separately from private dairy producers. For each trial, 200g of cow and buffalo milk was used separately for dahi preparation. A.R. grade chemicals manufacture by Glaxo Laboratories (India) Ltd., Mumbai, were used for chemical analysis. The milk samples were heated to 85–90°C for 15 min. and then cooled to 30±1°C before inoculating.

Sodium alginate was used as a stabilizer to increase the consistency and firmness of dahi. It is dissolved in small quantity of distilled water and then added to each sample under optimal stirring. Three levels of stabilizer *i.e.* control (S_1), 0.01 per cent (S_2) and 0.2 per cent (S_3) were included in the experimental trials. After setting dahi, samples were transferred and maintained at 5±1°C, till they were taken out for examination. The curd samples were brought at ambient temperature for their analysis.

The chemical analysis of fat, titratable acidity, pH and total solids were carried out as per IS: 1224 (Part–I), 1977; IS: 1479 (part–I), 1960; IS: 1479 (part–II), 1961, by Elico pH meter and IS: 1479 (part–II), 1961, by gravimetric method, respectively.

The Randomized Block Design with 3 replications was adopted to study the different aspects. The day on which the various observations were recorded was considered as one replication.

Results and Discussion

It was seen from the results that there was significantly ($P < 0.05$) higher acidity in cow (0.942 per cent L.A.) and buffalo (0.988 per cent L.A.) milk dahi prepared without sodium alginate. Incorporation of sodium alginate had resulted in significant decrease in the total titritable acidity with its increasing levels. These results were corroborated with the findings of Baisya and Bose (1974) and Jogdand *et al.* (1991). It might be either due to interference of this stabilizer with the metabolic path ways of the micro-organism or due to higher levels affecting the growth of starter culture resulting low production of these acids.

The results depicted in Table 46.1 indicate that the effect of addition of sodium alginate on pH of dahi was also highly significant. Dahi obtained from cow milk without addition of sodium alginate had significantly lower pH (4.80) than the dahi with 0.1 per cent sodium alginate (4.86) and with 0.2 per cent sodium alginate (4.92). Similarly, the dahi prepared from buffalo milk without sodium alginate had significantly lower pH (4.76) than the dahi prepared with 0.1 per cent sodium alginate (4.80) and dahi with 0.2 per cent sodium alginate (4.84).

Table 46.1: Effect of Levels of Sodium Alginate on Chemical Quality of Cow and Buffalo Milk Dahi

Chemical Qualities	*Level of Sodium Alginate*					
	S_1	S_2	S_3	*Mean*	*SE*	*C.D. at 5%*
(A) Cow milk						
Titratable acidity	0.942	0.884	0.834	0.887	0.006	0.018**
pH	4.803	4.860	4.928	4.864	0.005	0.016**
Fat %	3.779	3.768	3.773	3.773	0.006	N.S.
Total solids %	12.42	12.75	13.05	12.74	0.033	0.096**
(B) Buffalo milk						
Titratable acidity	0.988	0.940	0.897	0.942	0.009	0.025**
pH	4.76	4.80	4.84	4.80	0.013	0.039**
Fat %	5.273	5.274	5.278	5.275	0.003	N.S.
Total Solids %	**15.30**	**15.78**	**16.15**	**15.74**	**0.026**	**0.096****

The addition of sodium alginate resulted non-significant difference in fat content of dahi prepared from cow milk as well as buffalo milk.

From the result of present study it was also found that there was a significantly higher total solids in dahi prepared from cow milk (13.05 per cent) and buffalo milk (16.15 per cent) by addition of 0.2 per cent sodium alginate. Incorporation of sodium alginate has resulted in significant increase in total solids content of dahi with increasing rate of addition.

Total solids content in dahi increases with the concentration of sodium alginate. This might be due to the swelling action of stabilizers, which binds free water into "bound water" thus, concentrating the solids in the mixture into less water.

References

Baisya, R.K. and Bose, A.N. 1974. Effect of different salts and chemical additives upon quality of curd. *J. Food Sci. Technol.*, 11: 70–73.

Gupta, A. and Prasad, D.N. 2000. Use of stabilizers in cultured milk products. *Indian Dairyman*, 52(9): 19–24.

IS: 1224. 1977. *Determination of Fat by Gerber Method*, Part–I. Indian Standard Institute, Manak Bhawan, New Delhi.

IS: 1479. 1960. *Methods of Test of Dairy Industry*, Part–I: Determination of acidity in milk. Indian Standard Institute, Manak Bhawan, New Delhi, pp. 29.

IS: 1479. 1961. *Methods of Test for Dairy Industry*, Part–II. Indian Standard Institute, Manak Bhawan, New Delhi, pp. 25.

IS: 1479. 1961. *Methods of Test of Dairy Industry*, Part–II. Determination of pH in milk, Indian Standard Institute, Manak Bhawan, New Delhi.

Jogdand, S.B., Lembhe, A.F. and Ambedkar, R. K. 1991. Incorporation of additives to improve the quality of dahi. *Indian J. Dairy Sci.*, 44(7): 459–460.

Shreshta, R.G. and Gupta, S.K. 2001. Dahi from sweet cream buttermilk. In: *Processing Technology of Milk Products, Dairy Yearbook*, 2nd edition, p. 143–144.

Sinha, P.R. and Sinha, R.N. 2000. Importance of good quality dahi in food. *Indian Dairyman*, 52(7): 45–48.

Chapter 47

Potential Difference Across the Cell of Skin around the Transmission Tower

Vijay Kumar, Sachin Goyal, R.P. Vats and Tripti Johri

Department of Physics, M.S. College, Saharanpur – 247 001, Uttar Pradesh, India

ABSTRACT

The variation of potential difference across the spherical cells of a biological body due to external electric field is being studied. It is found that this potential difference caused by radiation field of transmission towers may be hazardous for the people nearby.

***Keywords:** Potential difference, Electromagnetic field (EMF), Spherical cells of body.*

Introduction

Our atmosphere has natural and manmade both types electromagnetic field (EMF). The effects of manmade EMF, which is transmitted from the transmission towers of radio, TV and mobile communication systems, are studied in this article. A biological body is homogeneous, whose macroscopic electrical properties are described by the complex permittivity. Interaction of EMF with biological tissues or cells is complex function of numerous parameters. The distribution of EMF inside the body depends upon the dielectric constant, conductivity, and geometrical factors that describe the tissue structure, of biological material. Potential difference across every cell or tissue is different due to different shapes and sizes of tissue of cell. When high frequency EMF penetrates inside the body, it decreases and a voltage gradient is produced inside the body. Thus, the potential difference across every spherical cell of same radius is different. Olsen (1990), Gajsek *et al.* (2001), Litwak *et al.* (2002), Kumar *et al.* (2003) and Pathak *et al.* (2003) make earlier studies on the effects of RF radiation on biological body.

Materials and Methods

High frequency electric field is present around the transmission tower. When the electric field penetrates inside the body, a potential difference is produced across the body-cells. The power of Massorie and Delhi TV transmission tower is 10 kW. They transmit at frequencies of 41 MHz and 202 MHz. These radiations produce an electric field around the tower. If this induced electric field penetrates inside the body, it decreases exponentially (Pathak *et al.*, 2003). The penetrated electric field at a depth z inside the body is given by:

$$E = E_0 \exp(-z/\delta) \quad (1)$$

This electric field produces a potential difference across the body cells. Body contains cells of many shapes and sizes. Here we calculate the potential difference across a spherical cell.

The potential difference across the spherical cell is given by (Berg, 1988)

$$\Delta U = 1.5\, r\, E \quad (2)$$

where,

r is radius of cell ($r = 5\ \mu m$) and E is induced electric field inside the body.

Results and Discussion

The potential difference across the spherical cell of skin and bone are given in Table 47.1 and 47.2 for both the radiation 41 MHz and 202 MHz. The values show that the potential difference across the cell is higher near the tower. The potential difference across the cell is higher near the tower. The potential difference is lower in higher depth of the body.

Table 47.1: Variation of Potential Difference Across the Spheriace Cell of Skin and Muscles

Frequency of EMF (MHz)	*Distance from the Tower (m)*	*Potential Difference Inside the Body at Depth*				
		1 mm	*2 mm*	*3 mm*	*4 mm*	*5 mm*
41	100	57.57	57.06	56.55	56.05	55.55
	500	11.51	11.4	11.31	11.2	11.10
	1000	5.75	5.70	5.65	5.6	5.55
202	100	56.89	55.71	54.56	53.43	52.33
	500	11.37	11.13	10.91	10.68	10.46
	1000	5.68	5.57	5.45	5.34	5.23

Table 47.2: Variation of Potential Difference Across the Spheriace Cell of Bone and Fat

Frequency of EMF (MHz)	*Distance from the Tower (m)*	*Potential Difference Inside the Body at Depth*				
		1 mm	*2 mm*	*3 mm*	*4 mm*	*5 mm*
41	100	58.04	57.99	57.94	57.89	57.84
	500	11.6	11.59	11.58	11.57	11.56
	1000	5.8	5.79	5.79	5.79	5.78
202	100	57.94	57.79	57.64	57.5	57.35
	500	11.58	11.55	11.52	11.49	11.46
	1000	5.79	5.77	5.76	5.75	5.73

If the applied electric field is 1.33×10^5 V/m, then for the cell of radius 5 μm, the potential difference across the cell become 1V which is the critical transmembrane breakdown voltage. Tables 47.1 and 47.2 also show that for higher frequency, potential difference is higher. Thus, if the frequency of radiation is increased, the induced potential difference of the cell win also increase which is harmful for the life of skin and bone cells.

Conclusion

Thus, from above analysis it can be seen that the frequency transmission tower radiate high frequency radiation, which is higher harmful for the cell life of the human body. Hence it is suggested that the transmission tower should be located far away from the populated area.

References

Berg, H. 1988. *Electrofusion of cells in modern Bioelectrodes*, (Ed.) A.A. Marino. Marcel Dekker, New York.

Olsen, R.G. 1990. *R.F. Energy for Warming Dives, Hands and Feet in Emerging Electromagnetic Medicine*, (Eds.) O. Connor, M.E. Bentall, R.H.C. and Monaha, J.C. Springer-Verlag, New York, p. 1–35.

Gajsek, P., Hurt, W.D., Senior Member, IEEE, Ziriax, J.M., and Mason, P.A. 2001. Parametric dependence of SAR on permittivity values in a man model. *IEEE Transaction on Biological Engineering*, 48: 1169–1177.

Litwak, E., Foster, K.R., and Repacholi, M.H. 2002. Health and safety implications exposure to electromagnetic fields in the frequency range 300 Hz to 10 MHz. *Bioelectromagnetic*, 23: 69–82.

Kumar, V., Maheswari, M., and Pathak, P.P. 2003. Variation of high and frequency electric fields in tissues. *Curr. Sci.*, 3(2): 281–284.

Pathak, P.P., Kumar, V. and Vats, R.P. 2003. Harmful electromagnetic environment near transmission tower. *Indian J. Radio and Space Phys.*, 32: 238–241.

Chapter 48

Antifertility Effect of Neem (*Azadirachta indica*) Bark in Male Albino Mice

Anju Puri Chakravarty and Bir Hans*

Department of Zoology and Fisheries, COBS & H, Punjab Agricultural University, Ludhiana – 141 004, Punjab, India

ABSTRACT

The effect of oral administration of alcoholic extract of neem (*Azadirachta indica*) bark was evaluated on fertility of male albino mice. Bark extract when administered at the dose level of 200 and 300 mg/kg body weight/day for 24 days caused remarkable changes in cauda epididymal sperm parameters and induced antifertility effect.

Keywords: *Antifertility, Azadirachta indica.*

Introduction

Earlier work revealed antifertility effect of neem leaves in male rodents (Joshi *et al.*, 1996; Parshad *et al.*, 1997). The present study was carried out to establish the antifertility properties of neem bark extract in male albino mice for its future use as a fertility regulating agent in male animal species.

Materials and Methods

Bark collected from neem plant was powdered. The powder was extract by percolation at room temperature with 70 per cent ethanol. The extract was finally concentrated under reduced pressure

* Present Address: Lecturer in Zoology, Baring Union Christian College, Batala – 143 505, Punjab, E-mail: c_drsumit@yahoo.com

and finally dried in vacuum desiccator. The bark residue was dissolved in propylene glycol (vehicle) at the rate of 100 mg/ml and was used for present experimental study. Colony bred adult male albino mice (8–10 weeks old with average 30 gm body weight) were maintained under controlled conditions. Standard rat feed and water were provided ad libitum. Female mice showing normal estrous cycle were also maintained. Acclimatized mice were divided into three different groups (8 in each group). Group I served as control and received vehicle only, while Groups II and III received bark extract at the dose level of 200 and 300 mg/kg bw/day for 24 days. After completion of each treatment 3 male mice were kept separate for the study of fertility index. Rests were sacrificed 24 hours after the administration of last dose. From the sacrificed mice epididymides were dissected out, blotted free of mucus and weighed. Cauda epididymal fluid was obtained by incising cauda epididymis in 0.5 per cent glucose saline for the study of various sperm parameters such as sperm count, sperm motility, normal sperms (per cent) and 1 : 2 abnormal sperms; (per cent). Sperm count was determined by haemocytometric method (Salisbury *et al.*, 1978). The percentage of motile sperms was calculated per unit area and expressed as per cent motility (Prasad *et al.*, 1972). Air dried smears of cauda epididymal fluid were stained with 2 per cent Giemsa stain for assessing the per cent normal and abnormal sperms. For the study of fertility index treated male mice of each group were caged separately with pro-estrous/ estrous females in the ratio of 1 : 2. Successful mating was confirmed by the presence of vaginal plug and by the presence of sperms in vaginal smear. This was taken as day 1 of pregnancy. Total number of pregnant females was recorded. From each group half of pregnant females were dissected on day 16th of gestation and number of implantations and resorptions (if any) were recorded while others were allowed to complete gestation period. Fertility index was calculated as:

$$\text{Fertility Index (\% fertility)} = \frac{\text{Number of Pregnant Females}}{\text{Total Number of Females Allowed to Mate}} \times 100$$

Results and Discussion

A significant reduction in the weight of cauda and caput epididymides was observed in Group III mice (Table 48.1). Androgen deficiency had been the cause of reduced epididymal weight in animal species treated with plant extracts (Khanna *et al.*, 1986; Akbarsha *et al.*, 1990). Treated groups revealed significant change in all sperm parameters (Table 48.1). Sperm count decreased significantly in Groups II and III. Anti-androgenic effect of plant product may be responsible for decreased sperm count (Shaikh *et al.*, 1993). Per cent sperm motility decreased in Groups II and III. Reduced sperm motility has also been reported in rodents administered with different plant extracts (Sarkar *et al.*, 2000). Reduced sperm motility in the present study would have resulted from alteration in epididymal milieu (Chinoy *et al.*, 1997). The percentage of abnormal sperms increased significantly in mice after treatment with bark extract. Morphological changes have also been observed in rats administered with neem leaf extract (Aladakatti and Ahamed, 1999).

Administration of bark extract in mice resulted in reduction in fertility index (Table 48.2). Control mice showed 100 per cent fertility while groups treated with bark extracts revealed reduced per cent fertility. Number of implantations reduced in females mated with males treated with bark extracts. In one of the females mated with males fed with 300 mg/kg bw of bark extracts, there were two resorptions. Reduced implantation sites and litter size have also been reported in females mated with males fed with neem leaf extract (Choudhary *et al.*, 1990) and male rats fed with steroidal extract of neem leaves failed to impregnate females (Parshad *et al.*, 1997). Sperm count is considered to be one of the most important parameters that affect fertility (Murugavel *et al.*, 1989) and decrease in sperm count may lead to reduced fertility/infertility in males (Safe *et al.*, 2001). Decline in sperm motility in males might have affected fertilization and implantation (Choudhary *et al.*, 1990) as only forward progressive motile spermatozoa are capable of penetrating and ascending the female reproductive tract.

Table 48.1: Effect of Oral and Administration of Neem Bark Extract (200 and 300 mg/kg body weight/day for 24 days) on Epididymal Weight and Various Sperm Parameters of Cauda epididymis

Parameters	*Group I Control*	*Group II 200 mg/kg bw*	*Group III 300 mg/kg bw*
Wt. of caput epididymis (g/100 g bw)	0.066±0.003	0.063±0.003	0.053±0.003*
Wt. of cauda epididymis (g/100 g bw)	0.038±0.004	0.034±0.008	0.031±0.002*
Sperm count (millions/ml)	43.32±1.78	20.66±1.44*	23.33±1.44*
Sperm motility (%)	70.0±0.47	12.33±3.53*	12.66±2.688
Normal sperms (%)	82.33±1.51	52.66±1.44*	42.66±2.17*
Abnormal sperms (%)	17.66±1.51	47.30±1.44*	57.33±2.17*

Values are mean ± S.E.

$P \leq 0.01$; *: Indicates significant change as compared to control.

Table 48.2: Female Mice with Implantation Sites and Litter Size Following Mating with Male Mice Administered with Different Doses of Neem Bark Extract (200 and 300 mg/kg bw/day, orally for 24 days)

Group	*No. of Males and Females*	*Male : Female for Breeding*	*Proportion of Pregnant Females*	*Pregnant Females (%) Dissected on Day 16th of Gestation*	*Implantation*		*% Female Delivered*	*Litter Range*	*Fertility Index (%)*	*Qualitative Fertility*
					No. of Foetus	*Resorption*				
Group I										
Control	3+6	1 : 2	6/6	50	7–9	Nil	50	7–9	100	Excellent
Group II										
200 mg/kg bw/day	3+6	1 : 2	4/6	50	2–3	Nil	50	2–4	66	Poor
Group III										
300 mg/kg bw/day	3+6	1 : 2	3/6	33	3	2	66	3–4	50	Poor

References

Akbarsha, M.A., Manivannam, B., Shahul, H.K. and, Vijayan, B. 1990. Antifertility effect of *Andrographis paniculata* (Nees) in male albino rat. *Indian J. Exp. Biol.*, 28: 421.

Aladakatti, R.H. and Ahamed, N.R. 1999. Effect of *Azadirachta indica* leaves on rat spermatozoa. *Indian J. Exp. Biol.*, 137: 1251–1254.

Chinoy, N.J., Patel, K.G. and Chawla, S. 1997. Reversible effects of aqueous extractor papaya seed on micro-environment and sperm metabolism of cauda epididymis of rat. *J. Medi. Arom. Plant Sci.*, 19: 717–723.

Choudhary, D.N., Singh, J.N., Verma, S.K. and Singh, B.P. 1990. Antifertility effects of leaf extracts of some plants in male rats. *Indian J. Exp. Biol.*, 28: 714–716.

Joshi, A.R., Ahamed, N.R., Pathan, K.M. and Manivannam, B. 1996. Effect of *Azadirachta indica* leaves on testis and its recovery in albino rats. *Indian J. Exp. Biol.*, 34: 1091–1094.

Khanna, S., Gupta, S.R. and Grover, J.K. 1986. Effect of long term feeding of tulsi (*Ocimum sanctum* Linn) on reproductive performance of adult albino rats. *Indian J. Exp. Biol.*, 24: 302–304.

Murugavel, T., Ruknudin, A., Thangavelu, S. and Akbarsha, M.A. 1989. Antifertility effect of *Vinca rosea* Linn leaf extract on male albino mice: A sperm parametric study. *Curr. Sci.*, 58: 1102–1103.

Parshad, O., Gardner, M.T., Williams, L.A.D. and Fletcher, C.K. 1997. Antifertility effects of aqueous and steroidal extracts of neem leaf (*Azadirachta indica*) in male wistar rats. *Phytother. Res.*, 11: 168–170.

Prasad, M.R.N., Chinoy, N.J. and Kadam, K.M. 1972. Changes in succinate dehydrogenase levels in rat epididymis under normal and altered, physiological conditions. *Fertil. Steril.*, 23: 186–190.

Safe, S.H., Pallaroni, L., Yoon, K., Gaido, K., Ross, S., Saville, B. and McDonnell, D. 2001. Toxicology of environmental estrogens. *Reprod. Fertil. Dev.*, 13: 307–315.

Salisbury, G.W., Hart, R.G. and Lodge, J.R. 1978. The spermatozoan genome and fertility. *Am. J. Obstet. Gynecol.*, 128: 342–350.

Sarkar, M., Gangopadhyay, P., Basak, B., Chakraporty, K., Banerji, J., Adhikary, P. and Chatterjee, A. 2000. The reversible antifertility effect of *Piper betle* Linn. on swiss albino mice. *Contraception*, 62: 271–274.

Shaikh, P.D., Manivannan, B., Pathan, K.M., Kasturi, M. and Ahamed, N.R. 1993. Antispermatic activity of *Azadirachta inidica* leaves in albino rats. *Curr. Sci.*, 64: 688–689.

Chapter 49

Palaeocurrent Analysis from Current Beddings and Pebbles of Tura Sandstone Group of Rocks Occurring in and Around Rangram Area, Meghalaya

Mahananda Borah, Anima Gogoi and P.K Das

Department of Geological Sciences, Gauhati University, Guwahati – 781 014, Assam

ABSTRACT

Cross-bedding forest direction and orientation of the long axes of the pebbles of Tura Formation occurring in and around Rangram area, Meghalaya, have been used for palaeocurrent analysis. The general palaeocurrent direction was from NW to SE as indicated by cross bedding and pebble analysis. The arithmetic mean of the resultant vectors of the pebbles at different locations is 132° which coincides fairly well with the cross-bedding direction (123°). The marine beach pebbles as indicated by their mean inclination are well rounded in nature. It indicates littoral environment of deposition.

Keywords: *Palaeocurrent, Current bedding, Pebble, Long axes, Rangram area, Meghalaya.*

Introduction

The current system responsible for transportation of sediments often leave record of their movement direction within the body of the sediment. This information can be used to decipher the palaeocurrent directions are of two types, directional and scalar. The directional elements (also called vectorial elements) are indicators of the palaeocurrent direction at the point of observation. The scalar properties show gradual change in the direction of transportation, and the palaeocurrent direction is known by systematically mapping this change over an area.

The directional elements useful for palaeocurrent analysis are mainly of two types, Planer and linear, but more complex structures are also known.

Current bedding in sandstone is one of the most useful planer structures. Cross bedding (q.v.) of various sizes are the commonest tool for studying palaeocurrcnt. The direction of inclination of a cross bedding forest (the azimuth of cross-bedding) indicates the direction of sediment transport. High and Davis (1984) reviewed the reliability of various types of cross-strata as palaeocurrent indicator.

Orientation and imbrication of long axes of sand grains are the linear structures which can indicate palaeocurrent direction. More commonly used for palaeocurrent study are the orientation and imbrication of pebbles and boulders. The long axes of pebbles generally lies in the direction of current and they dip upstream. But perpendicular arrangement of long axes of pebbles with respect to current direction are not unknown.

In the present study cross-bedding fore set direction and orientation of the long axes of the pebbles have been utilized for palaeocurrent analysis.

Geology of the Area

The Tertiary rocks occur as a north-south linear outcrop around the Rangram area. The rocks belonging to the Archean gneissic complex provided the basement for deposition of the Tertiary sediments. The deposition of the Tertiary sedimentary rocks give rise to Tura Formation. This formation is almost wholly made of dark brown, highly férruginous, hard, coarse to medium grained, gritty, poorly sorted sandstone. The traverse revealed the following succession which is consistent only near Rangram area.

Table 49.1

Age	*Description of Rocks*
Tertiary (Eocene)	Sedimentary Rock (Tura Formation)
Unconformity	
Jurassic	Basic Intrusive
Unconformity	
Precambrian	Granitic Gnesses and Schist, Pegmatites and vein Quartz Acid and Basic Intensive

Method of Study

Directional elements can be treated as vectors because they have both direction as well as magnitude. In the case of planner elements like cross-bedding, direction is given in three dimensions by the azimuth and inclination of the forest. (Direction of a linear structure is given by the attitude or orientation of the property concerned). The magnitude of the vector is determined arbitrarily, by assigning unit weight to each observation. When the datas are spread on either side of the origin, the usual method of computation of arithmetic mean gives absurd results (for example, the arithmetic mean of 340° and 20° is 180°), and the usual standard deviation of measurement can't also be used as a measure of dispersion.

But a meaningful average of a population directional data can be obtained computing the vector resultant as follows:

Each observation, be it a cross-bed or pebble, long axis orientation, is considered a vector with direction and magnitude. The magnitude is generally considered unity.

The north-south and east-west component of each observation vector are computed by multiplying the magnitude by the cosin and sine of the azimuth, respectively. These components are summed to give the components of the resultant vector, which again has a direction and a magnitude. The magnitude is reduced to 0–100 per cent scale by dividing by the sum of all observation vectormagnitudes and multiplying by 100. (If as most practical applications, the computations are performed on the percentage of observations in each sector or group the magnitude comes out in per cent without the last step). The calculations are outlined here.

N–S component = Σ_n COSθ

E–W component = Σ_n SINθ

$$\text{Ten}\,\bar{\theta} = \frac{\Sigma_n \text{SIN}\theta}{\Sigma_n \text{COS}\theta}, \quad \bar{\theta} = \tan^{-1}\frac{\Sigma_n \text{SIN}\theta}{\Sigma_n \text{COS}\theta}$$

$$R = [(\Sigma_n \text{SIN}\theta)^2 + (\Sigma_n \text{COS}\theta)^2]$$

$$L = \frac{R}{\Sigma_n} \times 100$$

where,

θ: Azimuth from 0° to 360° of each observation or group of observation.

$\bar{\theta}$: Azimuth of resultant vector

n: Observation vectormagnitude or in the case of grouped data of unit vectors, it is the number of observation in each group.

R: Magnitude of resultant vector.

L: Magnitude of resultant vector in per cent.

For determination of Σ_n COSθ and Σ_n SINθ value easily, a table proposed by Pincus (1956) has been utilized (Table 49.2).

The value of $\bar{\theta}$ (*i.e.*, azimuth of resultant vector) is the direction of palacocurrent.

Field Investigation and Interpretations

The pebble beds and the current beddings which are found in many places in the worked area has been used to find out the palaeocurrent direction.

Table 49.2: Table for Vector Calculation (After Pincus 1956)

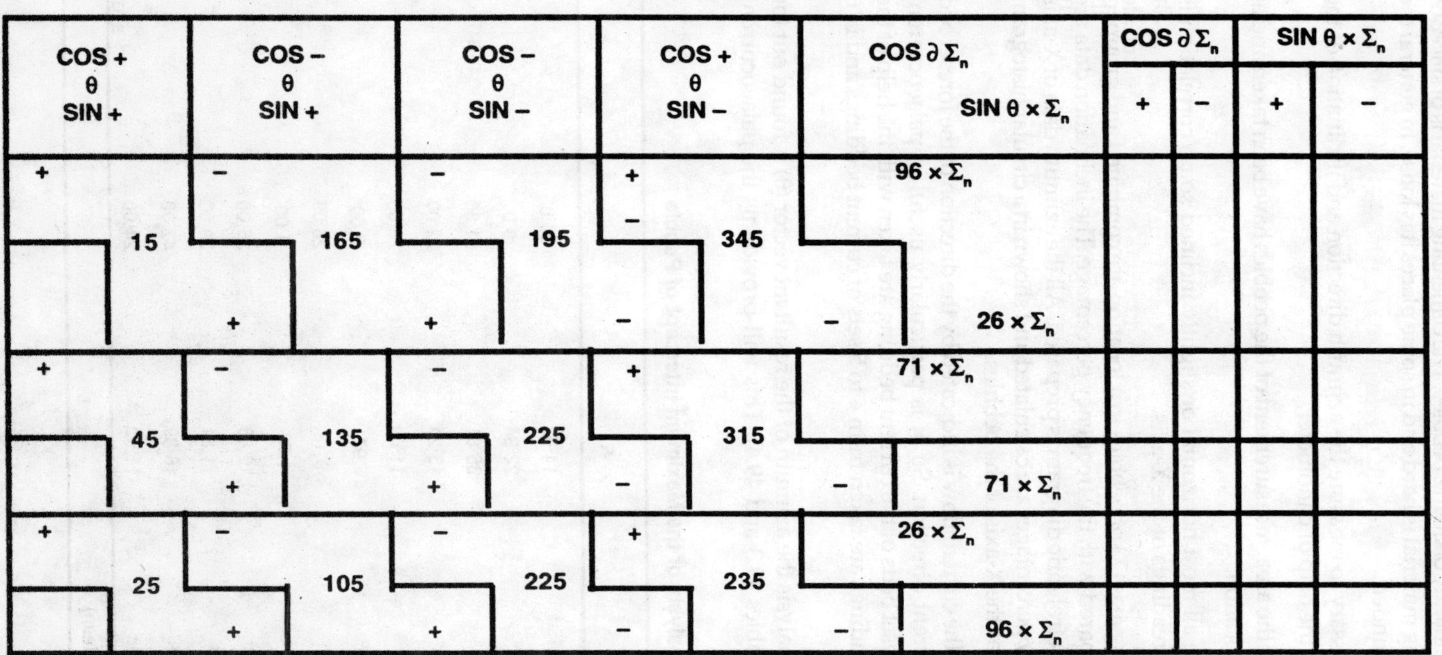

COS + θ SIN +	COS − θ SIN +	COS − θ SIN −	COS + θ SIN −	COS ∂ Σ_n / SIN θ × Σ_n	COS ∂ Σ_n +	COS ∂ Σ_n −	SIN θ × Σ_n +	SIN θ × Σ_n −
+ 15	− 165 +	− 195 +	+ − 345 −	96 × Σ_n − 26 × Σ_n				
+ 45	− 135 +	− 225 +	+ 315 −	71 × Σ_n − 71 × Σ_n				
+ 25	− 105 +	− 225 +	+ 235 −	26 × Σ_n − 96 × Σ_n				

The pebble beds are composed dominantly of quartz pebbles. The cementing material is of siliceous sediments. In some places this cementing material is hard and in some places it is loose. In general the pebbles are flate, elongated and well rounded.

To find the palaeocurrent it is necessary to measure the azimuth direction and inclination of the X–axis *i.e.* long axis of the pebbles with the help of clinometer.

In 13 locations at different places of the area, measurement of the pebbles have been taken and at least 30 pebbles are measured in each location.

As the pebble beds in the field are all most horizontal or slightly inclined so no correlation is needed in the azimuth and inclination readings of the X-axis.

Now, the angle of inclination of long axis (X) of pebbles in a location are combined and grouped into 10° interval and a histogram is prepared with the frequency percentage. The inclination data are also plotted on an equal area net and a petrofebric diagram is prepared. All the azimuth data of X-axis are grouped into 30° intervals, frequency percentages are calculated and shown in a circular histogram which shows the azimuth distribution or the X-axis of the pebbles.

In current bedding the direction of the current flow is indicated by the direction of the foreset bed which always inclined toward the current direction. So, it is particularly useful in palaeocurrent analysis. In the field the readings of foreset beds of the current bedding are taken with the help of the clinometer. In one location nearly 20 readings are taken from 4 to 5 sets of current beddings and in 6 locations the readings are taken.

Now with the help of the vector analysis the azimuth of the resultant vector (θ) is found out for both current beddings and pebbles (Tables 49.3 and 49.4). This will provide us the palaeocurrent direction.

Table 49.3: Vector Analysis of the Maximum Intercept of Pebble

Location	$\bar{\theta}$	*R*	*L*
P_1	141°75°	14°92	$30_{0}0$
P_2	171°47°	22°94	$9_{0}0$
P_3	134°71°	36°81	$81_{0}80$
P_4	133°98°	13°24	$22_{0}00$
P_5	149°64°	11°69	$28_{0}00$
P_6	174°29°	$_{0}97$	$3_{0}00$
P_7	143°50°	$9_{0}18$	$26_{0}00$
P_8	99°78°	$1_{0}77$	$5_{0}00$
P_9	133°06°	$10_{0}83$	$36_{0}00$
P_{10}	93°88°	$11_{0}81$	$29_{0}00$
P_{11}	126°89°	$16_{0}89$	$42_{0}00$
P_{12}	93°63°	$11_{0}90$	$24_{0}00$
	132°13° (Mean)		

Table 49.4: Vector Analysis of Azimuth of the Foreset Beds of Current Bedding

Location	$\bar{\theta}$	*R*	*L*
C_1	132°75°	19.94	87.00
C_2	116°47°	17.62	84.00
C_3	114°82°	16.85	94.00
C_4	124°02°	10.82	47.00
C_5	139°97°	16.38	66.00
C_6	113°96°	16.25	81.00
	123°67 (Mean)		

Table 49.5: Mean Angle Inclination of 'X'

Location No.	*No. of Reading*	*Mean Angle in Degree*
1	50	18.62
2	32	16.25
3	45	13.64
4	60	16.08
5	42	20.69
6	35	17.20
7	36	18.50
8	34	15.15
9	30	15.90
10	41	14.27
11	40	20.63
12	50	16.55
	495 (Total)	**14.96 (Grand Mean)**

Discussion and Conclusion

The inclination of the long axes of the pebbles have been found in up current and down current directions as generally expected in marine formation (Nicholson, 1965), but higher percentage of pebbles are inclined towards the down current direction. The mean inclination of the pebbles are generally less than 20° (Table 49.5) which indicates them to be of marine beach pebbles (Pettijohn, 1975, Ahmed, 1983).

The general palaeocurrent direction was from NW to SE as indicated by pebble and cross bedding analysis. The arithmetic mean of the resultant vectors of the pebbles at different locations is 132° which coincides fairly well with the cross-bedding direction (123°).

Geological observation indicate that long axes of the pebbles oriented in the direction of current flow (White, 1952) or transverse to the same (Twenhofel, 1950). The present analysis reveals that most of the pebbles are oriented along the current direction and only few are transverse. The maximum of the petrofabric diagram are in the NW and SE quadrant showing preferred orientation of the long axes of the pebbles.

The trend of the pebble bed is roughly NW-SE. The ancient shore line might have been running in the NW-SE direction.

The pebbles are fairly of large size and they are well rounded. The rounded nature or the pebbles is in favour of littoral environment for their deposition (Das and Borah, 1993).

The average size suggested that the current which transported the pebbles were or high velocities.

Acknowledgement

The authors are thankful to the Head of the Department of Geological Sciences of Gauhati University, Guwahati for providing the departmental facilities to undertake the work.

References

Ahmed. 1983. Depositional environment of the basal conglomerate of the Barapani formation, Shillong group, Khasi Hills, Meghalaya. *Quart. Jour. Geol. Min. Mat. Soc., India*, 55(2): 62–68.

Das, P.K. and Borah, M. 1993. Palaeocurrent analysis from current bedding of quartzite and pebbles of congloerate Mawryngkneng area, East Khasi Hills District, Meghalaya, 12F(1–2): 19–24.

Davis, G.H. 1984. *Structural Geology of Rocks and Regions*. John Willey and Sons Inc.

Pettijohn, F.J. 1975. *Sedimentary Rocks*. Harpar and Brothers, New York.

Pincus, H.J. 1956. Some vector and arithmetic operation on two dimensional orientation varieties, with application to geological data. *J. Geol.*, 1: 64.

Twenhofel, W.H. 1950. *Principles of Sedimentation*, 2nd Ed., New York, Mc-Graw Hills.

White, C.M. 1940. The Equilibrium of Grains on the Bed of Stream, Proc. Roy. Soc., London, 174A: 322–323.

Chapter 50
Isolation and Characterization of Microorganisms Found in Gums and Adhesives from Post Offices of Madurai City

*G. Jayavarshni, A. Angelin, P. Mourin Rosee, M. Rajalakshmi, Mary Esther Rani**

Department of Botany, Lady Doak College, Madurai – 625 002, Tamil Nadu, India
**E-mail: ranimam@yahoo.com*

ABSTRACT

A study was undertaken to isolate and characterize the microflora present in gums and adhesives collected from Post Offices in Madurai city, Tamil Nadu, India. 16 bacterial isolates and 21 fungal isolates were obtained from the collected samples. The bacterial isolates partially characterized through staining and biochemical tests included *Escherichia coli, Alcaligenes feacalis, Micrococcus leutus, Clostridium* sp., *Corynebacterium glutamicum, Leuconostoc mesenteroides, Enterococcus feacalis, Proteus vulgaris, Propionibacterium acne, Lactobacillus* sp, *Staphylococcus aureus, Aeromonas hydrophila* etc. The fungal strains identified through staining and microscopic examination included species belonging to the genera *Aspergillus,* and *Rhizopus.* The experimental findings show a high microbial load in the gums and adhesive samples.

***Keywords:** Gums and adhesives, Bacteria, Fungi.*

Introduction

Numerous kinds of gums and adhesives are available in the market nowadays. The preparation and use of adhesives has always been very inexact. The art of preparation and use of adhesives is centuries old although the science is very young. To define adhesives simply in the functional sense "Adhesives are substances or preparations of gummy or gelatinous character used for the purpose of joining together or affecting the mutual adhesion of surfaces of bodies."

Our study is limited to adhesives made from starch–a group of adhesives that has become extremely varied both in composition and application. Post offices are one among the various places where these gums and adhesives have a wide usage. Starch adhesives mainly used in such places are polysaccharides of corn, wheat flour, potato (Hull and Bangert, 1952) and tapioca.

Microorganisms are ubiquitous in nature and have adapted to more diverse living conditions than any other organisms. Hence it is no doubt that the starch adhesives would form an enriching medium and would attract a number of microorganisms. The different species of bacteria are differentiated by many factors including their morphology, chemical composition etc. that are detected by their staining reaction, nutritional requirements and biochemical activities. The various fungal species are differentiated based on their morphology and pigments produced.

Materials and Methods

The samples of the adhesive pastes and gums were collected from different post offices in Madurai city. Nutrient agar (NA) and Potato dextrose agar (PDA) plates were prepared for the cultivation of bacteria and fungi respectively. To inhibit the growth of bacteria, Streptomycin sulfate (0.5 mg/ml) was added to PDA plates. From serially diluted samples of the collected gums and adhesives 0.1 ml was added to each plate for the isolation of fungi and bacteria. For this pour plate method of pure culture was employed. NA plates were then incubated for 24 hours at room temperature, while the PDA plates were incubated at room temperature for 6–7 days. After isolating, the colonies of bacteria were identified by using morphological and bio-chemical characteristics (Sneath Peter, 1994). The isolated fungi were characterized based on their morphological characters (Onion *et al.*, 1981).

Results and Discussion

16 bacterial isolates and 21 fungal isolates were obtained from the adhesive samples (Tables 50.1–50.3). The experimental findings show a high microbial load in the gums and adhesive samples. This could be because there is sufficient free water available to encourage the propagation of stray molds and bacteria, that may enter it from the air, vessels or materials used in the preparation of gums and from the hands of the people who access the gum. Also microbial growth is favored due to the presence of sugars like maltose in the adhesives. A further comparison of the different Post Office samples demonstrated a higher microbial load in Head Post Offices, which is visited by a huge number of people everyday.

Gram-positive bacteria dominated samples from all the Post Offices (Table 50.1) and most of them were identified to be opportunistic pathogens and hence pose a threat to the health of the people who access the gum. *Aspergillus* species are also opportunistic pathogens and form a very large proportion of all the molds encountered in industrial work. These are found almost everywhere on any conceivable substratum due to the large number of enzymes they produce. Other fungal species identified like *Penicillium* sp., *Rhizopus* sp. are also serious agents of destruction and cause a number of opportunistic diseases.

Table 50.1: Partially Characterized Bacteria from Post Offices

Sl.No.	*Isolate No.*	*Bacteria*
1.	1	*Lactobacillus* sp.
2.	2	*Staphylococcus aureus*
3.	3	*Alcaligenes feacalis*
4.	4	*Aeromonas hydrophila*
5	5	*Leuconostoc lactis*
6.	6	*Escherichia coli*
7.	7	*Propionibacterium acne*
8.	8	*Micrococcus leutus*
9.	9	*Corynebacterium glutamicum*
10.	10	*Proteus vulgaris*
11.	11	*Enterococcus feacalis*
12.	12	*Leuconostoc mesenteroides*

Table 50.2: Partial Characterization of Bacteria Obtained from Post Offices

Biochemical Tests	*Bacterial Isolates*										
	1	*2*	*3*	*4*	*5*	*6*	*7*	*8*	*9*	*10*	*11*
Colony colour	White	Yellow	White	White	White	White	White	Yellow	White	Black	White
Gram's reaction	+	+	+	+	+	–	+	+	+	–	+
Shape	Rod	Cocci	Rod	Cocci	Cocci	Rod	Rod	Cocci	Rod	Rod	Cocci
Indole	–	–	–	–	–	+	+	–	–	+	–
Methyl red	–	–	–	+	–	+	–	–	–	+	+
Voges proskauer	+	+	–	–	+	–	–	+	–	–	+
Citrate utilization	+	+	–	–	+	–	–	+	–	–	+
H_2S production	+	+	–	–	–	–	–	–	–	+	+
Starch hydrolysis	–	–	–	–	–	–	–	–	–	–	–
Nitrate reduction	–	+	–	+	–	–	+	+	–	+	–
Ureas	+	+	–	–	–	–	–	–	+	+	–
Gelatin hydrolysis	–	–	–	+	–	+	–	–	–	+	–
Acid from glucose	–	+	–	–	+	–	+	+	+	+	+
Acid from fructose	–	+	–	–	+	+	+	–	+	–	+
Acid from sucrose	+	+	–	+	+	–	–	–	+	+	+
Acid from lactose	+	+	–	+	+	–	+	–	+	–	+
Acid from maltose	–	+	–	–	+	–	–	–	+	–	+

Table 50.3: Fungi Identified from Book Binding Shops

Sl.No.	Isolate No.	Fungi
1.	1	*Aspergillus* sp.
2.	2	*Rhizopus* sp.
3.	3	*Penicillium* sp.
4.	4	*Cordona* sp.

References

Alexpolous, C.J. and Mims, C.W. 1993. *Introduction to Mycology*, 3rd edition. Wiley Eastern Ltd. New Delhi, pp. 291.

Cappuccino, J.G. and Sherman, N. 1992. *Microbiology: Lab Manual*, 3rd Edition. Benjamin Publishing Co., California, pp. 77–80

Dikshit, V.R. and Desai, P.D. 2000. An agar plate assy to detect cell wall active antifungal agents. *Curr. Sci.*, 78: 615–617.

Hull, Bangert. 1952. Animal glue. *Ind. Eng. Chem.*, 44: 2275.

Khobragade, C.N., Bhasarkar, R.K. and Gacche R.N. 2001. Isolation and characterization of microbial strains found in the gums and adhesives from book binding shops. *Asian J. Microbiol. Biotech. & Env. Sci.*, 63: 223–238.

Onion, A.H.S., Allosopp, D.P. and Eggihs, H.O.W. 1981. *Industrial Mycology*. Edward Arnold Pub. Ltd., p. 283–320.

Saxena, R.K., Davidson, S.W. and Gupta, R. 1999. The fungistatic action of oleic acid. *Curr. Sci.*, 76: 1137–1139.

Sneath Peter, H.A. 1994. *Bergy's Manual of Systematic Bacteriology*. Williams and Wilkins, 2: 1104–1207, 1261–1418.

Chapter 51

Investigation of Physico-chemical Condition of Major Effluent Points of Nambul River, Manipur, India

R.K Rajeshwari Devi

Department of Zoology, Oriental College, Imphal –795 001, Manipur

ABSTRACT

The study was conducted to know the sources of pollution in Nambul river. Nambul river is the most polluted river in Manipur flowing through the heart of the city. Physio-chemical analysis of major effluent points in the greater Imphal area were studied for one year from January to December 2003 to assess the level of pollution in three major different effluent points of Nambul river *i.e.*, Samushang effluent point, Naga Nallah effluent point and Keishamthong effluent point These three influent points are mainly responsible for collection of sewage and solid waste from in and around the greater Imphal area and pour it out into the Nambul rivers. From the studies, it reveals that the distance of 3 kms river course flowing into the heart of the city received the maximum loads of pollutant from the above effluent points and it may consider to be the most polluted area in the Nambul river course, even the level of DO was found only 0.2 ppm in these 3 different effluent points in the month of October.

***Keywords**: Effluent, Nallah, BOD, Nambul river.*

Introduction

Many rivers of the world receive heavy flux of sewage, domestic waste, industrial effluents and agricultural waste etc. which contain substances varying from simple nutrients to highly toxic chemicals. In order to save the water quality of the river there is a need to create data base report of the

water quality for the future. This paper presents Physio-chemical characteristics of the major effluents in Nambul river. Nambul river arises from the Kangchup hill and passes through the heart of the Imphal city by collecting all the municipal and domestic sewage and waste garbages of the town. Nambul river also received excess water from many paddy fields and collect the residues of fertilizer and pesticides finally falls into the Loktak lake which is one of the lake of International Importance *i.e.*, Ramsar site. The Nambul river may contribute maximum pollutant to the Loktak lake.

Material and Methods

In the Nambul river four sites were selected for the water quality analysis. Sampling programme was conducted from January to December 2003. The four spots are Iroisemba (clean water upstream no effluents point, Samusang effluent points discharge many type of waste containing Hospital and Biomedical waste, Naga Nallah effluents points discharge all the market, Industrial and municipal waste and lastly Keishamthong Kabui Khull Maning spot discharge maximum sanitary waste, domestic waste and residues of local liquor. The parameters which have studied are Temp., PH, DO, CO_2, Turbidity, Hardness, BOD, Na, K and Cl etc. Parameters like Temp., PH and Turbidity was taken at the spot. For DO analysis sample was collected in a 250 ml reagent bottle and fixed at the spot and analysis were done in the laboratory. Sample collection was made for every first week of the month. The analysis was adopted the methods of APHA (1995).

Results and Discussion

Temperature varied from 18.1°C to 25.9°C in the year round. There was no mark difference in temperature in the three effluent points. pH ranges from 6.9 to 9.4 highest PH value was noted at the Naga Nallah mouth point during the month of October. pH is the scale of intensity or alkalinity and measures the concentration of iron in water value of DO was different in three effluent points. Comparatively higher value of DO was in Keishamthong then to Samushang, lowest DO was found in the Naga Nallah Spot. It was as low as 0.2PPm in the month of Oct. Highest value of DO was in the Keishamthong spot during the month of July. Decrease in DO is an index of increased organic pollution which are mainly due to addition of waste through point and non-point sources from its catchments area. These organic matters undergo degradation by microbial activity in the presence of DO resulting in the deoxygenating process and swift depletion of dissolve oxygen. Hardness ranges from 2.2 to 71 ppm in these three spot. The maximum value of hardness 71 ppm was. ound in the Naga Nallah spot during the month of May. Hardness of water recorded which is due to the high concentration of bicarbonates. Turbidity value varies from 49 ppm to 600 ppm. Minimum was in Samushang during the month of November and maximum was in June and August of Naga Nallah. BOD value falls from 2.6 ppm to 39.3 ppm. In comparison with Samushang and Keishamthong the BOD value was found to be very high in the Naga Nallah point reaches upto 39.3 ppm in the month of August indicating higher load of biodegradable accumulation in the effluent point. In case of Na high value was found in Naga Nallah and Keishamthong in the month of February and March. Keishamthong recorded the highest value of Cl than Samushang and Naga Nallah.

Table 51.4 shows the water quality of Iroisemba some miles away from the Kangchup origin of the river which is in the south of Nambul river. In this river course no efiluent has received except Field waste. The quality of water was almost normal in its DO content CO_2, PH, Turbidity and BOD. Turbidity was little high in some month *i.e.*, May, June and July but it may be due to the rainy season.

Thus, the result shows that Nambul river particularly in greater Imphal area are highly polluted. The three effluent points in Nambul river are mainly responsible for pollution of Nambul river even though they are different in degree of pollution. Nambul river water in these spots are not suitable in any purposes like Fishery and domestic use and Agriculture etc. Generally water quality was worse during winter season when the level of water in the river recedes. Among effluent point which we have studied Naga Nallah point which is in the heart of city mainly associated with Municipal and domestic waste have found to be the most polluted effluent point of Nambul river. So for proper maintenance of Nambul river at least these effluent points have to be treated before falling to main stream of the Nambul river.

Table 51.1: Physio-chemical Analysis of Nambul river (Samushang effluent point) During 2003

Month	*Temp. (°C)*	*pH*	*DO ppm*	*CO_2 ppm*	*Turbidity ppm*	*Hardness ppm*	*BOD ppm*	*Na ppm*	*K ppm*	*Cl ppm*
January	18.1	7.8	7.4	8.8	80	46	7.26	12	5	14.2
February	20.6	8.2	5.91	9.7	89	50	7.16	21	9	22.12
March	23.1	8.2	5.89	9.4	80	48	6.1	21	11	22.76
April	25.3	7.5	5.04	8.1	129	40	6.32	13.5	8.3	17.00
May	22.0	7.2	6.6	4.4	400	22	3.9	11.6	3.8	10.04
June	23.6	7.2	6.71	1.1	487	26	2.6	10.0	5	11.02
July	23.1	7.25	7.2	2.8	320	26	2.1	7.0	4	9.24
August	24.5	7.1	6.6	3.8	400	20.2	6.1	7.0	2.2	12.2
September	24.1	7.1	8.8	7.3	340	22.9	5.5	9.9	2.3	9.8
October	24.	07.2	6.8	8.3	128	29.2	6.9	10.1	2.9	10.9
November	20.1	7.2	7.4	7.0	49	38	7.0	9.0	5	13.3
December	18.6	7.6	8.2	12.1	50	48	8.2	3.0	5	151.6

Table 51.2: Physio-chemical Analysis of Nambul River (Naga Nallah effluent point) During 2003

Month	*Temp. (°C)*	*pH*	*DO ppm*	*CO_2 ppm*	*Turbidity ppm*	*Hardness ppm*	*BOD ppm*	*Na ppm*	*K ppm*	*Cl ppm*
January	18.1	7.4	3.6	7.4	90	66	12.06	18	9	36
February	20.6	7.7	3.91	16	72	68	12.63	38	17	32.1
March	23.3	9.1	3.6	21.2	95	62	12.6	46	21	34.7
April	25	8.9	2.41	12.6	234	62	11.6	27	73	28.6
May	21.7	5.5	4.8	18.6	430	28.8	6.2	19	10.8	17.2
June	24.7	6.6	5.4	2.8	567	26.0	4.6	24	9	17.6
July	23.2	8.5	7.41	3.3	452	29.0	3.5	12	7.1	18.0
August	25	9.6	3.8	8.7	600	24.9	39.3	18	13	87.1
September	24.6	2	0.2	11	385	30.6	25.7	14	10	31.2
October	24.6	4	0.2	11	170	32.1	23.3	16	8	20.5
November	20.7	7.9	3.0	24	68	38	12.1	15	14	21
December	20.2	8.2	24	6	57	71	11.7	17	9	27

Table 51.3: Physio-chemical Analysis of Nambul River (Keishamthong Khull Maning effluent point) During 2003

Month	*Temp. (°C)*	*pH*	*DO ppm*	*CO_2 ppm*	*Turbidity ppm*	*Hardness ppm*	*BOD ppm*	*Na ppm*	*K ppm*	*Cl ppm*
January	18.3	7.4	3.6	19.8	60	66	12.6	14	5	34.14
February	21.6	8.7	4.6	22	75	60	10.61	29	11	22.72
March	23.6	8.9	4.3	18.2	85	50	10.02	35	17	26.49
April	25.2	8.5	4.9	9.4	141	48	10	19	9	20.00
May	22.3	7.1	5.7	3.3	400	24	8.3	18	9	11.36
June	24.2	7.3	3.8	4.4	463	28	3.2	17	7	10.68
July	23.5	7.4	5.8	3.4	420	30	3.6	14	6	11.18
August	25.4	7.8	4.5	9.0	550	23.5	32.2	17.2	9.8	51.3
September	25.9	7.2	9.9	19.8	376	31.9	11.8	15.7	7.6	72.3
October	24.9	7.8	4.9	19	165	31.9	20.1	16.1	7.8	48.9
November	20.2	7.7	5.1	17.6	59	46.0	11.2	15	12	20.76
December	19.2	7.8	5.0	19.2	56	56	12.8	18	11	32.39

Table 51.4: Physico-chemical Analysis of Nambul River (Eroisemba upstream) During 2003

Month	*Temp. (°C)*	*pH*	*DO ppm*	*CO_2 ppm*	*Turbidity ppm*	*Hardness ppm*	*BOD ppm*	*Na ppm*	*K ppm*	*Cl ppm*
January	18	7.8	9.6	6.6	70	50	2.5	7	3	12.8
February	20.6	8.1	5.8	9.2	97	48	5.3	15	7	19.6
March	22.6	8.6	5.9	9.2	59	34	5.12	11	7.9	15
April	25	7.4	8.6	7.2	125	34	5.12	11	7.9	15
May	21.1	7.3	7.5	2.0	240	21	1.3	9	3	9.9
June	22.6	7.0	7.7	1.1	320	24	1.2	8	4	7.4
July	23.2	7.2	7.5	2.7	310.	23	2.0	7	4	7.2
August	23	7.1	7.8	3.4	300	21	2.1	7.2	4	7.5
September	22.1	7.0	8.1	3.2	85	25	7.2	9.3	4	8.6
October	21	7.2	8.0	6.1	80	28	8.1	7.9	4	10.8
November	20.1	7.2	7.4	7.0	49	32	7.2	8	4	12.6
December	18.6	7.4	6.9	11	46	42	8.1	11	4	14.7

References

APHA. 1995. *Standard Methods for the Examination of Water and Waste Water*, 19th Edition. American public health Association. American water works. Association and water pollution control federation, Washington. D.C., pp. 1134.

Baruah, B.K. and D. Baruah. 2003. Study on water quality of Subansiri river in Assam: An EIA approach for a proposed hydroelectric power project. *Indian J. Environ and Ecoplan.*, 7(L): 381–384.

Das, P.K. and Ranjana Borah. 2003. Geochemistry river sediments. *Indian J. Environ and Ecoplan.*, 8(1): 103–106.

Dinesh Kumar, Samiksh. 2004. A study on the characteristics of polluted water of Amanishah Nallah and neighbouring ground water sources with reference to heavy metal contents. *Indian J. Environ. and Ecoplan.*, 8(2): 371–374.

Goel, P.K. 1997. *Water Pollution: Causes, Effects and Control*. New Age Int. Pub., New Delhi, p. 97–115.

Gross, M.G. 1978. Effects of water disposal operation in estuaries and the coastal ocean. *Annual Review for Earth and Planetary Science*, 6: 127–143.

Kannan, K. 1996. *Fundamental of Environmental Pollution*. S. Chand and Co. Ltd., New Delhi.

Sengupta, Raw Kurseshly, T.W. 1999. Marine pollution bows and potential to the Indian marine environment. In: *Water Pollution: Conservation and Management*, (Eds.) Sinha A.K., Ramboojan and Viswanathan, P.N. Gyanodaya Prakashan, Nainital, India, p. 165–181.

Reginna, B. and B. Nabi. 2003. Physico-chemical spectrum of the Bhavani river water collected from the Kalingarayan Dam, Tamil Nadu. *Indian J. Environ and Ecoplan.*, 7(3): 633–636.

Tombi, H. 1996. Progress Report *Environmental Monitoring Laboratory*. Department of Life Sciences, Manipur University.

Trivedi, R.K. and Gopal, P.K. 1986. *Chemical and Biological Methods for Water Pollution Studies*. Environmental Publication, Kerala.

WHO. 1984. *Guidelines for Drinking Water Quality*. WHO Geneva, 1984 Vol. 182. Recommend WHO, Geneva.

WHO. 1992. *World Health Organisation's Standards of Drinking Water*, Geneva, Switzerland.

Chapter 52

Economic Appraisal of Biogas for Cooking and Electricity Generation: A Case Study

Er. Sarbjit Singh Sooch[**], *Er. Deepak Gupta*[*] *and Indervir Singh*[**]

[*]Assistant Professor, Department of Civil Engineering, College of Agricultural Engineering, PAU, Ludhiana

[**]Ex. B. Tech Student of College of Agricultural Engineering, PAU, Ludhiana

ABSTRACT

Biogas produced from anaerobic digestion is the best renewable energy source for cooking needs. This energy source can also be used for electricity generation. In this article economic appraisal of biogas has been carried out for cooking needs and electricity generation from non-functional community/institution biogas plants.

Keywords: *Biogas, Biogas plant, Cooking needs, Electricity generation, Economic appraisal.*

Introduction

Biogas is the best renewable energy source produced by anaerobic digestion of organic wastes. Thus, a number of community biogas plants were installed all over the state of Punjab (India). The major objective of these plants were to digest whole organic waste of a community, to provide proper sanitation in the community and the biogas thus produced be used as a by product, to fulfill the cooking needs of the community. But it has been found during survey in the villages that all such plants are lying junk. Hardly one or two plants are functional. This too is due to personal interest of the village panchyat. These plants failed due to mismanagement, wrong selection of sites and faulty

[*] E-mail: sarbjit_sooch@sify.com

laying of pipelines. The biogas so produced in the plant was not reaching to the individual home, either it was too less to meet the demands or the pressure of gas was too low (Chawla, 1986 and Khandelwal, 1989).

Thus a lot of money and land is blocked due to the installation of these plants which in other wards we can say, a wastage of lot of economy of the country due to non-functional of these biogas plants. The authors in this article have tried to compare that the use of biogas instead of for cooking be used for the electricity generation which would make the plant operational with good margin of profit, because the consumer in this case shall be only the state electricity board and generation of electricity shall be possible right at the biogas plant site.

Materials and Methods

It is proposed to use some alternative methods to revive the sick (non-functional) biogas plants. A case study of an Institutional biogas plant installed at Punjab Agricultural University, Ludhiana was considered. This plant (4 × 70 cubic meter capacity) was installed with the help of Ministry of Non-Conventional Energy Sources, Govt. of India, New Delhi. The raw material (cow dung) for feeding these biogas plants was collected from diary farm of PAU, Ludhiana. The gas produced from this biogas plant was used through underground gas pipeline system in boy's hostels. The plant was very effective for cooking in hostel mess but later on this plant was stopped due to management and some other problems (Singh, 2004).

This plant can be revived again by investing small amount of money for cooking as well as electricity generation. So to study this plant, firstly we have to design this institutional plant to get its present cost and then the economics (on the basis of Ludhiana market rates in the month of March 2005) of this plant will be compared both for cooking as well as electricity generation and finally we will see the viability of running this biogas plant for any of these two applications.

Results and Discussion

Cost of biogas plant is calculated as follows:

1. Cost of civil construction =	Rs. 5,71,900.00
2. Cost of steel gas holders =	Rs. 2,17,400.00
3. Cost of gas pipe line and other appliances =	Rs. 1,62,200.00
Total =	**Rs. 9,51,500.00**

Economics of Community Biogas Plant

The economics of community biogas plant is calculated on the basis of its application for cooking purpose and electricity generation (Singh, 2004).

Economics for Cooking Purpose

Operating Cost

The cattle dung available from PAU diary complex is used as FYM (Farm Yard Manure) only. The operating cost of the biogas plant is explained in Table 52.1.

Table 52.1: Operating Cost of Biogas Plant for Cooking Purpose

Sl.No.	Annual Working Cost (Rs.)	Use of Dung as Manure (Rs.)
1.	Civil construction work (50 yr. Life) Depreciation @ 2%	11,440
	Interest on Capital @ 8%	45,752
2.	Gas holder and tube well (10 yr. Life) Depreciation @ 10%	21,740
	Interest on Capital @ 8%	17,392
3.	Gas pipe line and appliances (20 yr. Life) Depreciation @ 50%	8,110
	Interest on Capital @ 8%	12,976
4.	Maintenance (1% of total cost)	9,515
5.	Cost of 1 labour/annum	31,200
6.	Transportation cost/annum	54,000
7.*	Cost of dung as manure (766T) at Rs. 250/T	1,91,500
	Total Cost	4,03,625

* Total fresh dung required for biogas plant in one year = 280 × 25 × 365 = 2555 ton

About 30 per cent manure is collected from fresh dung

∴ Quantity of manure = 2555 × 0.30 = 766 ton

Annual Income

Income from Biogas in Terms of LPG

Total amount of biogas produced in one year = 280 × 365 = 1,02,200 m^3

About 20 per cent of biogas is deducted from total production (which is due to operational faults, downfall in temperature in winter season etc.)

Net amount of biogas produced = 0.80 × 102200 = 81760 m^3

Amount of biogas in terms of LPG = 81760 × 0.43 = 35156.80

∴ Annual income from biogas in terms of LPG = 35156.8 × 20 = 703136

Income from Biogested Slurry = 191500

Total annual income = 703136 + 191500 = 894636.00

Profitability and Percent Return on Investment on Biogas Plant.

It is explained in Table 52.2.

Table 52.2: Profitability and Percent Return on Investment on Biogas Plants

Sl.No.	Use of Dung as	Income (Rs.)	Cost (Rs.)	Annual Profit (Rs.)	%Return on Investment
1.	Farmyard manure	894636	403625	491011	$\frac{491011 \times 100}{951500} = 51.60$

Payback period = Initial investment/Annual profit = 951500/51.60

= 1.91 years Ë 2 years (Approx.)

Economics for Electricity Generation

Initial Investment on the Plant

1.	Cost of civil construction work	=	Rs. 571900.00
2.	Cost of steel gas holder	=	Rs. 217400.00
3.	T and P construction cost and cost of room for generator (Assume 10 per cent of cost of biogas plant)	=	Rs. 78930.00
4.	Cost of 34 HP engine alteration	=	Rs. 170000.00
5.	Cost of appliances used with generator is 5 per cent of the total cost of biogas plant	=	Rs. 39465.00
	Total initial investment	=	**Rs. 1077695.00**

Operating Cost for Electricity Generation

It is explained in Table 52.3.

Table 52.3: Operating Cost of Biogas Plant for Electricity Generation

Sl.No.	*Annual Working Cost (Rs.)*	*Use of Dung as Manure (Rs.)*
1.	Civil construction work (50 yr life) Depreciation @ 2%	11,440
	Interest on Capital @ 8% per year	45,752
2.	Gas holder and tube well (life 10 yrs.) Depreciation @ 10%	21,740
	Interest on Capital @ 8% per year	17,392
3.	Generator (life 10 yrs.) Depreciation @ 10%	17,000
4.	T&P construction and generator room (20 yrs. life) Depreciation @ 5%	3,947
5.	Maintenance @ 1% of total initial investment	10,777
6.	Labour cost	31,200
7.	Transportation cost	54,000
8.	*Total diesel cost required for running 34 HP engine for 24 hrs/day where 25 per cent diesel is required @ 5.1 lit/hr/day	2,90,394
9.	Cost of dung as manure	1,91,500
	Total Running Cost (Rs.)	**6,95,142**

* Cost of Diesel required

Cost of Diesel required in engine when 25 per cent of Diesel and 75 per cent of Biogas is used for electricity generation from generator. In winter Diesel : Biogas is 30 : 70 and in summer Diesel : Biogas is 20 : 80 for dual fuel engine so we take average of these two *i.e.*, Diesel : Biogas is 25 : 75 (Bajpai, 1967 and Pander, 1949).

1.	Diesel consumption of 34 HP, 25 KVA generator	=	5.1 lit/hr/day
2.	Number of days for which engine is operated	=	365 days
3.	Cost of Diesel	=	Rs. 26/litre

Therefore Total Cost of diesel = 5.1 × 24 × 365 × 26 × 25 = Rs. 2,90,394.

Annual Income

Income from Biogas in Term of Electricity

Cost of electricity/unit = Rs. 4.00

Units of electricity = 183916 units

∴ Income = 183916 × 4.00 = Rs. 735664.00

Income from Biodigested Slurry

Cost as manure @ Rs. 250/ton = Rs. 176180

Total annual income = 191500 + 735664 = 927164.00

Profitability and Percent Return on Investment on Biogas Plant

It is explained in Table 52.4.

Table 52.4: Profitability and Per cent Return on Investment on Biogas Plant for Electricity Generation

Sl.No.	*Use of Dung as*	*Income (Rs.)*	*Cost (Rs.)*	*Annual Profit (Rs.)*	*%Return on Investment*
1.	Farmyard manure	927164	695142	232022	$\frac{232022 \times 100}{911844} = 51.60$

Pay Back Period of Plant for Electricity Generation

= Initial Investment/Average Annual Income

= 10,77,695/2,32,022 = 4.65 years

The economics of this plant for the purpose of cooking and electricity generation was compared and the payback periods of these plants using for above mentioned purposes has been calculated and is shown in Table 52.5.

Table 52.5: Payback Periods of Community Biogas Plant for Cooking Purpose and Electricity Generation

Sl.No.	*Use of Biogas*	*Payback Period (Years)*
1.	Cooking	2.00
	Electricity generation	4.65

The payback period of the plant for cooking purposes is less (2.00 years) as compared to payback of the plant for electricity generation as 4.65 years. But the present status of almost all the community biogas plants is non-functional in Punjab and the different reasons for this has already been discussed. The payback period calculated for community/institutional biogas plant for electricity generation is on the basis of new installed biogas plants. However this payback period can be highly reduced if the present study has to be implemented for the existing non-functional community/institutional biogas plants already installed for cooking purposes (Singh, 2004).

Conclusions

On the basis of the results obtained from this case study, it can be concluded that a lot of money and land, wasted for installation of community biogas plants in rural areas for cooking purpose can be re-utilized by operating community biogas plants for electricity generation. With the–use of electricity generation from biogas plants, the burden on power produced by conventional methods can be supplemented, in terms of economics etc. and also the dead community plants can be revived.

References

Bajpai, M.N. 1967. *Electrical Estimating and Costing*. Saroj Prakashan.

Chawla, O.P. 1986. *Advances in Biogas Technology*. Indian Council of Agricultural Research, New Delhi.

Khandelwal, K.C. and Maithini, P.C. 1989. *Guide to Biogas Plants*. Ministry of Non-conventional Energy Sources, New Delhi.

Mital, K.M. 1996. *Biogas Systems: Principles and Applications*. New Age International (P) Limited Publishers, New Delhi.

Singh, Indervir. 2004. *Economics of Community/Institution Biogas Plants for Cooking and Electricity Generation*. B. Tech Project Report, COAE, PAU, Ludhiana.

Pender, Harold and William A. DelMar. 1949. *Electrical Engineering Hand Book: Electrical Powers*. John Wiley and Sons, Inc., New York and London.

Chapter 53

Comparative Study of Community Biogas Plant Versus Family Size Biogas Plant: A Case Study

Er. Sarbjit Singh Sooch[**], *Er. Deepak Gupta*[*] *and Sukhwinder Singh*[**]

[*]Assistant Professor, Department of Civil Engineering, College of Agricultural Engineering, PAU, Ludhiana
[**] B. Tech Student of College of Agricultural Engineering, PAU, Ludhiana

ABSTRACT

Large number of biogas plants of family size and community size were installed in the State of Punjab. Some of the family size and all community type biogas plants failed due to technical, management and operational faults. Authors in this article have worked out the economics of small plants and big plants to show of the cost involved per person per year for the biogas supply for cooking.

Keywords: *Biogas plants, community biogas plant, family size biogas plant, economics, pollution control.*

Introduction

Family size biogas plants were installed all over the Sate of Punjab. In late seventies all were of KVIC type. Later drumless type of biogas plants of Janta and Deenbandhu type plants were also installed. The cost of these plants was high and was beyond the scope of many families although substantial subsidy was given. Government of India, thought of large size biogas plants to make use of the organic matter produced by the whole community. This would have served as a sanitation programme, solid waste treatment, water pollution control by anaerobic digestion of organic matter from human and animal waste and to serve poor people by supplying biogas for cooking either free of

[#] E-mail: sarbjit_sooch@sify.com

charge or on nominal cost. More than 100 such plants were installed all over the State but large number failed due to mismanagement, faulty design and high operative cost. Cost-wise also these plants were not cheaper as compared to the family size biogas plants. The authors in this article have endeavor to compare the family size biogas plants with community biogas plant by conducting survey of a village (Mittal, 1996).

The cost of the biogas plants as per Ministry of Non Conventional Energy Sources, Government of India, 1995, is as under (Table 53.1) (Biogas: A Rural Energy Source, 1995).

Table 53.1: Cost of Installation of Biogas Plants

Family Size Biogas Plant (Deenbandhu Model)		*Community Biogas Plant*	
Capacity of Plant (m^3)	*Average Cost in (Rs.)*	*Capacity of Plant (m^3)*	*Average Cost in (Rs.)*
1	6400	15	1,00,000
2	7500	20	1,29,000
3	9000	25	1,54,500
4	11500	35	2,29,500
6	13000	45	3,46,700
		60	3,89,500
		85	4,59,500

Note: The cost of the plants do not include the cost of gas pipe line system, cost of cattle dung and maintenance cost.

Materials and Methods

The economics of community biogas and family size biogas plant is to be calculated on the basis of the data regarding number of cattle and population of the village Rattanheri near Khanna of District Ludhiana (Punjab, India). The detailed data regarding number of cattle and population of village was collected. On the basis of the available data, community biogas plant and family size biogas plants were proposed to be installed and the cost of biogas per person per year was to be calculated (Singh, 1999).

Installation of Community Biogas Plant

1. Total no. of families = 207
   Total population = 1046
   Biogas required for cooking purpose per person per day = 0.3 m^3.
   Total gas requirement = 315 m^3
2. Total no. of cattle = 548
   Average quantity of dung available per cattle per day = 15 kg
   Total dung available = 8220 kg
   Biogas production per kg fresh dung in H.R.T. of 40 days = 0.04 m^3 (BIS, 1991)
   Total gas production = $8220 \times 0.4 = 329$ m^3

It means the gas production of 329 m^3 is more than that of gas requirement of 315 m^3.

So, the community biogas plant of 315 m^3 capacity (5 plants of 63 m^3) capacity, each) will be installed. On the basis of Table 53.1, the installation cost of 60 m^3 size biogas plant is 3,89,500.00. Thus the installation cost of 63 m^3 size biogas plant will be approximately Rs. 3,97,000 (389500 + 7200 = 396,700).

So, installation cost of community biogas plant size of 315 m^3 (5 × 63 m^3) is approximately = Rs. 3,97,000 × 5 = Rs. 19,85,000.00

Now according to KVIC recommendation out of the total cost of biogas plant, 40 per cent is the cost of gas holder and remaining 60 per cent is of the structural and installation cost (Indian standard–Family Size Biogas Plant, 1991).

Initial Investment of the Plant

Initial Cost of the Plant

1. Structure cost	=	Rs. 19,85,00 × 60/100 = Rs. 11,91,000.00
2. Cost of gas holder	=	Rs. 19,85,000 × 40/100 = Rs. 7,94,000.00
3. Cost of gas pipe line and appliances	=	Rs. 172500 + its 10%
(assume 10 per cent of the installation cost)	=	Rs. 1,90,000.00
Total cost of plant	=	Rs. 21,75,000.00

Running Cost of Biogas Plant

Considering the life of structural part as approximately 50 years, life of gas holder as approximately 10 years and life of gas pipeline system as 20 years.

The dung will be provided by the villagers and the spent slurry will be used by the villagers. Thus the cost of cattle dung is assumed to be negligible (not to be considered for running the plant).

1. Depreciation cost of structural part of the plant per year	=	11,91,000/50
	=	Rs. 23800.00
2. Interest on capital @10 per cent per year	=	Rs. 119100.00
3. Depreciation cost of gas holder per year	=	794000/10
	=	Rs. 79400.00
Interest on capital @ 10 per cent per year	=	Rs. 79400.00
4. Depreciation cost of gas pipeline system	=	1,90,000/20
	=	Rs.9500.00
Interest on capital @ 10 per cent per year	=	Rs. 19000.00

Cost of Labour

For operating the plant, 4 labourers and one technical person is required. The salary of one labour is equal to Rs. 2000.00 per month and that of technical person is equal to Rs. 4000.00 per month.

Therefore, total labour cost per year	=	Rs. 1,44,000.00
It means, total running cost of biogas plant per year	=	23,800 + 1,19,100 + 79,400 + 79400 + 9,500 + 19,000 + 1,44,000 = Rs. 474200.00

So, the expenses (cost) of biogas per person per year = 4,74,200/1046

= Rs. 453.35.00

≈ Rs. 455.00

Thus by expending Rs. 455.00 per year, the cooking needs of a person, can be fulfilled satisfactory.

Installation of Family Size Biogas Plants

For a family size biogas plant minimum 2–3 cattles are required, so on the basis of available data, number of family size biogas plants to be installed which are tabulated in Table 53.2.

Table 53.2: Number of Family Size Biogas Plants to be Installed

Sl.No.	*No. of Cattle Available (m^3)*	*Size of B.P. to be Installed*	*Cooking for no. of Person B.P. to be Installed*	*Biogas Plants to be Installed*	*Cooking for Total*
1.	2–3	1	2–3	79	197
2.	4–5	2	4–5	25 (22 + 3*)	99
3.	6–7	3	7–8	13 (5 + 8*)	77
4	8–9	4	10–11	1 (0 + 1*)	8
5.	10–12	6	14–16	6 (0 + 6*)	52
6.	13–20	8 = 2 × 4	20–22	2 × 3 = 6 (0 + 3*)	26
7.	25	12 = 2 × 6	28–32	1 x 2 = 2 (0 + 1*)	13
	Total			132	472

* It shows the number of families having less number of family members as compared to number of cattle.

Table 53.2 shows that total 132 number of family size biogas plants can be installed. With the installation of these plants, 472 persons (45 per cent of total population) of the village can get full benefit of biogas for cooking purpose.

Now referring Table 53.1 regarding cost of different size of Deenbandhu biogas plants. The cost of the biogas plants varies in the range of Rs. 9800 (7400 + 2400) for 1 m^3 size biogas plant to Rs. 15400 (13000 + 2400) for 6 m^3 size biogas plant. Rs. 2400.00 is the cost of gas pipeline system and life of gas pipeline system is 20 years.

Now considering the life of Deenbandhu biogas plant as approximately 40 years, the running cost of Deenbandhu biogas plant per year varies from approximately Rs. 332.00 to Rs. 580.00. Thus expenses on biogas to be invested on per person per year = Rs. 39.00 to Rs. 133.00.

Results and Discussion

As per economic analysis mentioned above, cost of biogas per person per year is Rs. 455.00 in community biogas plant and Rs. 39 to Rs. 133 in family size per biogas plant depending on the size and design of biogas plant and which is in the favour of installing family size biogas plant instead of community biogas plant. It is therefore essential to modify the design, the site of community plant and its management to make it suitable to maintain good sanitary conditions in the community. The objective of community biogas plant was not only to provide biogas for cooking, but to handle solid and liquid waste of the community at one place so as to provide good digested fertilizer for the fields.

The management can be improved by treating it as a private industry which will look after all the needs. Community will get regular supply of biogas, good digested fertilizer and neat and clean environment in the village (Duggal, Vyas and Sandhar, 1987 and Khandelwal and Maithani, 1989).

On the other hand individual family size biogas plant be installed instead of community biogas plants, because they are more successful and are looked after better (Singh, 1999).

Conclusions

Comparison of community biogas plants with family size biogas plants shows that cost-wise there is no benefit in installing big plants compared to family size biogas plants. Community biogas plants on the paper looks very attractive, but much more difficult in the management, operation and distribution. These can be economical only when it is to be treated as industry run by individuals.

References

Biogas: A Rural Energy Source. (1985). Ministry of Non-Conventional Energy Sources Publication, New Delhi.

BIS (1991). *Indian Standard: Family Size Biogas Plant.* Code of Practice, Bureau of Indian Standards, New Delhi.

Duggal, K.N., Vyas, S.K. and Sandhar, N.S. 1987. *Biogas Technology.* USG Publishers and Distributors, Ludhiana.

Khandelwal, K.C. and. Maithani, P.C. 1989. *Guide to Biogas Plants.* Ministry of Non Conventional Energy Sources Publication, New Delhi.

Mittal, K.M. 1996. *Biogas Systems: Principles and Applications.* New Age International (P) Limited Publishers, New Delhi.

Singh, Sukhwinder. 1999. *Comparative Study of Community Biogas Plant and Family Size Biogas Plant.* B. Tech Project Report, COAE, PAU, Ludhiana.

Chapter 54

A Study of Arsenic Effect on Blood and Tissue Glucose Concentrations in a Fish Model

Nimai Chandra Saha, Trilochan Midya and Nirmal Kumar Sarkar

Postgraduate Department of Zoology, Presidency College, Kolkata – 700 073, West Bengal, India

ABSTRACT

Glucose concentrations of blood, liver, kidney and muscle were determined in the live-fish, *Clarias batrachus* (L.) exposed to sodium arsenite in water for one month. The values were found to be significantly lower in the arsenic-treated fishes, as compared to the controls maintained in arsenic-free water. The changes were moderate in blood and liver, most pronounced in kidney and least in muscle. The overall findings indicate an adverse influence of chronic arsenic toxicity on glucose metabolism.

Keywords: *Glucose, Blood, Liver, Kidney, Muscle, Clarias batrachus.*

Introduction

Chronic arsenic toxicity has variously been reported to cause different physiological abnormalities like anaemia, keratosis and pigmentation of skin, hepatic malfunction, renal tissue damage, degenerative changes of gonad and cancers of different organs in human subjects (Bates, 1992; Morton and Dunnette, 1994; Guha Mazumder *et al.*, 1997; Buchancova *et al.*, 1998), laboratory mammals (Flora *et al.*, 1997; Guha and Sarkar, 2000), farm animals (Galloway *et al.*, 1992) and fish models (Shukla and Pandey, 1984; Kotsanis and Iliopoulou-Georgudaki, 1999; Sarkar and Sarkar, 2001). However, the relationship between arsenic toxicity and blood or tissue glucose levels remains a controversial issue as yet. Acute arsenic treatment has been reported to adversely affect gluconeogenesis and blood glucose level in laboratory mammals (Reichl *et al.*, 1988; Szinicz and Forth, 1988). Another report indicates a marked decrease of glucose concentrations in liver, muscle; gill and brain of the freshwater fish, *Tilapia mossambica*, following its exposure to sodium arsenite in water for 24 to 96 hours (Shobha Rani *et al.*, 2000). In another study, normal blood sugar level was found in 156 persons who were drinking arsenic-contaminated tube-well water (0.05–3.2 mg of arsenic/litre) for years in different rural areas of West Bengal, India (Guha Mazumder *et al.*, 1997). On the other hand, a high incidence of diabetes mellitus has been reported amongst people who were ingesting arsenic with artesian well water in a locality of Taiwan (Lai *et al.*, 1994).

In the present study, we have estimated glucose concentrations of blood, liver, kidney and muscle of the freshwater live-fish, *Clarias batrachus* (Linnaeus), following its exposure to sodium arsenite in water for one month. The objective of the study is to add to the existing knowledge of arsenic effect on blood and tissue glucose concentrations. We have chosen a fish model for the study, because a possibility of exposure of fishes to arsenic in nature cannot be nullified in ponds which are filled up in the summer with arsenic-contaminated groundwater by using pumps. It may be recalled here that the arsenic concentration of groundwater samples from rural areas of West Bengal has already been reported to reach a highest value of 3.7 mg/litre (Roy Chowdhury *et al.*, 1997).

Materials and Methods

Adult, male specimens of *Clarias batrachus* (20–22 cm in length and 95–110 g in weight) were maintained in some aquaria for one month, with 10 mg of sodium arsenite ($NaAsO_2$) per litre of water (total volume of water was 10 litre per aquarium). $NaAsO_2$ was chosen for the study, because a major portion of arsenic detected in groundwater as well as surface water-bodies in nature occurs in a soluble arsenite form (Thomton and Farago, 1997). The water of the aquaria were periodically changed and with each change, $NaAsO_2$ was added afresh. The concentration of $NaAsO_2$ was kept same as that used by an earlier group of workers in case of *Tilapia mossambica* (Shobha Ralli *et al.*, 2000). Control fishes were maintained in absence of $NaAsO_2$ in separate aquaria containing chlorine-free tap water. The water of all aquaria used in either the experimental or the control study had a pH of 7.1 to 7.2, a dissolved oxygen concentration of 6.5 to 6.6 ppm and a free carbon dioxide concentration of 1.4 to 1.5 ppm. All fishes were provided with commercial fish pellet and *Tubifex* as food *ad libitum*. Moreover, all fishes were allowed to acclimatise to the laboratory conditions for one week prior to the beginning of the study.

Both experimental and control fishes (five specimens per group) were sacrificed by striking their heads on a laboratory table and subjected to a puncture at the caudal peduncle. Blood samples were allowed to drip into clean watch-glasses from the caudal vein of the fishes. The samples were prevented from clotting by allowing a thin film of sequestrene (0.1 ml of a 1.0 per cent solution) to dry on each watch-glass one hour before collecting the samples. Tissue samples (liver, kidney and muscle) were homogenised in presence of chilled phosphate buffer (pH 7.0) (2.0 ml of the buffer was added to 1.0 9 of each tissue). Glucose concentration of blood and tissue samples were then determined colorimetrically following a modified version (Nath and Nath, 1990) of the well-known arsenomolybdate method (Somogyi, 1952). In the adopted method, 5.0 per cent zinc sulphate and 0.5N sodium hydroxide solutions have been used for deproteinization of blood and tissue samples while 5.0 per cent zinc sulphate and 0.3 N barium hydroxide solutions were used in the original method.

All studies were carried out twice for confirmation of results, initially in October–November, 2004 and then, in November–December, 2004. The temperature of water at the mid-day varied between 26.5 to 28.5°C during the first study and between 24.5 to 27.2°C during the second study. The differences between the results obtained from the experimental and the control groups of fishes were analysed statistically by means of Fisher's two-tailed t-test (Goon *et al.*, 1981).

Results

In either the first study (October–November, 2004) or the second study (November–December, 2004), glucose concentrations of blood, liver, kidney as well as muscle were found to be significantly decreased in the experimental group of fishes exposed to sodium arsenite in water for one month, as compared to the corresponding values observed in the control group of fishes. The most remarkable reduction of glucose concentration was noted in case of kidney (35.3–39.3 per cent reduction) while the least reduction was noted in case of muscle (14.6–16.3 per cent reduction). Moderate reductions of glucose concentrations were noted in blood and liver (25.6–27.9 per cent and 22.7–24.1 per cent reductions, respectively) of the treated fishes, as compared to the controls. Glucose concentrations recorded in blood and different tissues of the experimental as well as the control groups of fishes in two of our successive studies are presented in Tables 54.1 and 54.2, respectively.

Table 54.1: Blood and Tissue Glucose Concentrations (mg/100 ml of blood and mg/g of tissue) in the Fish Model During the First Study (October–November, 2004)

Tissue	*Glucose Concentration*	
	Control Group	*Arsenic-treated Group*
Blood	127.4±8.98	94.8±7.17*
Liver	17.49±1.01	13.51±0.95*
Kidney	6.12±0.41	3.96±0.40*
Muscle	16.54±0.87	14.12±0.75*

* Significantly different from control value at 1 per cent level.

Values represent means ± S.D. of five fishes in each group.

Table 54.2: Blood and Tissue Glucose Concentrations (mg/100 ml of blood and mg/g of tissue) in the Fish Model During the Second Study (November–December, 2004)

Tissue	*Glucose Concentration*	
	Control Group	*Arsenic-treated Group*
Blood	123.8±6.33	89.2±5.72*
Liver	17.11±0.92	12.98±0.94*
Kidney	6.16±0.42	3.74±0.38*
Muscle	16.24±0.82	13.59±0.81*

* Significantly different from control value at 1 per cent level.

Values represent means ± S.D. of five fishes in each group.

Discussions

Our findings on tissue glucose concentrations in arsenic-treated and control groups of the fish model, *Clarias batrachus*, are in conformity with that reported in another fish model (*Tilapia mossambica*) by an earlier group of workers (Shobha Rani *et al.*, 2000) who, however, did not determine blood glucose concentration. In so far as blood glucose concentration is concerned, our finding in the fish model is comparable to that in laboratory mammals by some other earlier workers (Reichl *et al.*, 1988; Szinicz and Forth, 1988). Considering our findings in conjunction with that of all the aforesaid workers, we conclude that chronic arsenic toxicity may considerably lower blood glucose concentration, probably by adversely affecting the rate of gluconeogenesis (Reichl *et al.*, 1988; Szinicz and Forth, 1988) in liver and kidney, which are the main sites of gluconeogenesis in the vertebrate body (Martin *et al.*, 1981). Muscle is not an active site of gluconeogenesis (Martin *et al.*, 1981). Therefore, a decreased blood glucose concentration might have been responsible for a decreased supply of glucose to the tissue, resulting in a fall of muscle glucose concentration. It may be added here that the arsenic-treated fishes are likely to be weaker than the control fishes owing to decreased blood and tissue glucose concentrations.

Finally, it seems necessary to account for the apparent discrepancy between our finding on blood glucose concentration in an arsenic-treated fish model and that of others in arsenic-affected human subjects. On account of their gill-breathing habit, fishes are continually exposed to arsenic in water. Moreover, arsenic concentration of water was 10 mg/litre in our study in *Clarias batrachus* as well as another study in *Tilapia mossambica* by an earlier group of workers (Shobha Rani *et al.*, 2000). On the other hand, human subjects studied by some earlier workers (Guha Mazumder *et al.*, 1997) were drinking water with an arsenic concentration raning between 0.5 to 3.2 mg/litre. Therefore, it is likely that a relatively larger quantity of arsenic entered everyday into the body of a fish than into the body of a human subject. This is why the blood glucose concentration remained unaffected in the aforesaid human subjects. However, it is difficult to understand the relationship between ingestion of arsenic with artesian well water and a high incidence of diabetes mellitus (hyperglycaemia), as reported in case of some people in a locality of Taiwan by another group of workers (Lai *et al.*, 1994). In view of the facts that (*i*) a decrease of blood glucose concentration has repeatedly been reported in arsenic-treated mammalian models (Reichl *et al.*, 1988; Szinicz and Forth, 1998) and (*ii*) hyperglycaemia has not been found in any of the many studied human subjects drinking arsenic-contaminated water in our country (Guha Mazumder *et al.*, 1997), we express that whether arsenic or any other substance present in groundwater was responsible for diabetes mellitus in some human subjects of a particular locality requires to be re-examined by means of as many sensitive methods as possible.

References

Bates, M.N. 1992. Arsenic ingestion and internal cancers: A review. *Amer. J. Epidemiol.*, 135: 462–472.

Buchancova, J., Klimentova, G., Knizcova, M., Mesko, D., Gallikova, E., Kubik, J., Fabianova, E. and Jakubis, M. 1998. Health status of workers of a thermal power station, exposed for prolonged periods to arsenic and other elements from fuel. *Eur. J. Pub. Hlth.*, 6: 29–36.

Flora, S.J.S., Pant, S.C., Malhotra, P.R. and Kannan, G.M. 1997. Biochemical and histopathological changes in arsenic-intoxicated rats co-exposed to alcohol. *Alcohol.* 14(6): 563–568.

Galloway, D.D., Wright, P.J., de Krester, D. and Clarke, I.J. 1992. An outbreak of gonadal hypoplasia in a sheep flock: Clinical, pathological and endocrinological features, aetiological studies. *Veter. Rec.*, 131(22): 506–512.

Goon, A.M., Gupta, M.K. and Dagupta, B. 1981. *Basic Statistics*, 3rd Edn. The World Press, Kolkata, p. 306–310.

Guha, A. and Sarkar, N.K. 2000. Chronic arsenic toxicity as a causative agent of anaemia and echinocytosis: An experimental study in Swiss mice. *Ind. J. Env. Ecoplan.*, 3(1): 19–26.

Guha Mazumder, D.N., Dasgupta, J., Santra, A., Pal, A., Ghose, A., Sarkar, S., Chattopadhyaya, N. and Chakraborti, D. 1997. Non-cancer effects of chronic arsenicosis with special reference to liver damage. In: *Arsenic-Exposure and Health Effects*, (Eds.) Abernathy, C.O., Calderon, R.L. and Chappell, W.R. Chapman and Hall, London, p. 112–123.

Kotsanis, N. and Iliopoulou-Georgudaki, A. 1999. Arsenic induced liver hyperplasia and kidney fibrosis in rainbow trout (*Oncorhynchus mykiss*) by microinjection technique: A sensitive animal bioassay for environmental metal toxicity. *Bull. Environ. Contamin. Toxicol.*, 62(3):.169–178.

Lai, M.S., Hsueh, Y.M. and Chen, C.J. 1994. Ingested inorganic arsenic and prevalence of diabetes mellitus. *Amer. J. Epidemiol.*, 139: 484–492.

Martin, D.W., Mayes, P.A. and Rodwell, V.W. 1981. *Harper's Review of Biochemistry*, 18th Edn. Lange Medical Publications, California, p. 160–185.

Morton, W.E. and Dunnette, D.A. 1994. Health effects of environmental arsenic. In: *Advances of Environmental Science and Technology*, (Ed.) Nriagu, J.O. John Wiley, New York, 27: 17–34.

Nath, R.L. and Nath, R.K. 1990. *Practical Biochemistry in Clinical Medicine*, 2nd Edn. Academic Publishers, Kolkata, p. 38–41.

Reichl, F.X., Szinicz, L., Kreppel, H., Fichtl, B. and Forth, W. 1988. Effect of As_2O_3 on carbohydrate metabolism after single or repeated injection in guinea pigs. *Toxicologist*, 8(1): 20–23.

Roy Chowdhury, T., Mondal, B.K., Samanta, G., Basu, G.K., Chowdhury, P.P., Chanda, C.R., Karan, N.K., Lodh, D., Dhar, R.K., Das, D., Saha, K.C. and Chakraborti, D. 1997. Arsenic in groundwater in six districts of West Bengal, India: The biggest aresenic calamity in the world: The status report upto August, 1995. In: *Arsenic-Exposure and Health Effects*, (Eds.) Abernathy, C.O., Calderon, R.L. and Chappell, W.R. Chapman and Hall, London, p. 93–111.

Sarkar, J. and Sarkar, N.K. 2001. Haematological profile in the freshwater fish, *Channa punctatus* (Bloch) following chronic experimental exposure to sodium arsenite in water. *Trans. Zool. Soc. East. India.* 5(1): 10–21.

Shobha Rani, A., Sudarshan, R., Reddy, T.N., Reddy, P.U.M. and Raju, T.N. 2000. Effect of sodium arsenite on glucose and glycogen levels in freshwater teleost fish, *Tilapia mossambica*. *Poll. Res.*, 19(1): 129–131.

Shukla, J.P. and Pandey, K.O. 1994. Impaired spermatogenesis in arsenic-treated freshwater fish, *Colisa fasciatus* (Bl. and Sch.). *Toxicol. Lett.*, 21(2): 191–195.

Somogyi, M. 1952. In: *Practical Biochemistry in Clinical Medicine*, 2nd Edn. (Eds.) Nath, R.L. and Nath, R.K. Academic Publishers, Kolkata, p. 38–41.

Szinicz, L. and Forth, W. 1988. Effect of As_2O_3 on gluconeogenesis. *Arch. Toxicol.* 61: 444–449.

Thornton, I. and Farago, M. 1997. The geochemistry of arsenic. In: *Arsenic-Exposure and Health Effects*, (Eds.) Abernathy, C.O., Calderon, R.L. and Chappell, W.R. Chapman and Hall, London, p. 1–16.

Chapter 55
Season-induced Metabolic Alterations in *Abelmoschus esculentus* (L.) Moench (Okra)

Supatra Sen and S. Mukherji*

Plant Physiology Laboratory, Department of Botany, University of Calcutta, Kolkata – 700 019, W.B., India

ABSTRACT

Season-controlled metabolic changes were studied in *Abelmoschus esculentus* (L.) Moench (Okra) in two developmental stages–28 days after sowing (DAS)–pre flowering and 58 DAS–post flowering and three seasons-summer, rainy and winter respectively. Carotenoids (carotene and xanthophyll), pectic substances along with activities of IAA oxidase and polyphenol oxidase (PPO) were measured. Summer season environmental conditions recorded high amounts of carotenoids and pectic substances while IAA oxidase and PPO activities were low. The winter season recorded an opposite trend in the amounts and activities of these constituents and enzymes. These results suggest that summer season is the most favourable while winter season environmental conditions are the most stressful for the growth and survival of this plant.

Keywords: *Season, Abelmoschus esculentus, Okra, Carotenoids, IAA oxidase, Polyphenol oxidase, Pectic substances, Environmental stress.*

Introduction

Plants coordinate their metabolic activity with the surrounding fluctuating environments, and growth and metabolism are greatly altered to deal with the abiotic stressors. The environment around

* *Address for Correspondence*: 2, Gopal Ghosh Lane, Kidderpore, Kolkata – 700 023, India, Email: supatrasen@vsnl.net
Present Address and Designation: Lecturer in Botany, Uluberia College, Uluberia, Howrah – 711 315

plants fluctuates regularly and predictably over daily and seasonal cycles. Plant responses to different seasons are detectable both by external morphological changes and internal biochemical and physiological alterations.

Crops are limited to about 25 per cent of their potential yield by the impact of environmental stress (Boyer, 1982). Because plants are rooted in one place, they have a limited capacity to avoid unfavourable changes in their environment such as extremes of temperature, water shortage, insufficient or excessive light and shortage of mineral nutrients (Smirnoff, 1995). Previous studies on *Abelmoschus esculentus* (L.) Moench (Okra) reveal that changes in seasonal conditions lead to adjustment of the relative amounts and concentrations of the photosynthetic and biochemical constituents (Sen and Mukherji 1998a, 1998b, 1999a, 2004) which also lead to differences in growth, yield and yield quality (Sen and Mukherji 1998c, 1998d, 1999b, 2002).

In this article, certain biochemical constituents *viz.* carotene, xanthophyll and pectic substances along with the enzymes IAA-oxidase and polyphenol oxidase were evaluated to record the seasonal impact on their concentration and activity and hence their varying behaviour in plant metabolism.

Materials and Methods

The work was carried out in pot culture in the experimental garden of the University Campus at Kolkata under natural environmental conditions. The seasons under study were summer (March–June), rainy (July–September) and winter (November–February).

Seeds of *Abelmoschus esculentus* (L.) Moench (Okra) variety Parbani Kranti were obtained from National Seed Corporation and sown (8–10 seeds per pot of diameter 12 inches) in sandy loam soil and farmyard manure in the ratio of 3 : 1. Fully expanded, penultimate (second from top) leaf samples were collected at two stages of development–pre flowering (28 DAS) and post flowering (58 DAS) in all the three above mentioned seasons. For all three seasons, identical water managements were maintained along with a constant nutrient status of the soil.

The duration of the experimental period was from November 2002–October 2003–a period characterized by considerable variations in temperature, photoperiod, light intensity, relative humidity and rainfall.

The meteorplogical data (Table 55.1) were obtained from Regional Meteorological Centre, Kolkata. The data indicated the monthly mean values, which have been finally expressed as seasonal mean values for the suitability of this work.

Table 55.1: Meteorological Data from November 2002 to October 2003

	Summer	*Rain*	*Winter*
Temperature (°C)	34	30.5	14.5
Sunshine Hours (Hour)	13.5	12.5	10.5
Light Intensity (Lux)	75000	68000	34000
Relative Humidity (%)	79	86	64
Rainfall (mm)	114	337	11

Data presented are the seasonal mean values.

Carotene and xanthophyll were estimated according to the method of Davies (1965), IAA oxidase was assayed according to Malik and Singh (1980), polyphenol oxidase was assayed according to Mayer and Harel (1979) and pectic substances determined according to Ranganna (1977).

The data obtained from three replications were statistically analyzed. Standard Error (S.E.) and Critical Difference (C.D.) values of both season and stage at 5 per cent and 1 per cent levels were calculated from the respective analysis of variance (ANOVA) tables.

Results and Discussion

Carotene and xanthophyll (carotenoid) contents were the highest in summer which declined significantly in the rainy and winter seasons (Figures 55.1 and 55.2). Carotenoid content also declined in the older post flowering stages as compared to pre flowering in all seasons. Absorption of excess light by chlorophyll results in the formation of triplet chlorophyll which can then pass excitation energy to oxygen resulting in formation of singlet oxygen. This is more reactive than ground state oxygen and can lead to peroxidation and breakdown of thylakoid lipids. The carotenoids function both in energy dissipation and to detoxify chlorophyll triplets and singlet oxygen (Owen, 1994).

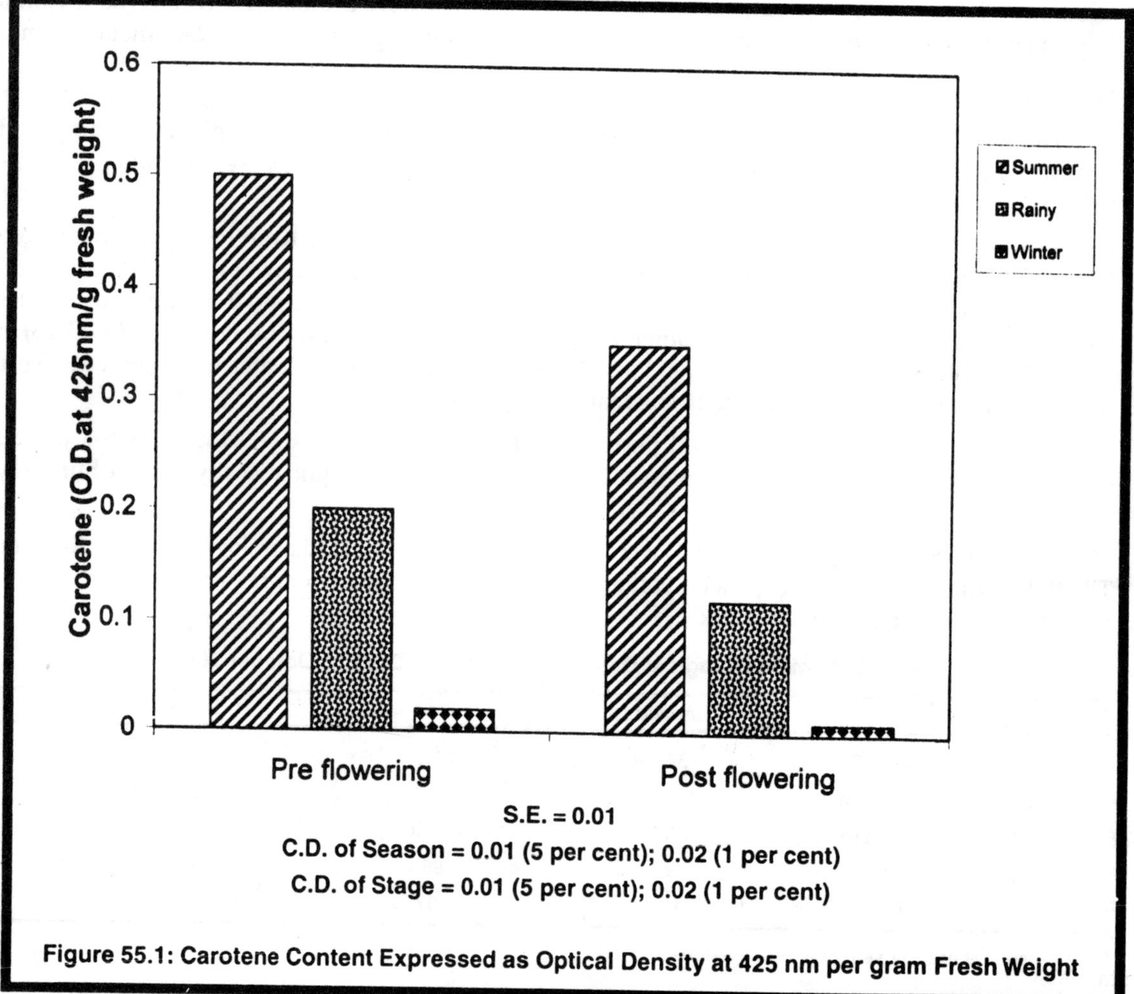

Figure 55.1: Carotene Content Expressed as Optical Density at 425 nm per gram Fresh Weight

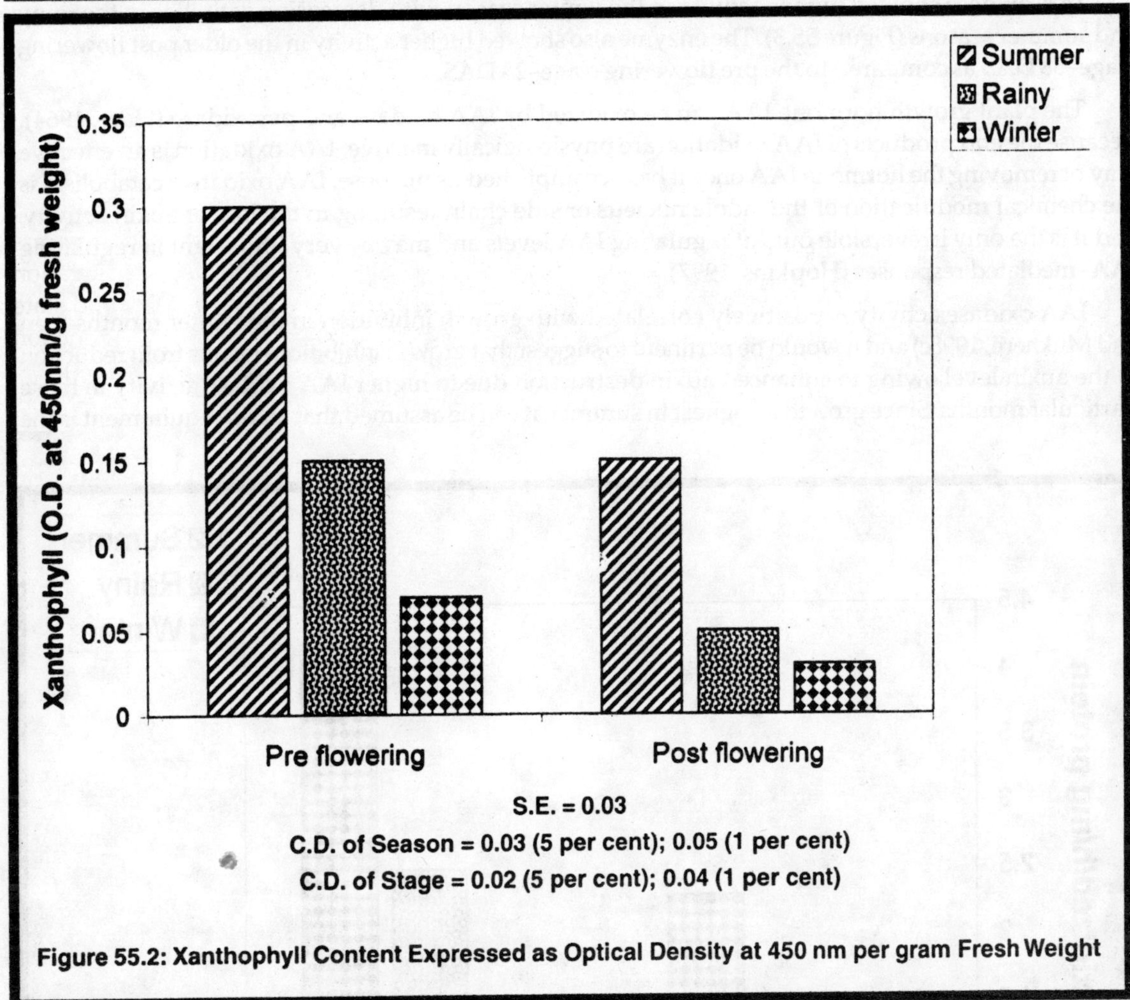

Figure 55.2: Xanthophyll Content Expressed as Optical Density at 450 nm per gram Fresh Weight

β-carotene prevents damage by singlet oxygen by acting as a quencher of triplet chlorophyll and singlet oxygen. Carotene-deficient mutants and plants treated with carotenoid synthesis inhibitors show rapid photo-oxidation of pigments and peroxidation of thylakoid lipids when exposed to light (Young and Britton, 1990).

Carotenoids help in maintaining the conformation of the chlorophyll-protein complexes (Hladik *et al.*, 1982). Besides they play an all important role in antioxidant defence (Larson, 1995) and can effectively inactivate electronically excited molecules such as $^1O_2$ and triplet chlorophyll (process termed 'quenching') to protect the chlorophyll and photosynthetic membrane against lethal oxidation and photo-oxidative damage (Goodwin and Mercer, 1983; Mayfield *et al.*, 1986).

Thus a decrease in carotenoid content in the rainy and winter seasons and the post flowering stages in *Abelmoschus* implies a reduced capacity to help maintain the conformation of the pigment–protein complexes as well as a decreased efficiency to scavenge the harmful free radicals (activated oxygen species).

IAA oxidase showed higher activity in the winter season with decreasing activities in the rainy and summer seasons (Figure 55.3). The enzyme also showed higher activity in the older post flowering stage–58 DAS as compared to the pre flowering stage–28 DAS.

The plant growth hormone IAA can be oxidized by IAA oxidase and peroxidase (Hare, 1964). Because the end products of IAA oxidation are physiologically inactive, IAA oxidation is an effective way of removing the hormone IAA once it has accomplished its purpose. IAA oxidative catabolism is the chemical modification of the indole nucleus or side chain resulting in the loss of auxin activity, and it is the only irreversible output regulating IAA levels and may be very important in regulating IAA–mediated responses (Hopkins, 1997).

IAA oxidase activity is positively correlated with growth inhibition in the winter months (Sen and Mukherji, 1998c) and it would be pertinent to suggest that growth inhibition–results from reduction in the auxin level owing to enhanced auxin destruction due to higher IAA oxidase activity in these particular months. Since growth is highest in summer, it can be assumed that auxin requirement is the

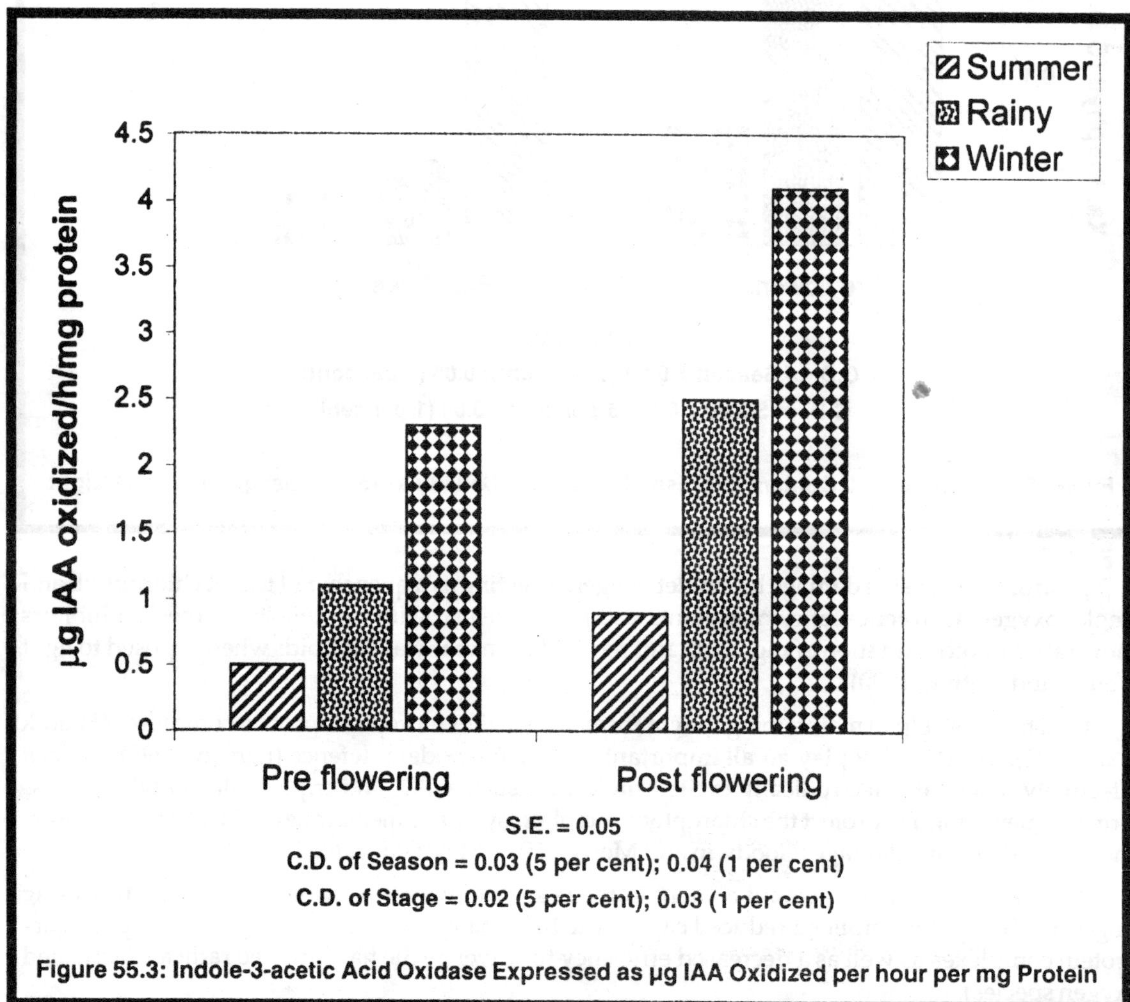

S.E. = 0.05

C.D. of Season = 0.03 (5 per cent); 0.04 (1 per cent)

C.D. of Stage = 0.02 (5 per cent); 0.03 (1 per cent)

Figure 55.3: Indole-3-acetic Acid Oxidase Expressed as µg IAA Oxidized per hour per mg Protein

highest in this season. In summer IAA oxidase activity was found to be the lowest indicating low auxin destruction. Pre flowering stages (28 DAS), when growth is higher, had higher auxin requirement in all seasons as compared to older post flowering (58 DAS) and this correlated to the lower IAA oxidase activity in the pre flowering stages.

p-diphenols, *o*-diphenols and polyphenols generally inhibit IAA oxidation. It has been suggested that these compounds serve a regulatory function in IAA peroxidative oxiaation. It was observed in this work that a high phenolic content could be correlated with low IAA oxidase activity in the summer season of *Abelmoschus* (Sen and Mukherji, 1998a.) An opposite trend was observed in the winter months. Similarly the pre flowering stage–28 DAS had higher phenolic content with low IAA oxidase activity and the post flowering stage 58 DAS showed the opposite trend.

The enzyme polyphenol oxidase (PPO) functions as a phenol oxidase in higher plants. PPO oxidizes phenolic compounds which have been associated with antioxidant activity (Larson, 1995). During periods of stress, this plastidial enzyme is released into the cytoplasm and it oxidizes phenols to lower the level to produce quinines which are quite toxic in nature (Mayer and Harel, 1979; Vaughn and Duke, 1984) and helps in prevention of chlorophyll bleaching.

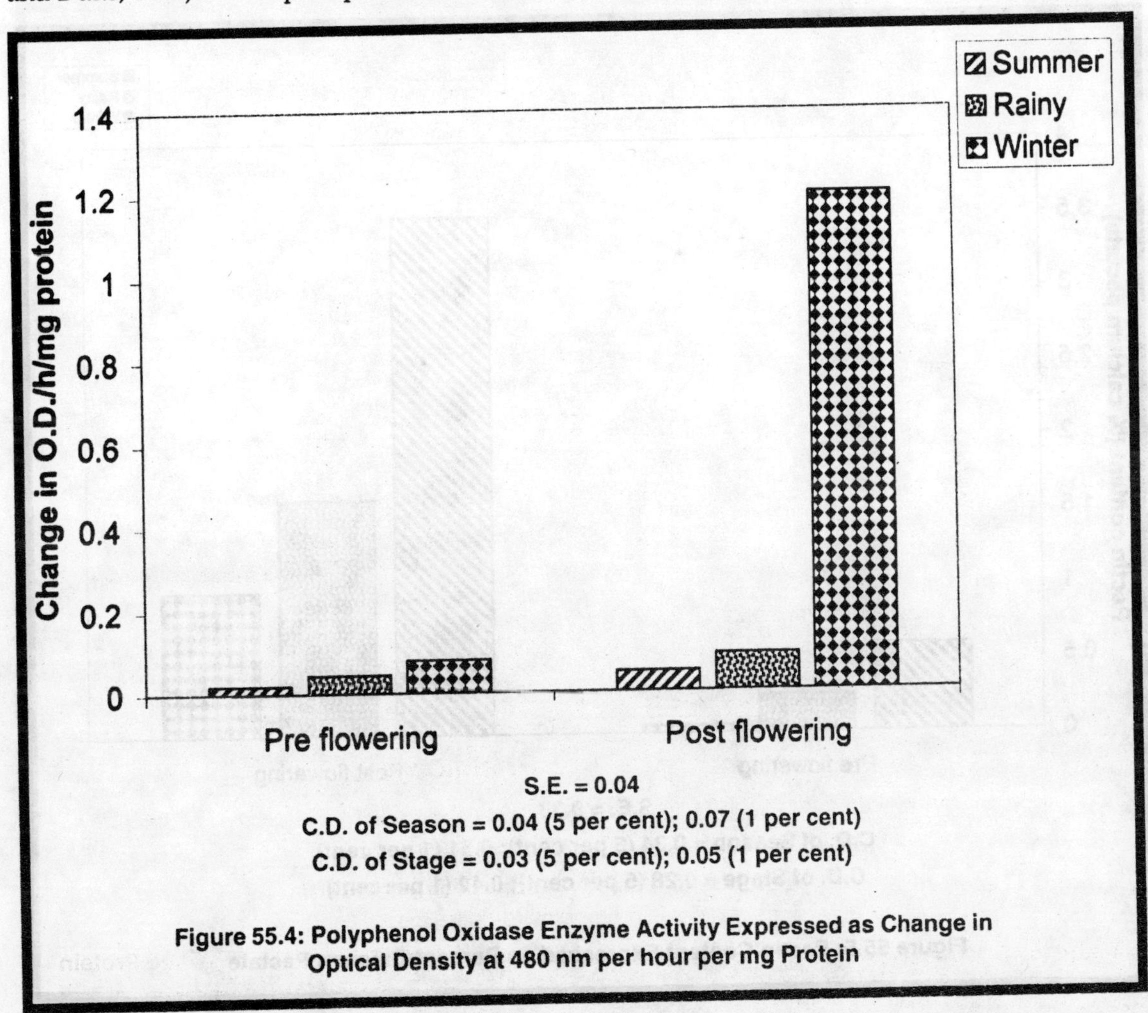

Figure 55.4: Polyphenol Oxidase Enzyme Activity Expressed as Change in Optical Density at 480 nm per hour per mg Protein

An increase in the activity of PPO in the winter and rainy seasons (Figure 55.4) may indicate a protective measure adopted by the plant in response to the prevailing conditions of environmental stress. From this work, it is observed that an increase in PPO activity always resulted in lowering of phenol content–winter and rainy seasons (Sen and Mukherji, 1998a). PPO was found to be more active in the post flowering 58 DAS stage in all seasons as compared to the young pre flowering 28 DAS stage thus indicating that older tissues are more susceptible to prevailing environmental stress conditions.

Pectic substances were higher in the summer season as compared to rainy and winter (Figure 55.5). The young pre flowering recorded a lower amount of pectic substances as compared to the older post flowering in all three seasons. Pectic substances are present in the middle lamella of cell walls and consist of pectic acids, pectin and protopectin. An earlier work on the Okra fruit reveal a similar trend (Sen and Mukherji, 1999a) as observed in the leaf samples.

From this work it is evident that *Abelmoschus* is well adapted to adjust to the summer season environmental conditions of Kolkata with high amounts of antioxidants–carotene and xanthophyll,

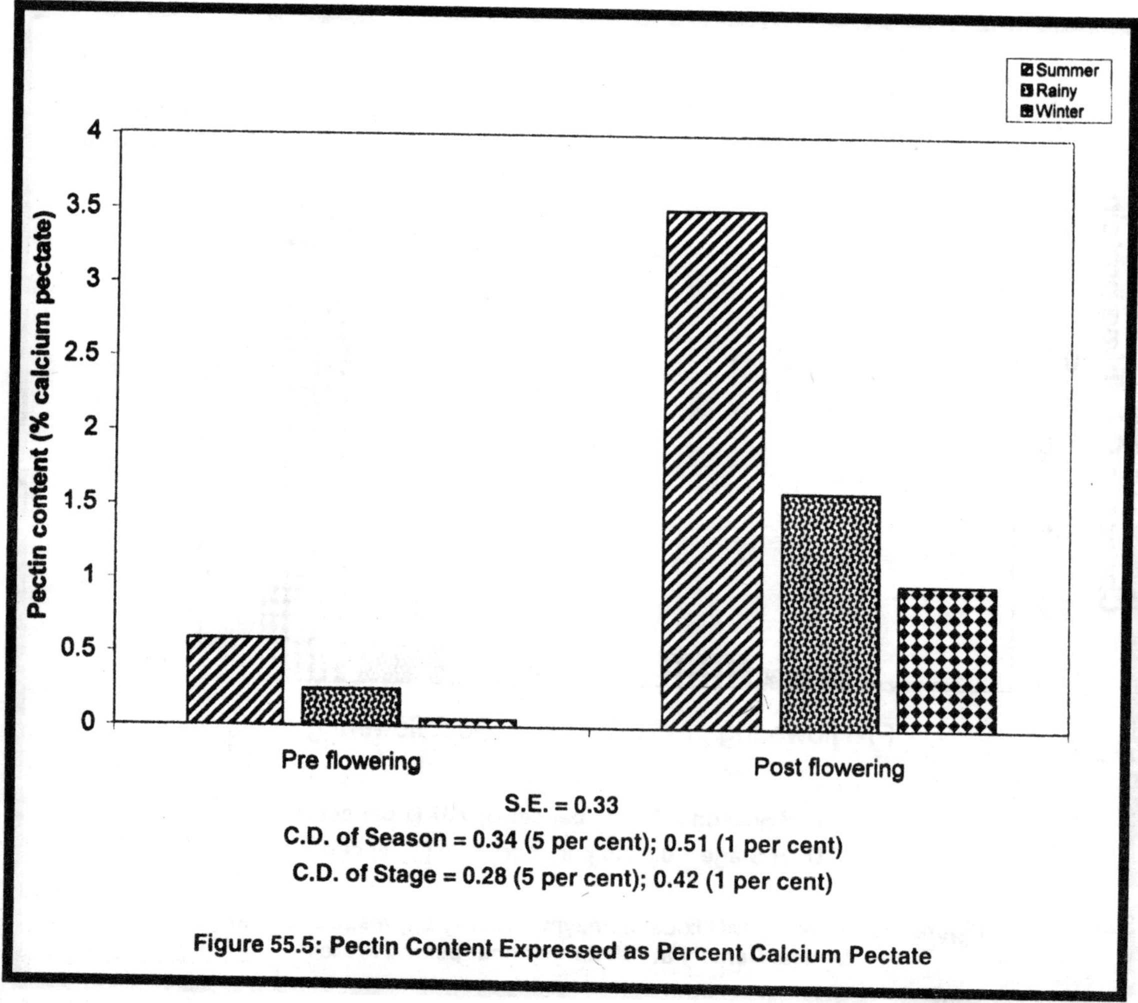

Figure 55.5: Pectin Content Expressed as Percent Calcium Pectate

low IAA oxidase activity ensuring greater auxin content, low PPO activity leading to high phenol content along with a high content of pectic substances. Rainy and winter are stressful seasons with low carotenoids and pectic substances and high IAA oxidase and PPO activities all of which have deleterious effects on the metabolism of this plant.

References

Boyer, J.S. 1982. Plant productivity and environment. *Science*, 218: 443–448.

Davies, B.H. 1965. Analysis of carotenoid pigments. In: *Chemistry and Biochemistry of Plant Pigments*, (Ed.) Goodwin, T.W. Academic Press, New York, p. 489–531.

Goodwin, T.W. and Mercer, E.I. 1983. In: *Introduction to Plant Biochemistry*. Pergamon Press, England, p. 116–118.

Hare, R.C. 1964. IAA oxidase. *Bot. Rev.*, 30: 129.

Hladik, J., Pancoska, P. and Sofrova, D. 1982. The influence of carotenoids on conformation of chlorophyll-protein complexes isolated from the cyanobacterium *Plectonema boryanum*, absorption and circulal dichroism study. *Biochem. Biophys. Acta.*, 681: 263–272.

Hopkins, W.G. 1997. *Introduction to Plant Physiology*, 2nd Edn. John Wiley and Sons, Inc.

Larson, R.A. 1995. Plant defences against oxidative stress. *Arch. Insect Biochem. Physiol.*, 29(2): 175–186.

Malik, C.P. and Singh, M.B. 1980. In: *Plant Enzymology and Histo-enzymology*. Kalyani Publishers, p. 51–53.

Mayer, A.M. and Harel, E. 1979. Polyphenol oxidases in plants. *Phytochemistry*, 18: 193–215.

Mayfield, S.P., Nelson, T. and Taylor, W.C. 1986. The fate of chloroplast proteins during photo-oxidation in carotenoid-deficient maize leaves. *Plant Physiol.*, 82: 760–764.

Owens, T.G. 1994. Excitation energy transferred between chlorophylls and carotenoid: A proposed molecular mechanism for non-photochemical quenching. In: *Photoinhibition of Photosynthesis, from Molecular Mechanisms to the Field*, (Eds.) Baker, N.R. and Boyer J.R. Bios Scientific Publishers, Oxford, p. 95–109.

Ranganna, S. 1997. In: *Manual of Analysis of Fruits and Vegetable Products*. Tata Mcgraw Hill Publ. Co. Ltd., New Delhi.

Sen, S. and Mukherji, S. 1998a. Seasonal variation in biochemical constituents of *Abelmoschus esculentus* (L.) Moench and *Lycopersicon esculentum* Mill. *J. Interacad.*, 2(3): 118–123.

Sen, S. and Mukherji, S.1998b. Seasonal effects on nitrogenous compounds in two crop plants. *Environ. Ecol.*, 16(4): 871–874.

Sen, S. and Mukherji, S.1998c. Seasonal changes in growth characteristics in *Abelmoschus esculentus* (L.) Moench and *Lycopersicon esculentum* Mill. *Ind Biol.*, 30 (2): 60–66.

Sen, S. and Mukherji, S. 1998d. Influence of seasons in determining the date of sowing and fruit quality of *Abelmoschus esculeritus* (L.) Moench (Okra) and *Lycopersicon esculentum* Mill. (Tomato). *Ind. Agric.*, 42(3): 161–166.

Sen, S. and Mukherji, S. 1999a. Changes in photosynthetic parameters in *Abelmoschus esculentus* (L.) Moench as affected by seasonal environmental conditions. *Asian J. Microbiol. Biotech. Env. Sc.*, 1(3–4): 157–161.

Sen, S. and Mukherji, S. 1999b. Biochemical evaluation of the Okra *Abelmoschus esculentus* (L.) Moench fruit under seasonal environmental changes. *Ecol. Environ. Cons.*, 5 (4): 381–384.

Sen, S. and Mukherji, S. 2002. Season–dependent mineral accumulation in fruits of Okra (*Abelmoschus esculentus*) and Tomato (*Lycopersicon esculentum*). *J. Env. Biol.*, 23(1): 47–50.

Sen, S. and Mukherji, S. 2004. Alternations in activities of acid phosphatase, alkaline phosphatase, ATPase and ATP content in response to seasonally varying Pi status in Okra (*Abelmoschus esculuntus*). *J. Env. Biol.*, 25(2): 181–185.

Smimoff, N. 1995. In: *Environment and Plant Metabolism: Flexibility and Acclimation,* (Ed.) Smimoff, N. Bios Scientific Publishers, Oxford, UK.

Vaughn, K.C. and Duke, S.O. 1984. Function of polyphenol oxidase in higher plants. *Physiol. Plant*, 60: 1016–1112.

Young, A. and Britton, G. 1990. Carotenoids and stress. In: *Stress Responses in Plants: Adaptation and Acclimation Mechanisms,* (Eds.) Alscher, R.G. and Cummings, J.R. Wiley-Liss, New York, p. 87–112.

Chapter 56

Air Borne Fungal Spores at Sharavathi Reservoir, Karnataka, India

K. Harish Kumar, B.R. Kiran, R. Purushotham, E.T. Puttaiah and S. Manjappa

Department of Environmental Science, Kuvempu University, Shankaraghatta – 577 451, Karnataka, India

ABSTRACT

The occurrence of fungal spores and hyphal fragments in the atmosphere of Sharavathi reservoir (Site 1) and its downstream stretch (Site 2) was, monitored for five months from August–December, 2004. Open plate method was employed for monitoring the mycoflora. During this study 11 types of fungal spores having diverse group were recorded *viz*., *Penicillium* sp., *Aspergillus* sp., *Cladosporium* sp., *Fusarium* sp., *Glomercularia* sp., *Helimanthosporium* sp., *Alternaria* sp., *Curvularia* sp., *Rhizopus* sp., *Cephalosporium* sp. and *Trichoderma* sp. While, *Penicillium* and *Aspergillus* species were common in both the sites. The highest fungal flora was found in August and lowest in December. The factors responsible for the occurrence of fungal spores are discussed in detail.

Keywords: Fungal spore, Reservoir.

Introduction

The main sources of viable airborne mycoflora are soil, water and vegetation. These are transported to the atmosphere and are transported to the atmosphere and are always associated with fine, dusty, particulate matter, survival of microflora depends on the morphological features, micro meteorological, climatological conditions and on the interaction between these parameters. From the atmosphere, microbes can reach the soil and waterbodies contaminate food or water supplies and or invades the people who may become healthy carriers or each develops a disease caused by them.

Fungi produce spores that becomes airborne, some also produce mycotoxins or volatile organic compounds. Some of the most common fungi that are found in the indoor and can cause health problems are *Penicillium* sp. *Aspergillus* sp. and *Alternaria* sp. The enumeration of fungi forms a major component of the bioparticles in the air. Nair *et al.* (1986) included the information obtained on the airborne fungal spores of different, regions of India. Sreeramulu (1967) and Chanda and Mandal (1978) reviewed the available literature and showed the diversity and richness of airborne fungal spores in Indian environment. In the present study the mycoflora in the air of Sharavathi reservoir site and its downstream stretch was monitored for a five months and the observations on the relative abundance and seasonal occurrence of different fungal spores type are present.

Materials and Methods

Air monitoring was carried out during August to December 2004 with open plate containing potato dextrose agar media exposing the plates for 5 minutes. Triplicate samples were collected at two sites (Site 1–Sharavathi reservoir itself, Site 2–down stream stretch of the reservoir) every month. The study area is located at latitude 14 ° 41' 24" N, longitude 74° 50' 54" E. The plates are incubates at 28° C till the appearance of growth. The fungal colonies were counted and identified by different literature (Gilman, 1949; Tandon, 1968; Ellis, 1976 and Subramanian, 1971)

Results and Discussion

The prevailing air temperature at Sharavathi reservoir varied between 28 and 30° C there was rainfall in most of the months (June to September).

The fairly high temperature prevailing throughout the year coupled with rainfall naturally promoted a good growth of vegetation and accumulation of organic matter supporting luxuriant growth of fungi. Table 56.1 shows the occurrence of fungal spores at Site 1 and 2 respectively. A total of 11 species of spores were identified *viz., Penicillium* sp., *Aspergillus* sp, *Alternanthara* sp. *Curvularia* sp, *Rhizopus* sp, *Helimanthosporium* sp., *Cladosporium* sp, *Fusarium* sp, *Glomerularia*, *Cephalosporium* sp. and *Trichoderma* sp. In this study, the dominant type was *Aspergilli,* which mostly consisted of the genera *Aspergillus* and *Penicillium* sp. In India, few studies which attempted to delineate the two genera in the aeromycoflora showed that *Aspergillus* is more prevalent than *Penicillium* (Agarwal *et al.*, 1969 and Rati and Ramalingam, 1976). It is therefore presumed that a similar situation may occur at Sharavathi reservoir stretch. This spore is also known to be more frequent in urban areas than in rural areas (Subba Reddi, 1970 and Bajaj, 1978). *Alternaria,* third in order of abundance has been reported as number one in the united states of America (Morrow *et al.*, 1964). In India its frequency is rather high in northern parts than in south (Nair *et al.*, 1986). A simultaneous survey using both cultural and visual means may throw light on this differential frequencies of *Alternaria* sp. Spores were encountered on all the days of the months trapping was done (Figure 56.1). The month of August witnessed the maximum incidence accounting to 35.08 and 38.84 per cent for Sites 1 and 2 respectively. At Site 1, the month of September recorded 26.32, per cent followed by October (19.30 per cent), November (12.28 per cent) and December (7.02 per cent). Similarity, in Site 2 the total count for spores were 24.79 per cent, 17.36 per cent 11.57 per cent and 7.44 per cent for the month of September, October–November and December respectively. The incidence of hyphal fragments (665–900/sq. cm) was considerably high. Most of the fragments were brown, septate and the hyaline ones rarely appeared. On several occasions they were trapped in association with the conidia of *Cladosporium, Cephalosporium, Curvularia* and *Alternaria.* They were more numerous during Aug–October since many of the air borne hyphal fragments remain viable and germinate in the manner of fungal spores (Pady and Gregory, 1963), they may be considered as an important constituent of the fungus air spora and as part of the dispersal mechanism of the parent fungus. Gregory (1973) reported that environmental factors are the most important which affect the total number of fungal population in the atmosphere. Hence, seasonal variation in the fungal concentration of Sharavathi reservoir and its downstream stretch area depends upon the climatic factors.

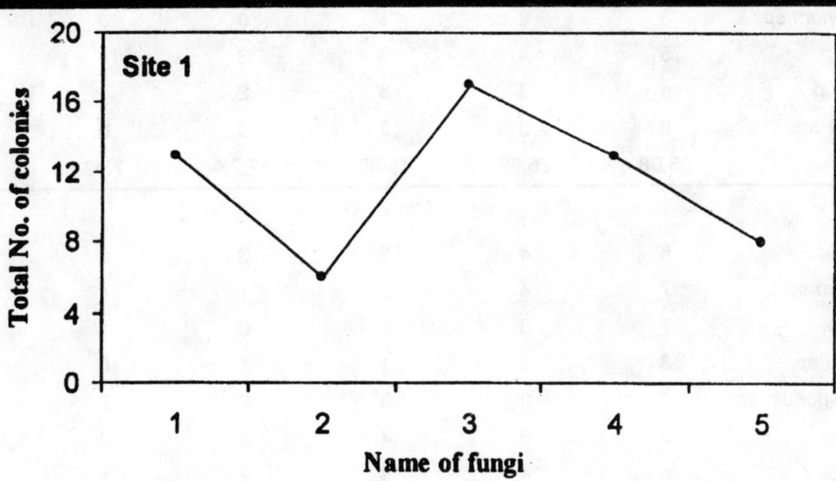

1. *Rhizopus* sp., 2: *Cephalosporium* sp., 3: *Penicillium* sp., 4: *Aspergillus* sp., 5: *Trichoderma* sp.

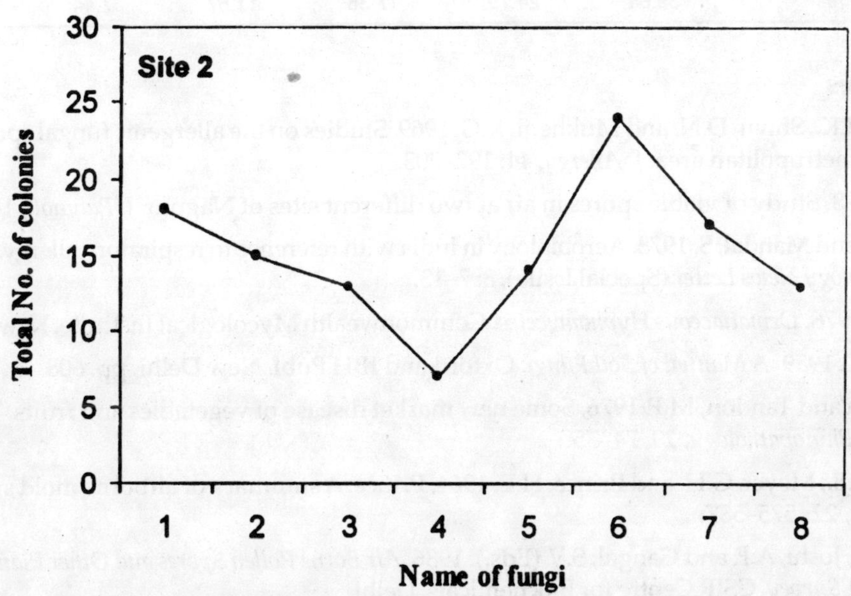

1. *Aspergillus* sp., 2: *Cladosporium* sp., 3: *Fusarium* sp., 4: *Glomerularia* sp., 5: *Helimanthosporium* sp., 6. *Penicillium* sp., 7: *Alternaria* sp., 8: *Curvularia* sp.

Figure 56.1: Total Number of Colonies of Different Fungal Species

Table 56.1: Distribution of Different Types of Fungal Spores at Sharavathi Reservoir (Site 1) and its Stretch (Site 2)

Name of the Species	*August*	*September*	*October*	*November*	*December*	*Total (%)*
			Site 1			
Rhizopus sp.	5	3	2	2	1	22.81
Cephalosporium sp.	3	2	1	0	0	10.52
Penicillium sp.	5	4	3	3	2	29.82
Aspergillus sp.	4	3	3	2	1	22.81
Trichoderma sp.	3	3	2	0	0	14.04
Percentage	35.08	26.32	19.30	12.28	7.02	
			Site 2			
Aspergillus sp.	6	4	3	3	2	14.88
Cladosporium sp.	7	4	2	1	1	12.40
Fusarium sp.	7	3	2	0	1	10.74
Glomerularia sp.	3	2	1	1	0	5.79
Helimanthosporium sp.	4	3	3	2	2	11.57
Penicillium sp.	9	6	4	3	2	19.83
Alternaria sp.	6	4	4	2	1	14.05
Curvularia sp.	5	4	2	2	0	10.74
Percentage	38.84	24.79	17.36	11.57	7.44	

References

Agarwal, M.K., Shivu, D.N. and Mukherji, K.G. 1969. Studies on the allergenic fungal spores of Delhi, India, metropolitan area. *J. Allergy*, 44: 192–203.

Bajaj, A. 1978. Study of viable spores in air at two different sites of Nagpur. *J. Palynol.*, 14: 136–149.

Chanda, S. and Mandal, S. 1978. Aerobiology in India with reference to respiratory allergy. *International Aerobiology News Letter* (Special Issue), 8: 7–13.

Ellis, M.B. 1976. *Dematiaceous Hyphomycetes*. Commonwealth Mycological Institute, Kew, England.

Gelman, J.C. 1959. *A Manual of Soil Fungi*. Oxford and IBH Publ. New Delhi, pp. 608.

Jamaluddin and Tandon, M.P. 1976. Some new market disease of vegetables and fruits. Tandon M.P. *Indian Phytopathology*, 29: 74–75.

Morrow, M.B., Meyer, G.H. and Prince, H.E. 1964. Prince: A summary of airborne mold surveys. *Ann. Allergy*, 22: 575–587.

Nair, P.K.K., Joshi, A.P. and Gangal, S.V. (Eds.). 1986. *Air Borne Pollen Spores and Other Plant Materials of India: A Survey*. CSIR Centre for Biochemicals, Delhi.

Pady, S.M. and Greogry, P.H. 1963. Numbers and viability of airborne hyphal fragments in England. *Trans. Br. Mycol. Soc.*, 46: 609–613.

Rati, E. and Ramalingam, A. 1976. Airborne *Aspergilli* at Mysore. *Aspects of Allergy and Applied*

Immunology, 9: 139–149.

Sreeramulu, T. 1967. Aerobiology in India. *J. Scient. Ind. Res.*, 26: 474–480.

Subba Reddi, C. 1970. A comparative survey of atmospheric pollen and fungus spores at two places twenty miles apart. *Acta. Allergol.*, 25: 189–215.

Subramanian, C.B. 1971. *Hypomycetes.* ICMR, New Delhi.

Chapter 57

Hydrological Studies for Dams in Ungauged Small Catchments: A Case Study

*Er. Sarbjit Singh Sooch**, Er. Deepak Gupta* and N.K. Khullar***

**Assistant Professor, **Professor-cum-Head,*
Department of Civil Engineering, College of Agricultural Engineering, PAU, Ludhiana

ABSTRACT

For the design of dams, spillways, location of outlets and to determine the life of reservoirs, it is necessary to make detailed hydrological calculations. These calculations will involve determination of design flood and rate of sedimentation and distribution of sediment in the reservoir. Dholbaha Dam and Maili Dam (earthen rockfill dams) in the Kandi area has been considered as representative cases for study as the hydrological, topographical, climatic and geological conditions are more or less indentical in the region. The conclusions derived from these dams for hydrological studies can be applied to other similar dams being provided in the area.

Keywords: *Hydrological studies, design flood, Sedimentation, Ungauged catchments, Flood control.*

Introduction

In the North-Eastern parts of Punjab State in Ropar, Gurdaspur and Hoshiarpur districts, there are number of flashy torrents called 'choes' which cause heavy damage to the adjoining areas during monsoons. After emerging from the hills, these choes fan out into several small torrents resulting in delta like formation in the plains. The abrupt and flashy discharges in these choes with steep bed slopes cause heavy damage to standing crops and property, whereas the flood during the falling river

* E-mail: sarbjit_sooch@sify.com

stages from these choes spread sandy deposits on the fertile lands. Numberous dam sites for storage of excess rain water exist in this region. By constructing a permanent structure of earth or masonry or a combination of both at suitable location in these ungauged catchments, it is not only possible to conserve the rain water and provide supplemental irrigation to the drought–stricken areas but also prevent soil erosion (Singh, 1991).

Hydraulic structures are designed for a flood which is likely to exceed once during the life of the structure. In gauged catchments, the design flood can be computed from the analysis of the flood data where as in the ungauged catchments this has to be computed by other methods.

Sedimentation studies have been carried out to determine the rate of sedimentation and the pattern of silt distribution in the reservoir in order to determine the safe level for fixing the irrigation outlet pipe and to assess the effective storage capacity available for irrigation supplies and flood control during various stages of the life of the reservoir. In this study average rate of sedimentation has been considered (Life of Reservoir, 1977; Mohammed, 1979 and Mutreja).

Methods and Materials

Various methods have been used for estimating design floods *viz.*, Empirical formula, Enveloping curves, Rational method, CWC short term report method, Standard Project Flood method, Synthetic hydrograph method, Comparison with meteorological homogenous catchments (Transposition) method and Flood frequency studies (Log-Normal distribution, Log-Pearson Type III distribution, Gumbel's method, Least Square method (Yen Te Chow method), Partial Duration Series method and Hazen's method) (Allen, 1975; Garg, 1977; Mccuen, 1977 and Mutreja).

Borland and Miller have presented six different methods for predicting the probable distribution of sediment in a reservoir. Of these methods, the empirical area reduction method and area incremental method seem to be more rational (Mohammed, 1979 and Singh, 1991).

From the sedimentation studies carried out on the basis of silt observations made at Dholbaha Dam and Maili Dam site, the annual rate of silt deposition in the reservoirs has been taken to be 8.0 hect-m and 7.34 hect-m respectively. This gives a rate of siltation of 0.1423 hect-m/year/km^2 of catchment for Dholbaha Dam and 0.43 hect-m/year/km^2 for Maili Dam.

Results and Discussion

A summary of results in respect of design flood and sedimentation distribution carried out for the Dholbaha and Maili Dams is given in Tables 57.1 and 57.2.

Table 57.1: Estimated Values of Design Flood Calculated by Different Methods

Sl.No.	*Method*	*Design Flood (cumecs)*	
		Dholbaha Dam	*Maili Dam*
1.	Empirical Flood formulae	1394	617
2.	Rational method	N.A.	635
3.	CWC Short Term Report method	1336	577
4.	Standard Project Flood method	2076	1120
5.	Synthetic Unit hydrograph	1621	611
6.	Transposition method	2261	918

Table 57.2: Peak Discharges

Year	Dholbaha Dam Annual Peak Discharge (cumecs) X	Dholbaha Dam X^2	Year	Maili Dam Annual Peak Discharge (cumecs) X	Maili Dam X^2
1959	198	38206	1975	201.90	40763.61
1960	100	10000	1976	76.90	5913.61
1961	322	108684	1977	184.85	34169.52
1975	472	222784	1978	220.17	48474.82
1976	222	49284	1979	94.35	8901.92
1977	421	177241	1980	228.95	52418.10
1978	334	111556	1981	47.27	2234.45
1979	47	2209	1982	138.70	19237.69

Design flood calculated by using different frequency analysis methods on the basis of available data of rain for the Dholbaha and Maili sites (Table 57.2) is given in Table 57.3.

Table 57.3: Design Flood by Frequency Analysis

Method	Peak Flood Values of Return Period (cumecs)									
	10		20		50		100		1000	
	Maili	Dholbaha	Maili	Dholbaha	Maili	Dholbaha	Maili	Dholbaha	Maili	Dholbaha
Log-Normal	245	635	330	734	365	832	403	926	542	1204
Log-Pearson III	215	518	309	599	354	673	389	735	–	–
Gumbel	263	398	305	564	361	667	396	747	486	1010
New Gumbel	200	367	216	399	236	440	252	470	303	572
Ven Te Chow's	270	514	335	650	384	750	432	849	591	917
Partial Duration Series	–	504	–	615	–	705	–	791	–	1078
Hazen's	232	461	256	508	279	582	282	604	295	637

From the perusal of the Table 57.1, it is found that the higher magnitude of floods peak is obtained from 'Standard Project Flood' method in case of Maili and Dholbaha Dams. In case of reservoir with storage less than 6000 hect-m, Standard Project Flood (SDF) method of a 100 years flood, whichever is higher is to be adopted as design flood. From the results of frequency analysis summarized in Table 57.3 it is seen that for Maili dam a 100 years flood has a magnitude of 432 cumecs. From Table 57.1 it is found that Standard Project Flood has a magnitude of 1120 cumecs. Therefore a design flood for determining the spillway capacity for the Maili dam may be taken as 1120 cumecs. Since the flood peaks determined from Standard Project Flood method are greatly dependent on the critical sequence of rainfall and the precise storm characteristics not available for the flood, it is desirable that the values of peak flood obtained by this method be cross checked. This cross-checking has been possible in this case by making use of the data available on Dholbaha catchment which has characteristics similar to Maili catchment. By transposition method the flood peak value for Maili dam is obtained as 918 cumecs (Table 57.1).

Keeping in view the above facts and average of values of flood peaks obtained by Standard Project Flood and Transposition method, 1019 cumecs is adopted as design flood value for the Maili dam. Similarly peak flood of 2168 cumecs is adopted as design flood value for the Dholbaha Dam.

Storage capacities calculated by both the methods *viz.*, Area incremental method and Empirical Area Reduction method for Maili dam and Dholbaha dam at the end of 15, 20 and 40 years and 15, 25 and 50 years respectively are not much of difference. Thus any of two methods may be adopted for the purpose of determining sediment distribution in a reservoir. Estimated useful life of Maili dam and Dholbaha dam has been computed as 37 years and 92.50 years respectively.

The elevations of sediment deposits and reduced capacities available in the reservoir resulting from sediment deposition at 15, 20 and 40 years age of the Maili reservoir and at 15, 25 and 50 years age of Dholbaha reservoir are indicated in the Table 57.4.

Table 57.4: Reduced Storage Capacities in Reservoirs after Sediment Deposition

Dam	*Period (Years)*	*Zero Level (metre)*	*Storage Capacities (hect-m)*		
			By Empirical Area Reduction Method	*By Area Incremental Method*	*Average*
Maili	Original	355.40	489.75	489.75	489.75
	After 15 years	359.00	415.334	416.45	415.89
	After 20 years	360.00	407.50	399.02	403.28
	After 40 years	364.00	328.84	331.47	330.15
Dholbaha	Original	391.67	1338.40	1338.40	1338.40
	After 15 years	396.85	1218.50	1216.46	1217.48
	After 25 years	398.70	1138.00	1133.22	1135.61
	After 50 years	402.19	937.50	935.46	936.48

Conclusion

On the basis of sedimentation studies carried out for Dholbaha and Maili Dams, the reservoirs are expected to silt upto elevations 396.85, 398.70 and 402.19 in 15, 25 and 50 years respectively for Dholbaha Dam and elevations 359.0 360.0 and 364.0 in 15, 20 and 40 years respectively for Maili Dam. The intake of the irrigation outlet pipe is therefore proposed to be provided at EL 403.50 for Dholbaha Dam, and at EL 365.00 for Maili Dam. A temporary intake (to be pludged later on) is proposed at EL 398.00 to enable utilization of additional live storage in the first 15 years life of the Dholbaha Dam when the level of sediment deposits is low. Similarly for the Maili Dam the temporary intake of the irrigation outlet pipe is proposed to be provided at EL 359.0 to enable utilization of maximum quantities of water in early phase of life of dam.

The live storage capacities available for storage of irrigation supplies and for flood absorption at various stages in the life of dams are given in Table 57.5.

Design flood of Maili Dam has been adopted as 1020 cumecs whereas the full reservoir level and irrigation outlet level has been recommended as EL 373.00 and EL 365.00 respectively. Similarly design flood for Dholbaha Dam has been adopted as 2170 cumecs whereas full reservoir level and irrigation outlet level has been recommended as EL 417.00 and EL 403.50 respectively. The useful life for Maili and Dholbaha reservoir has been determined as 37 years and 92.50 years respectively.

Table 57.5: Available Live Storage Capacity of Reservoir

Dam	*Life of Dam (yrs)*	*Irrigation Supplied*	
		Between Elevations (m)	*Storage Available (hect-m)*
Maili	15	359.00 to 373.00	415.89
	20	365.00 to 373.00	369.00
	40	365.00 to 373.00	300.25
Dholbaha	15	398.00 to 417.00	1200
	25	403.50 to 417.00	1035
	50	403.50 to 417.00	920

References

Allen, T., Hjemfelt J.R. and Cassidy, J.J. 1975. *Hydrology for Engineers and Planners*. Iowa Sate University Press, U.S.A.

Garg, Santosh Kumar. 1977. *Water Resources and Hydrology*, 2nd Edition. Khanna Publishers, New Delhi.

Hydrological Studies for the Dholbaha Dam. Irrigation Department, Puniab, Head Office, Chandigarh.

Hydrological Studies for the Maili Dam. Irrigation Department, Punjab, Head Office, Chandigarh.

Life of Reservoir. 1977. Technical Report No.19, Central Board of Irrigation and Power, New Delhi.

Mccuen, R.H., Rawls, W.J., Fisher, G.L. and Power, R. (1977) Flood flow frequency for ungauged watersheds: A literature evaluation. *Agricultural Research Service*, Betsvile, Maryland, ARS-NE-8.6, pp. 140.

Mohammed, Akram Gill. 1979. Sedimentation and useful life of reservoirs. *J. Hydrology*, 44: 89–95.

Mutreja, K.N. *Applied Hydrology*. Tata McGraw Hill Publishing Company Limited, New Delhi.

Singh, Sarbjit. 1991. Hydrological Studies for Dams in ungauged small water shed. *M.Tech Thesis*, Punjab Agricultural University, Ludhiana, India.

Chapter 58

Prediction of Nitrate Pollution of Groundwater: A Case Study

Sarbjit Singh Sooch, Baljeet S. Kapoor**, Bijay Singh*** and N.S. Grewal*****

**Assistant Professor, ****Professor-cum-Head (Retd.)*
Department of Civil Engineering, Punjab Agricultural University, Ludhiana, Punjab
***Director, Punjab Engineering College, Chandigarh*
****Senior Soil Scientist, Department of Soils,*
Punjab Agricultural University, Ludhiana, Punjab

ABSTRACT

Nitrate pollution of groundwater in the Bist Doab Tract in Punjab was studied as a function of anthropogenic activities on the soil surface and recharge of groundwater. Data pertaining to application of nitrogenous fertilizers to crops grown in 27 blocks of the study area and disposal of urban and industrial wastes on agricultural lands were collected. Data pertaining to nitrate-N content in groundwater samples were obtained from Central Groundwater Board. On an overall basis, during the last decade nitrate-N content in the study is slowly decreasing due to decreasing recharge and increasing water table depth. In four out of 27 blocks in the study area, average fertilizer-N use was more than 200 kg N/ha. As groundwater recharge increased with passage of time, increase in nitrate pollution of groundwater was parallel to increase in fertilizer use. In 15 blocks where farmers apply intermediate doses of fertilizer-N to crops, annual groundwater recharge is rapidly declining with time. Nitrate pollution of groundwater is thus not changing with time inspite of substantial application of fertilizer-N. In the remaining 8 blocks, fertilizer-N use was low and ranged from 43 to 71 kg N/ha and thus did not influence nitrate pollution of groundwater. Localized high nitrate-N concentration in groundwater was observed possibly due to irrigation with urban and industrial wastes in the vicinity of cities and presence of manure pits in the periphery of villages. Localized high nitrate pollution regions kept on shifting due to changing pattern in the use of waste waters for irrigation by farmers depending upon crops grown in the vicinity of cities.

Keywords: *Ground water quality, Nitrate pollution, Prediction.*

Introduction

Due to excessive fertilizer N use in developed countries during the last 3 to 4 decades, groundwater at many locations is already polluted with nitrates leaching from cultivated fields. Similar situations may also arise in states like Punjab where fertilizer N use in agriculture is among the highest in the country. The consumption of water, containing high levels of nitrate N can lead to methaemoglobinemia (blue baby syndrome) particularly in infants less than 6 months old (WHO, 1978). The critical nitrate level of 10 mg NO_3–N/L in the drinking water has been set by World Health Organization. Excessive ingestion of nitrates may also increase the risk of cancer in human population through *in vivo* formation of carcinogenic nitrosamines by reaction of ingested amines with nitrite in the human stomach.

Bajwa *et al.* (1993) investigated 236 samples from 21 to 38 m deep tube wells in different blocks of Punjab where fertilizer-N consumption ranged from 151 to 249 kg/ha/year. No significant correlation was observed between fertilizer consumption and nitrate content of well waters, possibly because samples were taken from deep wells. Nevertheless, percentage of groundwater samples containing more than 5 mg/L nitrate-N from tube wells located in vegetable growing areas was 17 per cent as compared to 3 and 6 per cent from wells located in regions where, respectively, rice-wheat and potato-wheat rotations were followed.

A feature of nitrate contamination of groundwater has been noted as inverse relationship between nitrate concentration and depth below the land surface or with in an aquifer (Freez and Cherry, 1979; Bijay Singh and Sekhon, 1978; Hallberg, 1986; Hallberg *et al.*, 1984; Libra *et al.*, 1984; Alfeldi, 1983; Hill, 1996, Ritter and Chimside, 1984). One controlling factor in this general relationship between depth and nitrate concentration is the depth of water table or conversely the thickness of the unsaturated zone and the properties of the soils and surface geologic materials. Groundwater contamination may not be evident for many years after excess fertilize-N use begins in areas with a thick, medium-to-fine-textured soil mantle or in areas with a substantial thickness of unsaturated zone. In sandy soils, if the water table is deep, some studies suggest it will take 10–50 years for the leaching of NO_3^-–N to reach the groundwater (Pratt *et al.*, 1972; Adriano *et al.*, 1972; Adelman *et al.*, 1985; Carey and Lloyd, 1985).

Several studies suggest that fertilizer alone do not contaminate groundwater under all conditions. Kolenbrander (1972) observed at various locations in Netherlands that concurrent with a total increase of 150 kg N/ha in the mean annual quantity of fertilizer N applied during 1920–1967, the nitrate content of groundwater increased only by 0.37 mg/L NO_3^-–N and this increase was confined to only 33 per cent of the wells. Kansal and Bajwa (1992) while reviewing the work on environmental pollution through fertilizer use did not find any report where nitrate concentration in groundwater had been associated with fertilizer use in the intensively cultivated areas of Northern India. Many other factors along with fertilizers contribute to groundwater enrichment with NO_3^-–N. Soil organic matter, animal wastes and plants also contribute to nitrate pollution but their inputs are difficult to determine. In general, there exists a remote possibility of nitrates originating from fertilizers to leach down to groundwater except where high rates of nitrogenous fertilizers are applied to heavily irrigated crops (Bijay Singh *et al.*, 1995). Thus, in regions where fertilizer nitrogen usage is high, soils are coarse, percolation losses are high and crops are shallow rooted, the possibility of nitrate-N leaching to groundwater bodies is there.

The objective of the present study is to predict scenarios with respect to nitrate pollution of groundwater in Bist Doab Tract (Figure 58.1) in the Punjab using records pertaining to use of nitrogen fertilizers and the irrigation water and nitrate content in ground water during last several years. This small area is bounded by Beas and Sutlej rivers and comprises of districts of Hoshiarpur, Jalandhar, Nawanshahar, Kapurthala and Nurpur Bedi block of Ropar district.

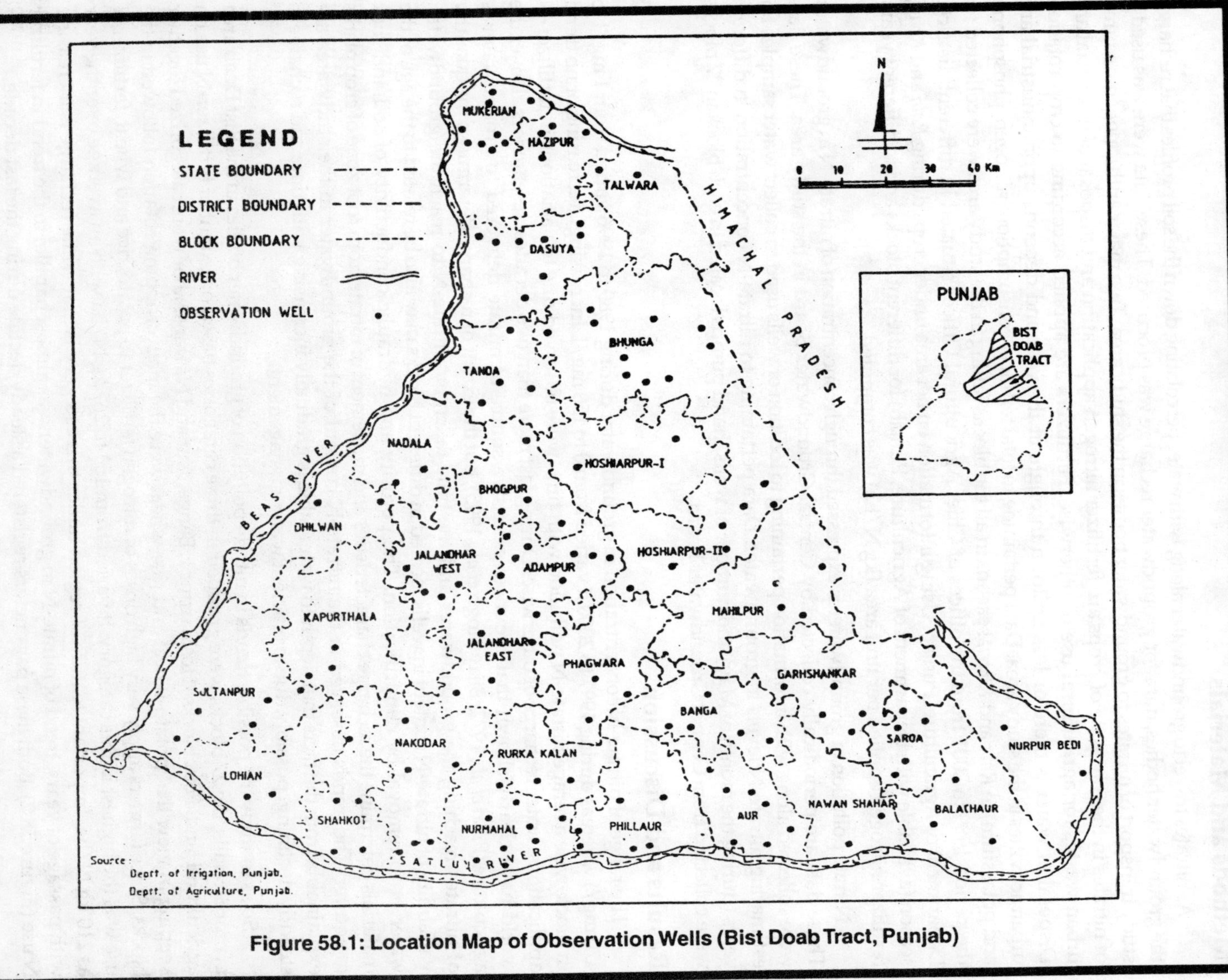

Figure 58.1: Location Map of Observation Wells (Bist Doab Tract, Punjab)

Methods and Materials

As nitrates travel to groundwater along with water percolating down the soil profiles that recharges the groundwater bodies, data for groundwater recharge was procured. These data were was used to study transport of nitrates to groundwater bodies in the study area. Two major anthropogenic sources of nitrates in the root zone of crops are fertilizer nitrogen application and disposal of industrial and urban wastes for agricultural use or otherwise. Fertilizers are applied according to crop rotation. Disposal of water is more or less uniform throughout the year and depends upon industrial and urban activities in the study area. Data pertaining to total fertilizer consumption (nitrogen, phosphorus and potassium) in *Rabi* and *Kharif* seasons in all the blocks falling in the study area were collected for the period 1989 to 2001 from the offices of Chief Agricultural Officer situated in different districts, Department of Agriculture, Punjab. Using information on net area under crops during *Rabi* and *Kharif* season (available with Department of Agriculture, Punjab) for different blocks and in different years, fertilizer nitrogen applied per unit area (kg N/ha) was computed.

Nitrate pollution of groundwater is expressed through concentration of nitrate-N in groundwater. These data are periodically collected by Central Groundwater Board in the study area. These data were collected along with information pertaining to location of wells used to collect water samples for estimating nitrate-N levels in groundwater. Data pertaining to nitrate-N concentration in different years (during the month of May) and number of wells/hand pumps located in each block in 27 blocks were collected from Central Groundwater Board.

Results and Discussion

The range of nitrate-N concentration in groundwater during 1989 to 1996 varied from 1 mg/l to 470 mg/l whereas during 1997 to 2000 it varied from 0 to 150 mg/l. Increasing, decreasing or no trend in block-wise average nitrate-N pollution with time were observed due to variations in fertilizer-N applications, water recharge or otherwise. During last more than two decades, researchers all over the world have clearly shown that the most extensive source of nitrate delivered to groundwater is agriculture (Pratt *et al.*, 1972; Bijay Singh *et al.*, 1995), but the extent of recharge determines the quantity of nitrates reaching the groundwater and the time taken by nitrate-N to reach the groundwater. Variability of nitrate-N content in wells located in different blocks may also be related to the age of the water or to removal by denitrification. The porosity and hydraulic conductivity of soil and rock materials determine the recharge characteristics and movement of nitrate from root zone of crop plants to aquifer. The trends observed in the nitrate-N content of the groundwater in the study area are explainable from the facts that depth to water table is gradually increasing with time and recharge is steadily declining, possibly due to efficient water management.

So as to study the possible trends in nitrate pollution of groundwater bodies in the Bist Doab area in more details, the 27 blocks were grouped in three categories based on extent of fertilizer-N use on agricultural land. Four blocks–Hoshiarpur 1, Banga, Aur and Nawanshahar were categorized together keeping in view that potato-wheat and rice-wheat were the main cropping rotation followed in the blocks. Since farmers apply very high doses of nitrogen fertilizer to potato, rice and wheat, fertilizer N use was the highest in this category. It varied from 153 to 267 kg N/ha with an average over the year as 210 kg N/ha. Important feature of region under this category is that while recharge is increasing with passage of years, trend of nitrate-N in groundwater was almost parallel to the trend in fertilizer N use (Figure 58.2). Both these parameters registered a steady decline during the last decade.

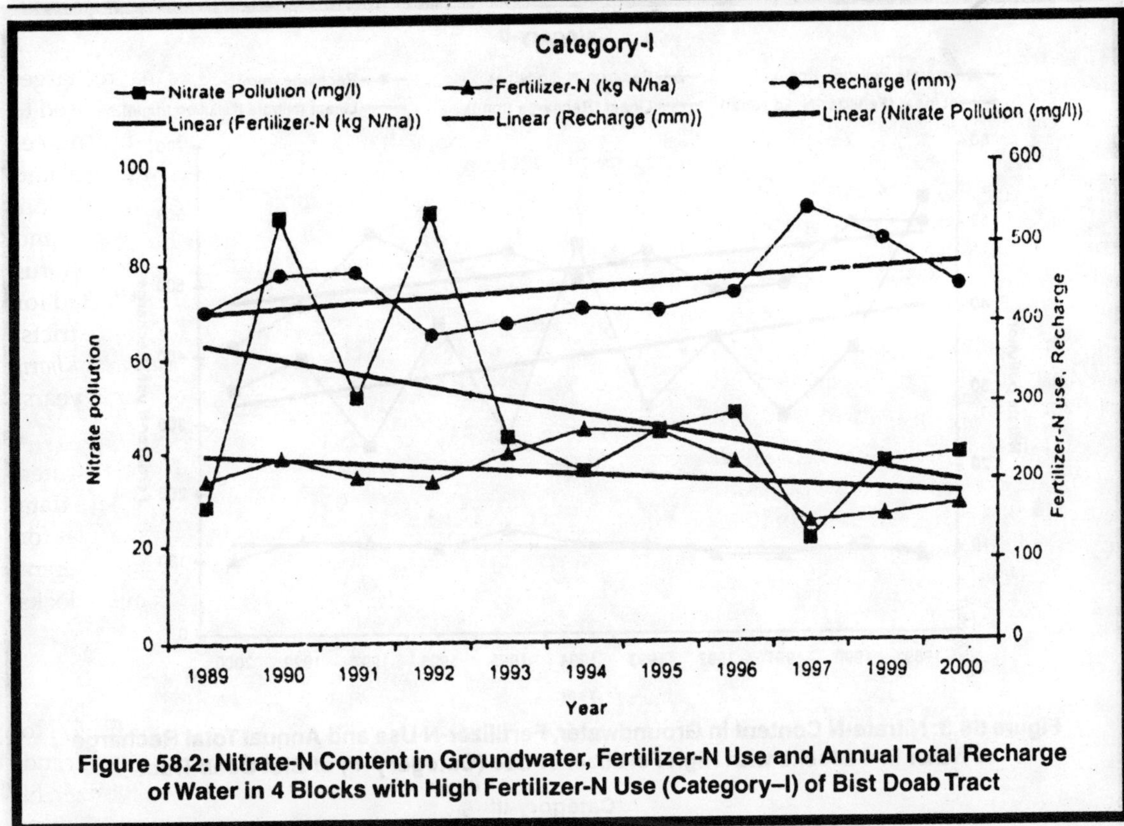

Figure 58.2: Nitrate-N Content in Groundwater, Fertilizer-N Use and Annual Total Recharge of Water in 4 Blocks with High Fertilizer-N Use (Category–I) of Bist Doab Tract

In the second category, there were 15 blocks (Garhshankar, Tanda, Jalandhar-East, Jalandhar-West, Adampur, Bhogpur, Kapurthala, Sultanpur, Shahpur, Phillaur, Nurmahal, Nakodar, Rurka, Phagwara and Nadala). Average fertilizer use in agricultural lands in this category was almost half of that recorded in first category blocks. During the last decade, average fertilizer N use varied from 108 to 146 kg N/ha. Most important feature of the blocks falling in this category was that annual recharge exhibited a rapid declining from more than 580 mm to around 400 mm in 2000–01. Thus fertilizer use did not strongly influence the nitrate-N content of groundwater and it was almost not changing with passage of time (Figure 58.3). Remaining 7 blocks were grouped in third category (Hoshiarpur-2, Bhunga, Malpur, Talwara, Dasuha, Mukerian, Hazipur and Balachaur). Fertilizer use in all the blocks in this category was very low. Average value ranged from 43 to 71 kg N/ha during different years, but it was steadily increasing with time. Along with fertilizer use, amount of recharge recorded annually was also found to be rapidly increasing with passage of years (Figure 58.4). It looks likes that inspite of increasing water recharge, nitrate-N content of groundwater was not increasing because fertilizer-N use on agricultural fields was very low. When data pertaining to 27 blocks are averaged and plotted against time (Figure 58.5), it is observed that although fertilizer use is very gradually increasing, due to decreasing recharge and increasing water table depth, nitrate-N content of groundwater in the study area is slowly decreasing. These trends should be viewed in the light of reports that even in sandy soils, if the water table is deep, it may take 10–50 years for the leaching of nitrate-N to reach groundwater (Pratt *et al.*, 1972; Adriano *et al.*, 1972; Carey and Lloyd, 1985; Adelman *et al.*, 1985).

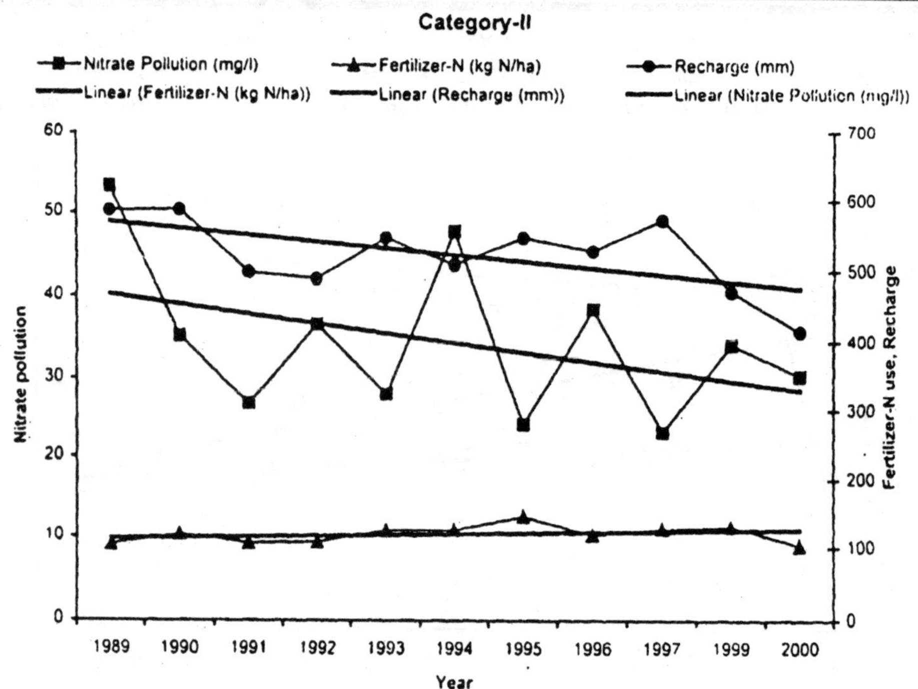

Figure 58.3: Nitrate-N Content in Groundwater, Fertilizer-N Use and Annual Total Recharge of Water in 4 Blocks with High Fertilizer-N Use (Category–II) of Bist Doab Tract

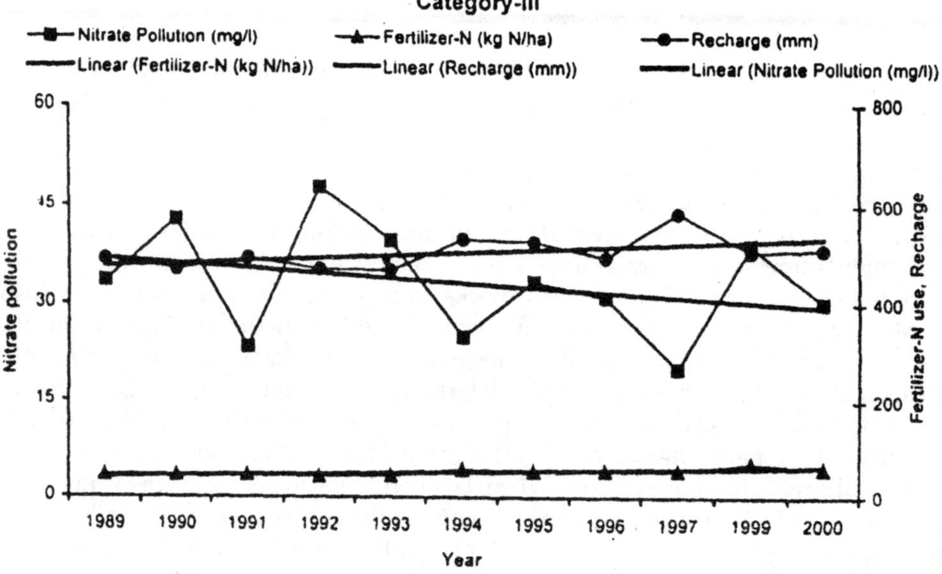

Figure 58.4: Nitrate-N Content in Groundwater, Fertilizer-N Use and Annual Total Recharge of Water in 4 Blocks with High Fertilizer-N Use (Category–III) of Bist Doab Tract

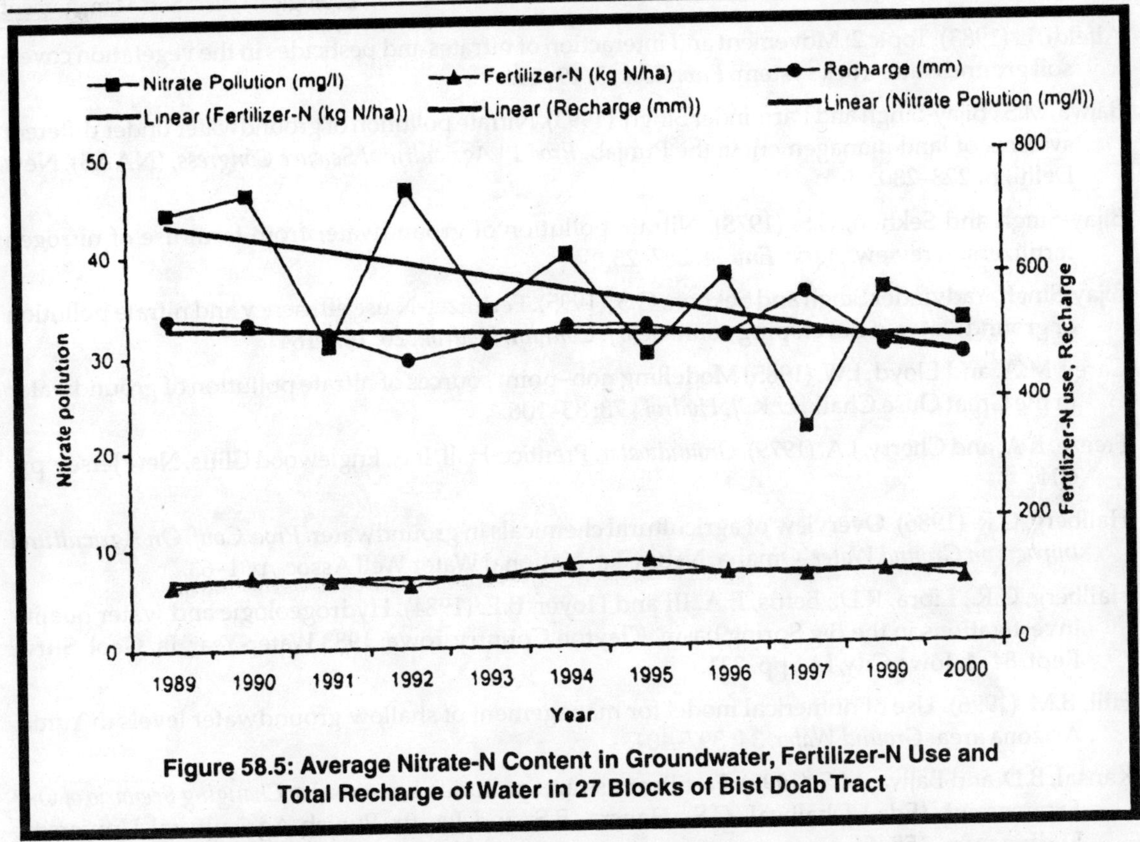

Figure 58.5: Average Nitrate-N Content in Groundwater, Fertilizer-N Use and Total Recharge of Water in 27 Blocks of Bist Doab Tract

On an overall basis nitrate-N content in the groundwater in the study area is not increasing during the last decade due to declining fertilizer use and relatively less recharge in high fertilizer use region, rapidly declining water recharge with passage of time in blocks where fertilizer use is intermediate and very low fertilizer use in remaining area.

Conclusion

Nitrate pollution of groundwater can be effectively controlled by efficient use of water and nitrogen fertilizer by crop plant. It seems that with increasing awareness among farmers to use fertilizer and water resources efficiently for increased profits from farming, already there is a stabilizing effect on nitrate pollution of groundwater in the Bist Doab Tract. Coupled with reducing annual recharge and resultant increasing depth of water table, possibility of increased nitrate pollution in the study is very less. However, farmers need to be educated about the possible harmful effects of nitrate pollution of good water as that they strictly follow the practices that lead to efficient management of fertilizer-N and water.

References

Adelman, D.D., Schroeder, W.J., Smaus, R.J. and Willin, G.R. (1985). Overview of nitrate in Nebraska's groundwater. *Trans. Nebr. Acad. Sci.*, 13: 75–81.

Adriano, D.C., Pratt, P.F. and Takatori, F.H. (1972) Nitrate in the unsaturated zone of an alluvial soil in relation to fertilizer nitrogen rate and irrigation level. *J. Environ Qual.*, 1: 418–422.

Alfeldi, L. (1983). Topic 2: Movement and interaction of nitrates and pesticides in the vegetation cover-soil groundwater-rock system. *Env. Geol.*, 5: 19–25.

Bajwa, M.S., Bijay-Singh and Parminder Singh. (1993). Nitrate pollution of groundwater under different systems of land management in the Punjab. *Proc 1st Agricultural Science Congress*, (NAAS), New Delhi, p. 223–230.

Bijay-Singh and Sekhon, G.S. (1978). Nitrate pollution of groundwater from farm use of nitrogen fertilizers: A review. *Agric. Env.*, 4: 207–25.

Bijay-Singh, Yadvinder Singh and Sekhon, G.S. (1995). Fertilizer-N use efficiency and nitrate pollution of groundwater in developing countries. *J. Contam. Hydrol.*, 20: 167–184.

Carey, M.A. and Lloyd, J.W. (1985) Modelling non–point sources of nitrate pollution of groundwater in the Great Ouse Chalk, U.K. *J. Hydrol.*, 78: 83–106.

Freeze, R.A. and Cherry, J.A. (1979). *Groundwater*. Prentice-Hall, Inc., Englewood Cliffs, New Jersey, pp. 604.

Hallberg, G.R. (1986). Overview of agricultural chemicals in groundwater. *Proc. Conf. On Agricultural Impacts on Ground Water*, Omaha, Nebraska, National Water Well Assoc., p. 1–63.

Hallberg, G.R., Libra, R.D., Bettis, E.A. III and Hoyer, B.E. (1984). Hydrogeologic and water quality investigations in the Big Spring basin, Clayton Country, Iowa: 1983 Water-Year. Ia. Geol. Surv., Rept. 84–4, Iowa City, IA., pp. 231.

Hill, B.M. (1996). Use of numerical model for management of shallow groundwater levels in Yuma, Arizona area. *Ground Water*, 34: 397–404.

Kansal, B.D. and Bajwa, M.S. (1992). Fertilizer use and pollution hazards. In: *Changing Scenario of Our Environment*, (Eds.) Dhaliwal, G.S., Hansra, B.S. and Jerath. Punjab Agricultural University, Ludhiana, p. 158–64.

Kolenbrander, G.J. (1972). Does leaching of fertilizer affect the quality of groundwater at the water works? *Stikstef*, 15: 8–15.

Libra, R.D., Hallberg, G.R., Ressmeyer, G.R. and Hoyer, B.E. (1984). Groundwater quality and hydro geology of Devonian-Carbonate aquifers in Floyd and Mitchell Counties, Iowa. Ia. Geol. Surv., Rept. 84–2, Iowa City, IA., pp. 106.

Pratt, P.F., Jones, W.W. and Hunsakes, V.E. (1972). Nitrate in deep soil profiles in relation to fertilizer rates and leaching volumes. *J. Env. Qual.*, 1: 97–102.

Ritter, W.F. and Chimside, A.E.M. (1984) Impact of land use on groundwater quality in southern Delaware. *Ground Water*, 22: 38–47.

World Health Organization (WHO) (1978). Nitrate, nitrites and N-Nitroso compounds. Environmental Health Criteria 5. World Health Organization, Geneva.

Chapter 59

Design Flood by Frequency Analysis for Relatively Ungauged Small Watersheds: A Case Study

Er. S.S. Sooch#, N.K. Khullar**, Er. Deepak Gupta**

*Assistant Professor, **Professor cum Head,
Department of Civil Engineering, College of Agricultural Engineering, PAU, Ludhiana,
#E-mail: sarbjit_sooch@sify.com

ABSTRACT

The hydrological variables are random in nature and it may be appropriate to use probabilistic approach to determine flood flows for different recurrence intervals. In water resources development and management, a study is normally carried out for identification of a probability function which best fits an annual series. The following probability functions were used for analysis of flood events (extreme values) in the Dholbaha catchment of Kandi watershed: (1) Log-normal distribution; (2) Pearson Type III distribution; (3) Gumbel's method; (4) Yen Te Chow method and (5) Hazen's method. Gumbel distribution (EYI) and log normal distribution seem to be the suitable choices for describing the set of annual flood series.

Keywords: *Hydrological studies, Design flood, Frequency analysis, Ungauged catchments, Flood control.*

Introduction

In the north-eastern parts of the Punjab State comprising of Ropar, Gurdaspur, Hoshiarpur and Nawanshahar districts (area is popularly known as Kandi area), there are number of streams/choes which cause heavy damage to neighbouring lands during monsoon. The Punjab Government with the aid of World Bank proposed the construction of series of earth and rockfill dams to store water for flood moderation and irrigation purposes. These dams are going to be provided with spillways to pass the flood water. Historical record of discharge in the streams in Kandi area is very limited, so much so that it is almost non-existent in case of some of the streams on which dams are proposed to be built. In planning stages, therefore, some gauging sites were selected and historical discharge data collected, naturally such data covers only few years. Dholbaha catchment in the Kandi area has been selected for the case study. An index map of Dholbana Dam Project is shown in Figure 59.1

Log-normal, Pearson Type-III, Gumbels method, Yen Te Chow's method and Hazen' s method were fitted to the empirical frequency distribution of annual flood peaks. Gumbel's distribution and Log-normal distribution seems to be fitting well the empirical annual flood series. These frequency curves can be utilized to determine the flood peaks for desired recurrence intervals. For the other catchments in the same meterological homogenous zone, the results of the above study can be transposed.

Methods and Materials

In the case of Dholbaha catchment, the flood peak data available for 8 years is given in Table 59.1 (Project Report, 1985).

Table 59.1: Peak Annual Floods at Dholbaha Dam Site

Year	*1959*	*1960*	*1961*	*1975*	*1976*	*1977*	*1978*	*1979*
Annual Peak Curnecs	198	100	322	472	222	421	334	47

It is seen that the length of record is not sufficient to lend itself to frequency study. It was, therefore, decided to extend the data, for number of years beyond the present data period.

Generation of data has to be done keeping in view the type of theoretical probability curve to be fitted to empirical frequency distribution. Random values are available in the tables for uniform and standard normal distribution. From these random numbers, log-normal, Gumbel and Log-Pearson type III distribution were worked out (Table 59.2) (Sarbjit, 1991 and Mutreja, 1986).

For Gumbel's method, generation of data is done in manner similar to generation of data for Log-Pearson type-III distribution. The data (original as well as generated) for Gumbel and Log-Pearson type III method would, therefore, be the same (Table 59.2).

The most commonly used distribution for frequency studies of annual flows and extreme value events are; two parameter log-normal distribution (LN2); Log–Pearson Type III distribution (LP III), Gumbel's distribution (EVI), VenTe Chow's method and Hazen's method.

Log Normal Distribution (LN2)

Many hydrological variables show a marked right skewness partly due to influence of the natural phenomenon having values greater than zero or some other lower limit and being unconstrained theoretically in the upper range. In such cases the variate will not follow the normal distribution and instead their logarithms follow a normal distribution (Stedinger, 1980). The probability density function of such a variable ($Log_e.x$).

$$f(x) = \frac{1}{x.\sigma_n.(2\pi)^{1/2}} \text{Exp.}\left[-1/2\left(\frac{\log_e x - \mu_n}{\sigma_n}\right)\right] \text{ for } x \geq 0 \quad (1)$$

and $f(x) = 0$ for $x < 0$

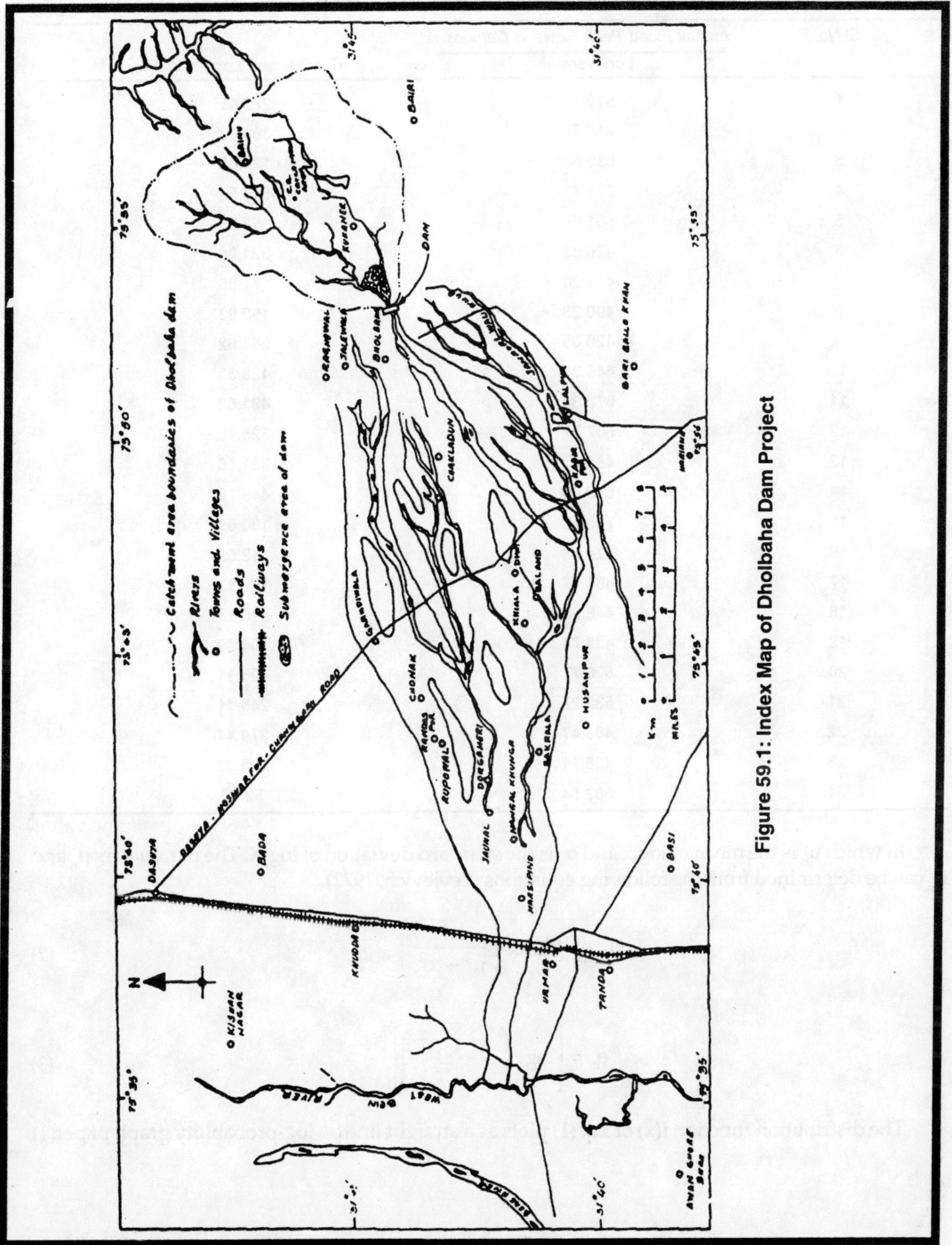

Figure 59.1: Index Map of Dholbaha Dam Project

Table 59.2: Generated Data for 24 Years

Sl.No.	Annual Flood Peak Flows in Cumecs	
	Log-Normal	Gumbel and Log-Pearson
1.	519.12	279.45
2.	418.26	388.52
3.	533.02	124.03
4.	703.75	518.53
5.	501.04	352.66
6.	426.33	321.28
7.	433.20	171.85
8.	490.28	150.93
9.	420.65	361.62
10.	446.35	439.33
11.	673.34	491.63
12.	507.17	325.76
13.	481.47	351.16
14.	510.46	449.79
15.	788.25	130.01
16.	424.83	512.55
17.	467.57	177.83
18.	448.00	218.17
19.	435.74	104.60
20.	520.77	496.11
21.	532.27	249.11
22.	468.47	279.44
23.	408.11	203.23
24.	502.54	52.30

In which μ_n is the mean of $\log_e x$ and σ_n is the standard deviation of $\log_e X$. The parameters μ_n and σ_n can be determined from the following equations (Yevjevich, 1972).

$$\mu_n = \tfrac{1}{2} \log_e \left(\frac{\mu^4}{\mu^2 + \sigma^2} \right) \text{ and} \qquad (2)$$

$$\sigma_n = \left[\log_e \left(\frac{\sigma^2 + \mu^2}{\mu^2} \right) \right]^{1/2} \qquad (3)$$

The distribution function f(x) of Eq. (1) plots as a straight line on log-probability graph paper.

Log-Pearson Type III Distribution

Log-Pearson Type III distribution is a special case of gamma distribution and is extensively used in hydrologic frequency analysis. It provides a skew adjustment. However, its properties remain obscure because of logarithmic transformations used (Hjelmfelt, 1975).

Gumbel Distribution

Gumbel distribution used in flood frequency analysis is an example of extreme value distribution law:

$$P = 1 - \exp[-\exp(-y)] \tag{4}$$

where,

P is the probability of a given flow being equalled or exceeded and *y* is the reduced variable which is a function of probability and is given by

$$Y = \frac{1}{0.78\,\sigma}^{(x - \bar{x} + 0.45\sigma)} \tag{5}$$

In Eq. (5), x is a flood magnitude with probability of exceedance P and e is the base of natural logarithm.

Ven Te Chow's Method

Chow made a modification of Gumbel's distribution (equation 4) by modifying the equation in the following form:

$$X = A + By \tag{6}$$

where,

X = magnitude of the flood

Y = $\log_{10}[\log_{10}T - \log_{10}(T-1)]$

A, B = regression constants

T = recurrence interval

Hazen's Method

According to Hazen, the frequency factor k = ϕ (T,Cs) is a function of both recurrence interval and skewness. Hazen' s formula is

$$P = (2M - 1)/2N$$

where,

M = Rank number

N = Total number of items

Calculate Coefficient of Variation

$$Cv = \sqrt{\frac{\sum x^2}{N-1}}$$

Calculate Coefficient of Skewness

$$C_s = \frac{\Sigma x^3}{(N-1)(Cv)^3}\text{; then } C_s = FC_s$$

Results and Discussion

The generated data for the three distribution is given in Table 59.2.

For the above distributions, the parameters were computed and are shown in Table 59.3.

Table 59.3: Parameters of Distributions

Parameters	*Log-normal*	*Log-Pearson Tpe-III X = log_{10}*	*Gumbel*
X	443.1	2.392	289.57
S_x	156.15	.0.28	125.81
$g_1{}^*$	1.1	– 0.88	– 0.40
$g_2{}^*$	4.25	2.60	3.16

$g_1{}^*$ and $g_2{}^*$ are skewness and kurtosis coefficients respectively for the sampled data.

Making use of estimated parameters given in Table 59.3 for different distributions and equation $x = \overline{X} + K.s_x$ flows have been determined with different probability levels and are shown in Tables 59.4–59.6.

Table 59.4: Flood Flows Using Log-normal Parameters

Probability, P (%)	*Recurrence Interval, T*	*Chow Frequency Factor, K*	*Flood Flow $x = \overline{X} + K.s_x$ (cumecs)*
99	0.0100	– 1.63	188.57
95	0.0105	– 1.31	238.54
80	0.0125	– 0.83	313.49
50	0.0200	– 0.16	418.11
20	0.0500	0.73	557.08
5	0.2000	1.86	733.53
1	1	3.09	925.60
0.1	10	4.87	1203.55
0.01	100	6.71	1490.86

Frequency curve data (Tables 59.4–59.8) for all the distribution was plotted. Log-Pearson Type III frequency distribution neither plotted as straight line on log-log paper nor on probability paper. However, Log-normal frequency distribution (Figure 59.2), Gumbel (EYl) frequency distribution and Yen Te Chow's method plotted as straight lines on log-probability (Figure 59.3), respectively. Upper and lower confidence limits for Gumbel's frequency curve at 95 per cent confidence probability have also been shown.

Table 59.5: Flood Flows Using Log-Pearson Type III Distribution $\overline{X}$ = 2.392; S_x = 0.28 and g_1 = – 0.88

Probability, P	*Recurrence Interval, T (yrs.)*	*Frequency Factor, K*	*Log Q = $\overline{X}$ + K.s_x*	*Flood Flow, Q (cumecs)*
0.99	1.010	– 2.94	1.568	36.98
0.90	1.111	– 1.34	2.180	104.23
0.50	2	0.114	2.432	270.39
0.10	10	1.150	2.714	517.60
0.02	50	1.560	2.828	672.97
0.01	100	1.674	2.860	724.43

Since the data fell on the straight line, there was no necessary of applying any numerical check of goodness of fit. Frequency curve for Hazen' s method is plotted as shown in (Figure 59.4) which is not a straight line.

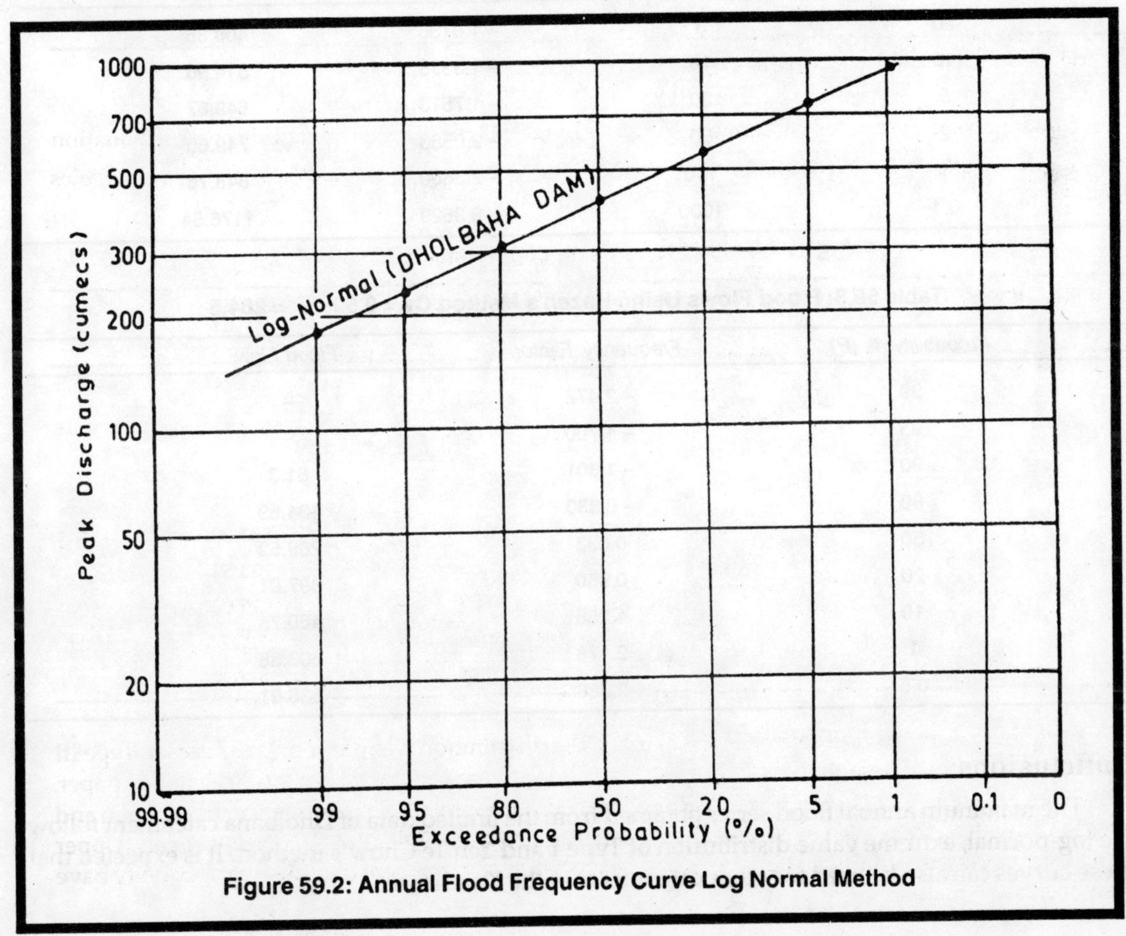

Figure 59.2: Annual Flood Frequency Curve Log Normal Method

Table 59.6: Flood Flows Using Gumbel's Distribution $\overline{X}$ = 289.57; S_x = 125.81

Probability, P (%)	*Recurrence Interval, (yrs.) T*	*Frequency Factor, K*	*Flood Flow $x = \overline{X} + K.s_x$ (cumecs)*
20	5	0.8616	397.96
10	10	1.531	482.18
5	20	2.1736	563.03
2	50	3.0072	667.90
1	100	3.6310	746.38
0.1	1000	5.727	1010.08

Table 59.7: Floods Flows Using Ven Te Chow's Method A = 76.80; B = – 327.11

Probability, P	*Recurrence Interval, (yrs) (T)*	*$Y = Log_{10}$ (log_{10} T/T–1)*	*Flood Flow X = A + B (cumecs)*
20	5	– 1.0136	408.36
10	10	– 1.3395	514.96
5	20	– 1.7513	649.67
2	50	– 2.0563	749.60
1	100	– 2.3600	848.78
0.1	1000	– 3.3629	1176.54

Table 59.8: Flood Flows Using Hazen's Method Cv = 0.59; $\overline{X}$ = 264.5

Probability, % (P)	*Frequency Factor*	*Flood Flow*
99	– 2.472	–
95	– 1.700	–
90	– 1.301	61.3
80	– 0.830	134.89
50	0.033	269.52
20	0.850	397.01
10	1.258	460.75
1	2.178	603.58
0.5	2.388	636.91

Conclusions

The maximum annual flood series obtained from the limited data of Dholbana catchment follow the log-normal, extreme value distribution of Type 1 and Yen Te Chow's method. It is expected that these curves can also be used for other catchments in the Kandi tract.

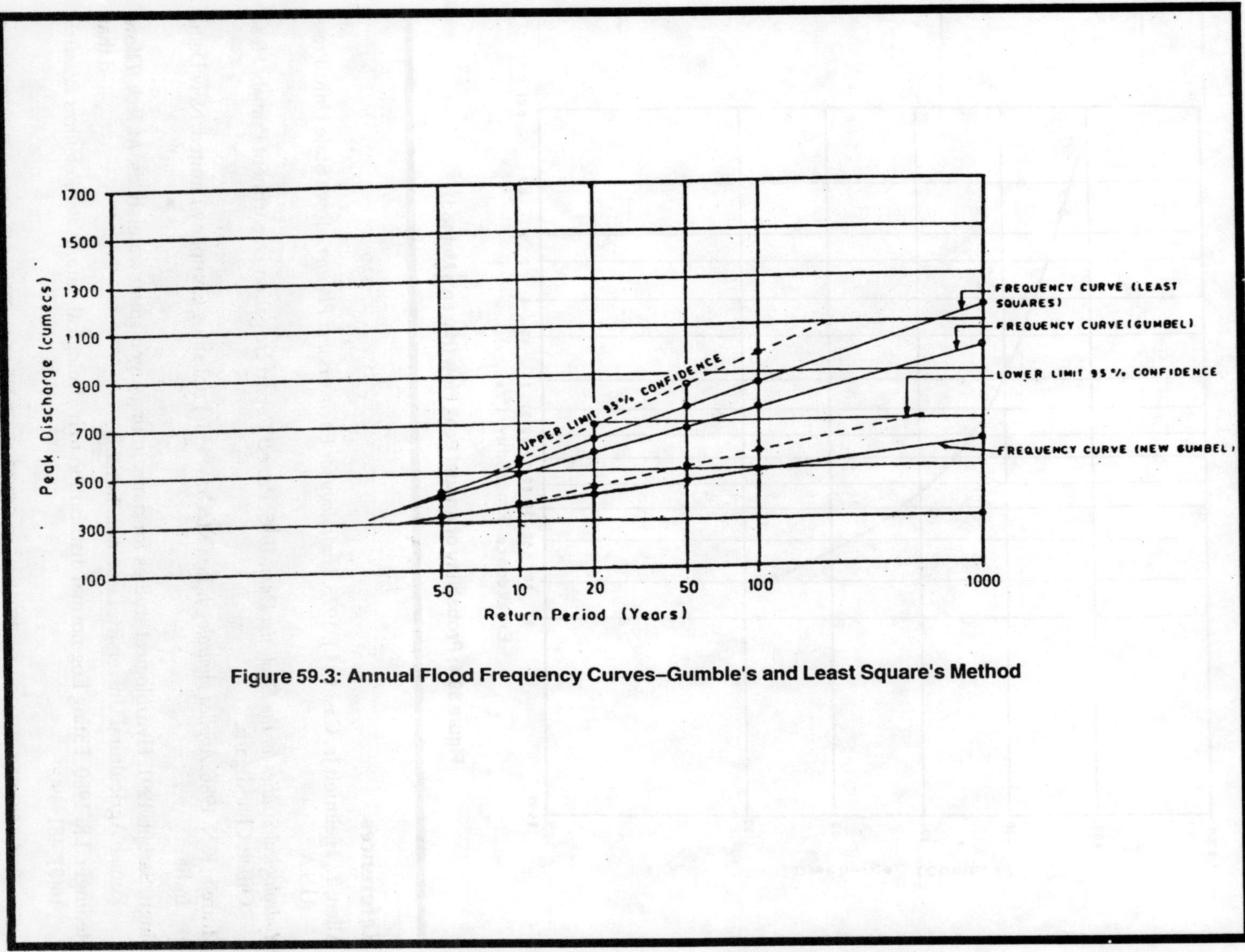

Figure 59.3: Annual Flood Frequency Curves–Gumble's and Least Square's Method

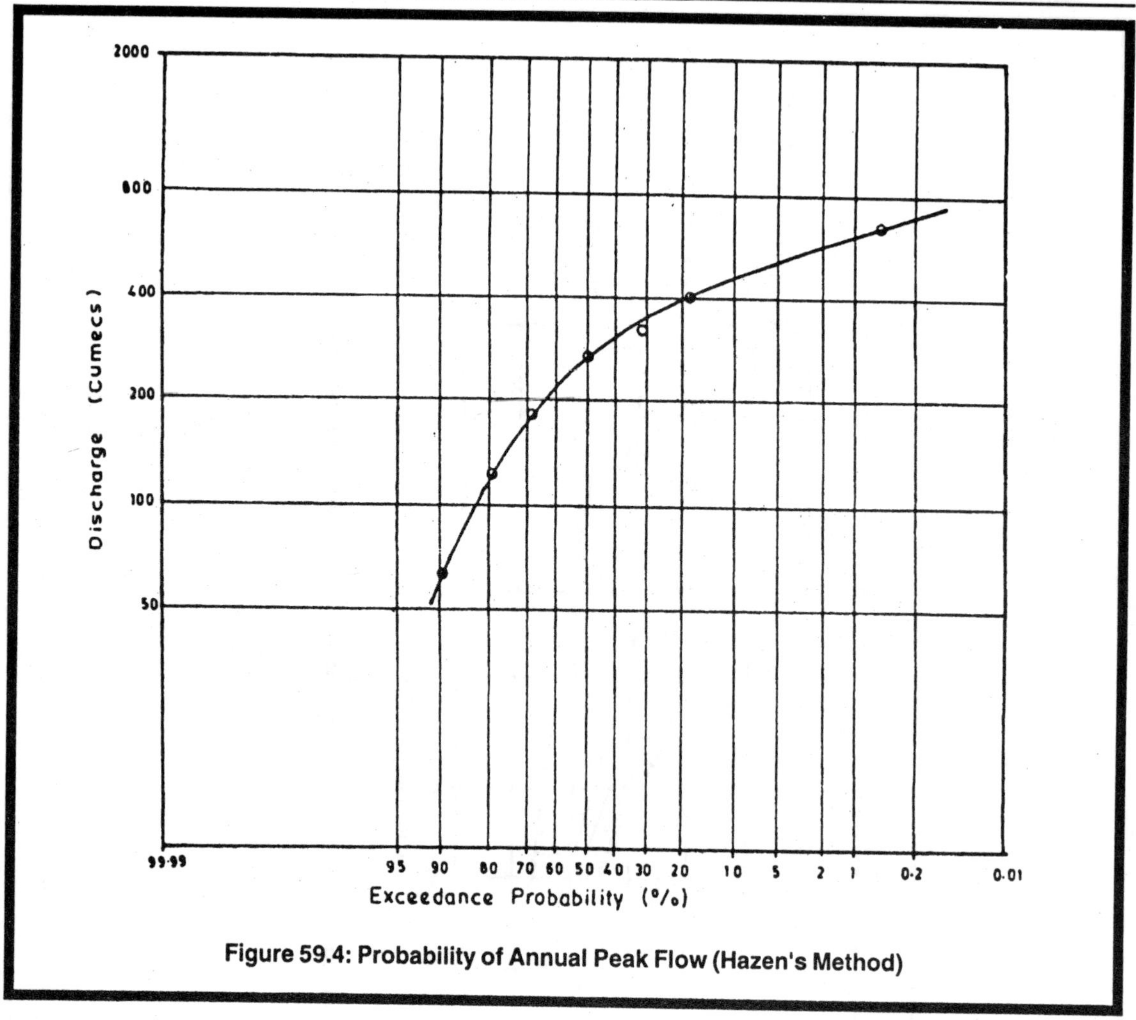

Figure 59.4: Probability of Annual Peak Flow (Hazen's Method)

References

Allen, T., Hjeimfelt Jr., Casidy, J.J. 1975. *Hydrology for Engineers and Planners.* Iowa State Univ. Press, U.S.A.

Hydrological Studies for the Dholbaha Dam. 1985. Project Report Irrigation Department Punjab, Head Office, Chandigarh.

Mutreja, K.N. 1986. *Applied Hydrology.* Tata McGraw Hill Publishing Company Limited, New Delhi, India.

Singh, Sarbjit. 1991. Hydrological studies for dams in ungauged small watersheds. *M.Tech. Thesis,* Punjab Agricultural University, Ludhiana, India.

Stedinger, J.R. 1980. Fitting Log-normal distribution to hydrological data. *Water Resources Research,* 16(3): 481–490.

Chapter 60
Effectiveness of Three Botanicals Against Dermatophytes

B.K. Dutta, I. Rahman and S. Karmakar

Defence Research Laboratory, Post Bag No. 2, Tezpur – 7840 011, Assam, India

ABSTRACT

Laboratory experiment was conducted to find out the antifungal property of *Terminalia chebula* dry fruit, *Punica granatum* fruit rind and *Piper betel* leaf aqueous extracts against five dermatophytes *viz., Trichophyton tonsurans, T. rubrum, T. mentagrophytes, Microsporum fulvum* and M. *gypseum.* Extract of *T chebula* was found more effective followed by *P. granatum* and *P. betel* against test pathogens. Among five test pathogens, *T. mentagrophytes* showed more susceptibility while *M. gypseum* registered more resistant to the extracts.

Keywords: *Dermatomycosis, Dermatophytes, Antifungal property, Aqueous extracts, Inhibition percentage.*

Introduction

Dermatomycosis is superficial skin infection in man and animals caused by a group of pathogenic fungi known as dermatophytes. They attack the outer horny layer of skin as well as hair and nails. According to the natural distribution, they are classified as anthropophilic, zoophilic and geophilic. Dermatomycosis is not fatal but highly contagious in favourable conditions which lead to discomfort with intense itching/scaling, ultimately decreases the work performance in both man and animals. A report of a family outbreak of tinea corporis by *Trichophyton mentagrophytes* initially caused scalp infection of eight years old boy was established (Otberg *et al.*, 2001). Another report highlighted that 98 individuals from a child care centre in Denmark were clinically examine for mycological observation and *Microsporum audounii* was cultured. Out of them, 12 were suffering from tinea capitis and tinea corporis (Haeder Sdal *et al.*, 2003). A report of tinea pedis among 58 medical students with a mean age of 23.9 yrs in Japan revealed, 13 (22.4 per cent) positive by direct microscopic examination and the causative organisms were *Trichophyton mentagrophytes* and *T. rubrum*. Similarly, among 37 research workers, 21 (56.8 per cent) showed positive by same observation and isolates were *T. mentagrophytes*. While in case of 31 females wearing boots (mean age 21 yrs), 7 (22.6 per cent) were infected by tinea pedis and the dominant species was *T. mentagrophytes* (Watanobe *et al.*, 2001). Dermatophyte infections on 227 subjects (112 were male and 115 were female) in Duzce western Black Sea region of Turkey were examined. Total 120 samples (47 nail fragments, 73 skin scales) were collected from 31 patients for mycological analysis on the basis of clinical evaluation. Collected samples were analyzed by direct microscopy and culture. The most frequently isolated aetiological agents were *T. rubrum* and *T. mentagrophytes*. In conclusion, prevalence of dermatophytic infections was attributed to traditional and religious habits such as cohabitations and performing ritual ablutions (Sahin *et al.*, 2004).

Hot and humid climate of NE India is most favourable to transmit this disease and it is presumed that this region is a natural incubator to spread dermatomycosis. On the other hand the brilliant biodiversity of this region have immense potential medicinal plants to overcome many human and animal ailments.

To avoid the toxic effects of chemotherapeutic agents, recent trend in research is emphasizes the use of botanicals for healing of many diseases. The World Health Organisation (WHO) estimated that four billion people *i.e.*, 80 per cent of the world population used herbal medicine for some aspect of primary health care (Farnsworth *et al.*, 1985). In *in vitro* study, *Azadirachta indica* leaf and seed extracts exhibited antimycotic property against *T. mentagrophytes* and other four dermatophytes (Natarajan *et al.*, 2002). Vonshak *et al.* (2003) screened 28 south Indian medicinal plants for their antifungal activity against six species of cutaneous fungi. Aqueous gall extract of *Terminalia chebula* showed inhibitory effect on three dermatophytes (*Trichophyton* spp.) and three yeasts (*Candida* spp.) and seed extract inhibited the growth of *T. glabrata*. Dutta *et al.* (2004) reported the effectiveness of *T. chebula* dry fruit aqueous extract and obtained 0.5 per cent MIC against six dermatophytes and 10 per cent MIC against a dominant species *Candida albicans*. *Punica granatum* bark and fruit rind aqueous extracts exhibited antifungal property against three species of *Trichophyton*, one species of *Trichosporon* and two species of *Microsporum* was at 5 per cent concentration except *Candida albicans* (Dutta *et al.*, 1998). *Piper betel* leaf extract showed fungicidal property at 5 per cent concentration and fungistatic property at 2 per cent concentration against *T. mentagrophytes* (Dutta *et al.*, 2004). In *in vitro* studies the aqueous extracts of *Allium sativum* bulbs showed antifungal property against *T. rubrum* isolated from infected patients (Samuel *et al.*, 2000).

In the present investigation, an attempt was made to evaluate the antifungal property of three botanicals such as *Terminalia chebula* dry fruit (small type), *Punica granatum* (fruit rind) and *Piper betel* leaf (type–satyavaram) aqueous extracts against five dermatophytes in the laboratory conditions.

Materials and Methods

Test dermatophytes were collected from School of Tropical Medicine, Kolkata for conducting the study. These test organisms are:

Trichophyton tonsurans (*Tt*)

As an anthropophilic species distributing in the nature causing chronic superficial infection on hairs.

T. rub rum (*Tr*)

Anthropophilic species commonly isolated from the glabrous skin infection of man.

***T. mentagrophytes* (*Tm*)**

A zoophilic species causing inflammatory ringworm of the foot, hands and glabrous skin particularly of the intertriginous areas and also causes tinea pedis and tinea cruris.

***Microsporum fulvum* (*Mf*)**

Geophilic in nature and occasionally found causing tinea pedis.

***M. gypseum* (*Mg*)**

Universally distributed as a geophilic species. It causes infection of the scalp in man, occasionally causes inflammatory infections.

Aqueous extracts were prepared with clean *T. chebula* dry fruit, *P. granatum* fresh fruit rind and fresh *P. betel* leaf. The materials were grounded and boiled with double distilled water at a ratio 1 : 4 (w/v) for 10 minutes separately. After cooling and filteration by Whatman filter paper No.1 the filtrates were considered for conducting the study.

Extracts of *T. chebula*, *P. granatum* and *P. betel* were incorporated in Sabouraud's Dextrose Agar (SDA) medium at 5, 2, 1, 0.5, 0.25 and 0.125 per cent concentrations respectively. After autoclaving at 151b pressure for 15 minutes, 25 ml media was poured aseptically into separate sterilized 90 mm diameter petriplate inside the laminar air flow chamber and allowed to solidify. The test pathogens were inoculated in the centre of each petriplate and incubated in inverted condition at a temperature of 28°C in BOD incubator for proper mycelial growth. Respective controls were also maintained containing plain SDA media. Data was recorded as a mean of five replications up to the period when each test pathogen completed full (90 mm) mycelial growth in the control plates. The inhibition percentages were estimated by applying the following formula (Mittal, *et al.*, 2002).

$$\text{Inhibition Percentage} = \frac{\text{Dia of Colony in Control j In Extract}}{\text{Dia in Colony in Control}} \times 100$$

Results and Discussion

In Table 60.1, full mycelial growth (90mm) in control plates recorded in 15 and 26 days both for *Tm*, *Mf* and *Tr*, *Mg* respectively but for *Tt* it was 35 days. No mycelial growth was recorded at 0.5 per cent concentration in *Tr*, *Tm* and *Mg* but for *Tt* and *Mf*, it was 1 and 2 per cent respectively. Inhibition percentage of *Tt* (7.56 per cent) was more than *Tr* (3.78 per cent) and *Mg* (6.67 per cent) at 0.125 per cent concentrations despite showing nil growth at 1 per cent concentration.

In Table 60.2 the same pathogens were tested against *P. granatum* extract, no mycelial growth was recorded at 0.5 per cent concentration in *Tm*, but for *Tt*, *Tr*, *Mg* it was 1 per cent and for *Mf* it was 5 per cent *Mf* was the highly resistant against the extract exhibited 3 mm mycelial growth at 2 per cent concentration but at 0.25 per cent concentration it was recorded highest inhibition percentage (20 per cent) in respect of others. At 0.125 per cent concentration, the highest mycelial growth 89.4 mm was recorded both *Tr* and *Tm* followed by 89 mm for *Mg*, 88.8 mm for *Mf* and 88 mm for *Tt*. The highest inhibition percentage recorded at 0.125 per cent concentration in *Tt* (2.22 per cent) followed by *Mf* (1.33 per cent), *Mg* (1.11 per cent) and both *Tr* and *Tm* (0.67 per cent) respectively. *Tm* was highly susceptible to the extract and recorded no mycelial growth even at 0.5 per cent concentration, but at 0.125 per cent concentration mycelial growth attained the maximum point along with *Tr* and revealed lowest inhibition percentage.

Table 60.1: Efficacy of *T. chebula* (dry fruit) Aqueous Extract

Test Dermatophytes	**MG in Control (days)*	*Parameters*	***Concentrations (%)*					
			5	*2*	*1*	*0.5*	*0.25*	*0.125*
T. tonsurans	35	MG (mm)	0	0	0	14.4	69.8	83.2
		MG (%)	0	0	0	16	7756	92.44
		Inh. (%)	0	0	0	84	22.44	7.56
T. rubrum	26	MG (mm)	0	0	0	0	67.8	86.8
		MG (%)	0	0	0	0	75.33	96.22
		Inh. (%)	0	0	0	0	24.67	3.78
T. mentagrophytes	15	MG (mm)	0	0	0	0	55.8	76.2
		MG (%)	0	0	0	0	62	84.67
		Inh. (%)	0	0	0	0	38	15.33
M. fulvum	15	MG (mm)	0	0	1.8	3.8	83.8	90
		MG (%)	0	0	2	4.22	93.11	100
		Inh. (%)	0	0	98	95.78	6.89	0
M. gypseum	26	MG (mm)	0	0	0	0	56.8	84
		MG (%)	0	0	0	0	63.11	93.33
		Inh. (%)	0	0	0	0	36.89	6.67

Table 60.2: Efficacy of *P. granatum* (fruit rind) Aqueous Extract

Test Dermatophytes	**MG in Control (days)*	*Parameters*	***Concentrations (%)*					
			5	*2*	*1*	*0.5*	*0.25*	*0.125*
T. tonsurans	35	MG (mm)	0	0	0	10.2	79	88
		MG (%)	0	0	0	11.33	87.78	97.78
		Inh. (%)	0	0	0	88.67	12.22	2.22
T. rubrum	26	MG (mm)	0	0	0	18	88	89.4
		MG (%)	0	0	0	20	97.78	99.33
		Inh. (%)	0	0	0	80	2.22	0.67
T. mentagrophytes	15	MG (mm)	0	0	0	0	81.4	89.4
		MG (%)	0	0	0	0	90.44	99.33
		Inh. (%)	0	0	0	0	9.56	0.67
M. fulvum	15	MG (mm)	0	3	4.8	5.8	72	88.8
		MG (%)	0	3.33	5.33	6.44	80	98.67
		Inh. (%)	0	96.67	94.67	93.55	20	1.33
M. gypseum	26	MG (mm)	0	0	0	3.6	79	89
		MG (%)	0	0	0	4	87.78	98.89
		Inh. (%)	0	0	0	96	12.22	1.11

MG: Mycelial growth; Inh.: Inhibition.

*: Complete MG in control plate (90 mm dia); **: Mean of 5 replications.

In Table 60.3, except *Mg* none of the test pathogens exhibited mycelial growth at 5 per cent concentration against *P. betel* extract. *Mg* showed high resistance towards the influence of the extract which attained 90 mm mycelial growth *i.e.* nil inhibition percentage at 0.25 per cent concentration. In case of other four test pathogens at 2 per cent concentration; highest inhibition percentage was recorded (90 per cent) in *Tr* and gradually decreased at 76, 49.56 and 45.56 per cent for *Tt*, *Tm* and *Mf* respectively. At 0.125 per cent concentration inhibition percentage of *Tr* recorded nil but highest (90 per cent) at 2 per cent concentration in respect of others. Highest mycelial growth at 0.125 per cent concentration was registered 90 mm for *Tr*, *Tm* and *Mg* followed by 87.6 mm for *Mf* and 73.4 mm for *Tt* respectively.

Table 60.3: Efficacy of *P. betel* (type–sathyavaram) Aqueous Extract

Test Dermatophytes	**MG in Control (days)*	*Parameters*	***Concentrations (%)*					
			5	*2*	*1*	*0.5*	*0.25*	*0.125*
T. tonsurans	35	MG (mm)	0	21.6	55.4	58.6	65.4	73.4
		MG (%)	0	24	61.56	65.11	72.67	81.56
		Inh. (%)	0	76	38.44	34.89	27.33	18.44
T. rubrum	26	MG (mm)	0	9	61	74.6	87.8	90
		MG (%)	0	10	67.78	82.89	97.56	100
		Inh. (%)	0	90	32.22	17.11	2.44	0
T. mentagrophytes	15	MG (mm)	0	45.4	78.4	85.4	89.4	90
		MG (%)	0	50.44	87.11	94.89	93.33	100
		Inh. (%)	0	49.56	12.89	5.11	0.67	0
M. fulvum	15	MG (mm)	0	49	73	75.4	82.2	87.6
		MG (%)	0	54.44	81.11	83.78	91.33	97.33
		Inh. (%)	0	45.56	18.89	16.22	8.67	2.67
M. gypseum	26	MG (mm)	35.80	61.8	80	88.8	90	90
		MG (%)	39.78	68.67	88.89	98.67	100	100
		Inh. (%)	60.22	31.33	11.11	1.33	0	0

MG: Mycelial growth; Inh.: Inhibition.

*: Complete MG in control plate (90 mm dia); **: Mean of 5 replications.

The comparative study of Tables 60.1–60.3 can be summarized as the minimum mycelial growth revealed against *T. chebula* extract in *Mf* at 1, *Tt* at 0.5 and remaining *Tr*, *Tm*, *Mg* at 0.25 per cent concentration respectively. In *P. granatum* extract it was 2 per cent for *Mf*, 0.5 per cent for *Tt*, *Tr*, *Mg* and 0.25 per cent for *Tm* but the influence of *P. betel* extract over the test pathogens were less where mycelial growth was started at 5 per cent concentration in *Mg* and. rest *Tt*, *Tr*, *Tm*, *Mf* were recorded at 2 per cent respectively. Among the extracts, *T. chebula* showed highest inhibition percentage at 0.125 per cent concentration against *Tr* (3.78 per cent), *Tm* (15.33 per cent) and *Mg* (6.67 per cent) whether in *P. betel* it was *Tt* (18.44 per cent) and *Mf* (2.67 per cent) respectively. *P. granatum* extract revealed no significant inhibition percentage against any of the test pathogens.

It can be concluded that *T. chebula* extract is more effective against test pathogens followed by *P. granatum* and *P. betel* extracts, on the other hand *Tm* showed more susceptibility and *Mg* recorded more resistant to the extracts. Though each plant is a chemical factory, further study is needed to exploit these botanicals to find out the active ingredients for preparing herbal alternatives to cure dermatomycosis.

Acknowledgement

Authors are grateful to Dr. S.N. Dube, Director, Defence Research Laboratory, Tezpur (Assam) for extending his full cooperation and encouragement throughout the study. Thanks are also due for Professor S. Basak, Professor and Head of Mycology Department, School of Tropical Medicine, Kolkata for providing the test pathogens.

References

Dutta, B.K., Rahman, I. and Das, T.K. 1998. Antifungal activity of Indian plant extracts. *Mycoses*, 41: 535–536.

Dutta, B.K., Rahman, I. and Das, T.K. 2004. *Piper betel:* An effective antidermatophytic plant. *Env. and Eco.*, 22: 480–482.

Dutta, B.K., Karmakar, S., Rahman, I., Das, J., Choudhury, K and Das, T.K. 2004. Antidermatophytic property of dry fruit extract of *Terminalia chebula*. *Geobios*, 31: 293–295.

Farnsworth, N.R., Akerele, A.O., Bingel, A.S., Sojarta, D.O. and EnD, Z. 1985. Medicinal plants in theraph. *Bull. World Health Organ.*, 63: 965–981.

Haeder Sdal, M., Stenderup, J., Mailer, B., Agner, T. and Svejgarrd, E.L. 2003. An outbreak of tinea capitis in a child care centre. *Den. Med. Bull.*, 50: 83–84.

Mittal, M., Sharma, R. and Paul M.S. 2002. Antifungal properties of some fern from north-west Himalayas. *Geobios*, 29: 244–246.

Natarajan, V., Prasad, P.V.S. and Ramasamy. K. 2002. *Azadirachta indica* in the treatment of dermatophytosis. *J. Ecobiol.*, 14: 201–204.

Otberg, N., Tietz, H.J., Henz, B.M. and Haas, N. 2001. Kerion due to *Trichophyton mentagrophytes:* responsiveness to fluconazole versus terbinafine in a child. *Acta Dermato-Vener.*, 81: 444–445.

Sahin Idris, Oksuz Sukru, Kaya Demet, Sencan Irfan and Cetinkaya Reyhan. 2004. Dermatophytes in the rural area of Duzce, Turkey. *Mycoses*, 47: 470–474.

Samuel, J.K., Andrews, B. and Jebashree, H.S. 2000. *In vitro* evaluation of the antifungal activity of *Allium sativum* bulb extract against *Trichophyton rubrum*, a human skin pathogen. *World J. Microb. & Biotechnol.*, 16: 617–620.

Vonshak, A., Barazani, O., Sathiyamoorthy, P., Shalev, R., Vardy, D., Golan-Gold Hirsh, A. 2003. Screening South Indian medicinal plants for antifungal activity against cutaneous pathogens. *Phytotherapy Res.*, 17: 1123–1125.

Watanabe, K., Taniguchi, H., Nishioka, K., Katoh, T., Ara, K. and Kayane, S. 2001. Epidemiological investigation of tinea pedis in groups of healthy students, research workers and female wearing boots. *Japanese J. of Med. Mycology*, 42: 253–258.

Index

A

Abelmoschus esculentus moench, 310-317
- carotene content expressed as optical density, 312
- indole-3-acetic acid oxidase expressed, 314
- introduction, 310
- materials and methods, 311
- meteorological data, 311
- pectin content expressed as percent calcium pectate, 316
- polyphenbol oxidase enzyme activity expressed as change in optical density, 315
- results and discussion, 312317
- xanthophyll content expressed as optical density, 313

Abelmoschus, 316

A. esculentus, 311

Absorption spectra, 224

Abusaria, Sharda, 39

Acridotheres tristis, 157-159

ACTH, 159, 257

Administration, 275

Agarkar, S.V., 193

Age, 69

Age-related changes in long bones of Indian toad, *Bufo melanostictus*, 58-69
- calcium and phosphorus, 59
- collagen, 59
- introduction, 58
- materials and methods, 59
- results, 59
  - Ca/P molar ratio, 62
  - correlation between body between body weight and phosphorus content, 62
  - – – – – and calcium content, 59, 623

Air borne fungal spores at Sharavathi reservoir, Karnataka, India, 319-322
- different types of fungal spores at Sharavathi reservoir, 322
- introduction, 319
- materials and methods, 320
- number of colonies of different fungal species, 321
- results and discussion, 320

Airborne mycoflora, 319

Akela, B.P., 157

Albino mice, 115

ALT, 249

Alternation in oxygen consumption in freshwater snail *Bellamya bengalensis* during pesticide exposure, 206-208
- introduction, 206
- material and methods, 207
- results and discussion, 207

Ambystome mexicanum, 110

Amena, 32

American College Dictionary, 188

Analysis and seasonal comparative study of Amanishah Nallah in Sanganer town, Jaipur, 187-192
- Biochemical Oxygen Demand (BOD), 190
- calcium and magnesium hardness, 191
- carbonate and non-carbonate hardness, 191
- causes of pollution, 188

chlorides, 191
fluorides, 192
introduction, 188
materials and methods, 188
nitrates, 192
pH, 190
physico-chemical parameters of various sampling stations, 189
results and discussion, 189
temperature, 190
total alkalinity, 191 Total Hardness (TH), 191
Total Dissolve Solid (TDS), 191
Total Solid (TS), 191
Total Suspended Solid (TSS), 191

Anand, J. Justin, 261
Ananthapadmanaban, S., 210
Anbarasu, K., 236
Andia, B.N., 58
Angelin, A., 285
Animal tissues, 62
Anitha, S., 152
Anopheles, 162
Antifertility effect of neem bark in male albino mice, 274-276
introduction, 274
materials and methods, 274
results and discussion, 275
Antimicrobial activity of *Premna tomentosa* willd. to chosen pathogenic bacteria, 79-82
activity index of solvent extracts of *P. tomentosa*, 81
inhibition zone of solvent extracts of *P. tomentosa*, 81
introduction, 79-80
materials and methods, 80
results and discussion, 81-82
Antimicrobial activity of *Syzygium aromaticum* and *Millingtonia hortensis* against human pathogens, 231-234
antimicrobial activity of
Millingtonia hortensis, 233
Syzygium aromaticum, 233
Candida albicans, 234
Escherichia coli, 234
Klebsiella pneumomae, 234
materials and methods, 232
Pseudomonas sp., 234
Staphylococcus aureus, 234
Anusuya, D., 127
APHA, 23, 106, 123, 227, 290
Arsenic affect on blood and tissue glucose concentrations in a fish model, 305-308
blood and tissue glucose concentration in the fish model, 307
introduction, 305
materials and methods, 306
results, 307
Artificial neural networks, 262, 263
Aspergillus, 320
Assessment of pollution soils near aluminium industry Warangal, Andhra Pradesh, 198-203
biological characteristics, 202
chemical characteristics, 201-202
enzyme characteristics, 203
introduction, 198
materials and methods, 198
physical characteristics, 199
results and discussion, 199
AST, 249
AWWA/WPCF, 106

B

Babu, S. Madhan, 210
Baccillus subtilis, 165-168
Bajwa, 330
Bakkialalakshmi, S., 221
Banerjee, Dhriti, 114
Barik, B.B., 84
Bayer LTD, Bombay, 170
Behera, H.N., 58
Bellamya bengalensis, 206-208
Belsare, D., 257
Beta vulgaris, 11, 12
Bhutra, M.K., 178
Bhuvaneshwar, N., 118, 169
Biochemical Oxygen Demand (BOD), 21
Biodegradation of tannery effluent using *Pseudomonas aeruginosa* and *Baccillus subtilis*, 165-168
biodegradation of tannery effluent using *Pseudomonas aeruginosa*, 167
introduction, 165
materials and methods, 166
Biological materials, 96
Biological treatment system, 165
Bist Doab Tract, 335
Blocks, 333
BOD, 24, 26, 40, 102, 125, 139, 166, 290
Bohra, Chandan, 1
Borah, Mahananda, 278
Bose, B., 20
BTL Cattha Factory, 51
Bufo melanostictus, 127-129
Bufo melanostictus, 58-69
Buprenorphine in mice, 114-117

C

Calcium levels, 62
Carassills auratus, 110
CCRAS, 80
Ceratophyllum demersum, 3
Certain herbal ethno-medicines used by the Karbi tribe of Assam, India, 252-255
introduction, 252-253
list of medicinal plants used by the Karbis, 254-255
methodology, 2 53
results and discussion, 253

Chakravarty, Anju Puri, 274
Chandrakala, D., 221
Changes in Changes in sodium and potassium ratio in pressmud with influence of earthworm *Eisenia foetida*, 46-49
  histogram showing the % increase of potassium and sodium, 48
  introduction, 46
  materials and method, 47
  results and discussion, 49
Channa orientalis: 104-111
C. striata, 169, 170
  discussion, 171
  introduction, 169
  material and methods, 170
  results, 170-171
  sublethal concentration of oxydemeton-methyl exposure on the blood glucose and ovarian glycogen, 170-171
Charya, M.A. Singara, 198
Chatterjee, S., 84
Chaudhari, P.G., 267
Chelladurai, V., 80
Chemical impact on the histologial studies thyroid in the freshwater fish *Channa orientalis*: 104-111
  assessment of thyroid activity, 107
  materials and methods, 106-107
  results and observations, 107
    effect of thiourea treatment on the thyroid gland of the fish, 108
    histology of thyroid exposed to low, 108
    thiourea treated fish thyroid, 107
Chemical properties, 200
Choes, 324
Christy, I., 127
Chronic kidney disease, 95
CK, 249
Clarias bastrachus, 105, 306, 307
Clinical importance of hypertension, diabetes mellitus and heart rate associated with acute myocardial infraction, 218-220
  diabetic status, 219
  heart rate, 219
  hypertension, 219
  introduction, 218-219
  latent diabetic, 219
  materials and methods, 219
  results and discussion, 219-220
COD, 102, 166
Comparative study of community biogas plant versus family size biogas plant, 300-304
Comparison of mosquito fauna in Srivilliputhus town and Krishnankovil village, Tamil Nadu, 161-163
  discussion, 162
  introduction, 161-162
  materials and methods, 162
  mosquito recorded in Srivilliputhus town and Krishnankovil village, 163
Consider, 82
Crude drugs, 232
Culex, 162
CWC, 325
Cycling of minerals, 1

D

D.S. College Aligarh, U.P., 47
Das, Ashutosh, 210
Das, P.K., 278
Decomposition processes, 21
Department of Animal Science and Dairy Science, 268
Deshmukh, S.V., 104
Design flood by frequency analysis for relatively ungauged small watersheds, 337-346
  annual flood frequency curve log normal method, 343
    - - - - Gumble's and least square's method, 345
  calculate coefficient of skewness, 342
    - - - variation, 341
  flood flows using log-normal parameters, 342
  generated data, 340
  gumble distribution, 341, 344
  Hazen's method, 341, 344
  index map of Dholbaha dam project, 339
  introduction, 337-338
  log normal distribution, 338
  log-Pearson type III distribution, 341
  methods and materials, 338
  parameters of distribution, 342
  peak annual floods at Dholbaha Dam Site, 338
  probability of annual peak flow, 346
  results and discussion, 342
  Ven te Chow's method, 341, 344
Devi, C.S. Shymala, 248
Devi, R.K. Rajeshwari, 289
Devi, Y. Nanda, 241
Devika, R., 118
Dhankhar, Monika, 39
Dhasarathan, P., 261
Dholbaha dam, 327
Diabetic coma, 95
Diel variation in waterfowl during winter at Sirpur Tank, Indore, 141-150
  Diel percentage composition of various groups of waterfowl at Sirpur Tank, 149

-- in cormorant and darter at Sirpur Tank, 147
--- gallinules and coot at Sirpur Tank, 148
--- geese and ducks at Sirpur Tank, 148
--- grebe, 147
--- herons and egrets at Sirpur Tank, 147
--- jacana at Sirpur Tank, 148
--- percent composition of various groups of waterfowl at Sirpur Tank, Indore, 146
--- shore birds at Sirpur Tank, 149
--- waterfowl group at Sirpur Tank, 149
group-wise Diel variation of waterfowl at Sirpur Tank, Indore, 146
introduction, 141-142
material and methods, 142
mean Diel winter count of waterfowl at Sirpur Tank, Indore, 145
morning, 150
noon, 150
numbers of species per family, 144
results and discussion, 142-150
waterfowls of Sirpur Tank, Indore, 143
Dimethyl sulfoxide (DMSO), 84, 85, 90
Dissolved Oxygen (DO), 21, 30, 166, 213, 290
DNA, 14, 16
DOK, 106
Drinking water quality of Basavanhole tank with reference to physico-chemical characteristics, 122-126
introduction, 123
materials and methods, 123
results and discussion, 123
physico-chemical parameters of basavanahole tank, 124
Durairajan, M.S., 261
Dust accumulation studies at Hyderabad, 32-38
composition of living matter in the air dust samples, 36
dust pollutants in different areas of twin cities of Hyderabad, 37
-- in Urban, Suburban, Rural and protected areas, 37
introduction, 32
list of trees and shrubs from twin cities of Hyderabad, 34-35
materials and methods, 33
microns of aggregate particles collected from the air samples, 37
result and discussion, 33
Dutta, B.K., 347

E

Ecological research, 17
Economic appraisal of biogas for cooking and electricity generation, 294-299
introduction, 294
materials and methods, 295
results and discussion, 295
cost of biogas plant, 295
economics of community biogas plant, 295
- for cooking purpose, 295
operating cost of biogas plant for electricity generation, 297
payback periods of community biogas plant for cooking purpose, 298
Effect of DAP toxicity on certain hematological parameters of *Acridotheres tristis*, 157-159
introduction, 157
results and discussion, 158
effect of DAP toxicity on blood parameters of *Acridotheres tristie*, 158
Effect of levels of sodium alginate on chemical quality of *dahi*, 267-269
introduction, 268
materials and methods, 268
results and discussion, 268-269
levels of sodium alginate on chemical quality of cow and buffalo milk dahi, 269
Effect of metal poisoning on total body ascorbic acid in *Sphaerodema rusticum*, 173-177
body ascorbic acid concentration in insects on treatment with
copper sulphate, 175
Mohr's salt, 177
cobalt nitrate, 174
copper sulphate, 174
experimental, 174
introduction, 173-174
Mohr's salt, 174
result and discussion, 175
body ascorbic acid concentration in insects on treatment with cobalt nitrate, 176
treatment with cobalt nitrate, 175
-- Mohr's salt, 176
total body ascorbic acid concentration of control insects, 175
treatment with copper sulphate, 175
Effect of multifloral honey on blood glucose profile of rabbits after induced lipidosis, 257-259
blood glucose level in rabbits, 258

discussion, 258
introduction, 257
materials and methods, 258
observations, 258
Effect of polyvinyl pyrrolidone and dimethyl sulphoxide on ethyl cellulose transdermal patches of verapamil hydrochloride, 85
Effect of scrotal heating on the reproductive organs of the laboratory rat, *Rattus norwegicus*, 91-93
discussion, 93
introduction, 91
materials and methods, 92
observations, 92
histology of epididymis, 93
-- the testis, 92
untreated controls, 92
Effectiveness of three botanicals against dermatophytes, 347-352
efficacy of *T. chebula* aqueous extract, 350
-- *P. granatum* aqueous extract, 350
-- *P. betel* aqueous extract, 351
introduction, 347-348
materials and methods, 348
Microsporum fulvum (Mf), 349
M. gypseum (Mg), 349
T. mentagrophytes (Tm), 349
T. rub rum (Tr), 348
Trichophyton tonsurans (Tt), 348
Ehteshamuddin, S., 173
Eisenia foetida, 47
Electromagnetic field (EMF), 271
Elodea canadensis, 2
Environmental and ecological impact of genetically modified organisms, 10-18
Environmental induced changes in the biomodal gas exchange and haematology of facultative air-breathing fish, *Mystus punctatus*, 118-121
discussion, 120-121
environmental induced changes in the bimodal respiration of *M. punctatus*, 120
introduction, 118
materials and methods, 119
one hour air-exposure on the haematology of *M. punctatus*, 120
results, 119
Environmental problem, 11
Environmental risk, 16
Environmental stress, 311
Errant vascular hydrophytes, 5
Estimating the oxygen requirement in a mass bathing tank
actual dissolved oxygen, 215
chlorine final Vs DO final, 214
estimation of typical retention time in the tank, 212
-- DO contributed by chlorine at the outlet, 212
---- contributed by atmosphere, 213
experimental procedure, 211
chemical modeling of dissolved oxygen, 211
chlorine dosage at the inlet, 211
introduction, 210-211
nomogram for saturation concentration of dissolved oxygen, 214
simplified line sketch of Mahamaham tank, 212
Ethyl cellulose (EC), 84
European Economic Community (EEC), 40

F

Facing problems, 197
Family hydrocharitaceae, 3
Farmers, 47
Floating, 7
Free fatty acid (FFA), 257
Fungal spores, 320

G

Genetic effects, 15
Genetic engineering, 11, 13
Genetically modified organisms (GMOs), 14
Geographic information, 17
Gogoi, Anima, 278
Gogoi, Sanjib kumar, 252
Gopalakrishnan, S., 79
Goyal, Sachin, 271
Grewal, N.S., 329
Growing field, 157
Growth hormone (GH), 15
GTF complex, 259
Gulati, Suman, 236
Gupta, Er. Deepak, 294, 300, 324, 337

H

Handique, Pratap J., 252
Hans, Bir, 274
Haryana, 40
Heavy metal pollution in river Tunga, Bhadra and Tungabhadra at Kudali, Karnataka, 100-103
Hegde, B.A. Kumara, 136
Heteropneustes fossilis, 110
Histological alterations in tadpoles of *Bufo melanostictus*, exposed to a sublethal concentration of chromium, 127-129
Histological approach, 128
Hoshiarpur, 330
Hydraulic structures, 325
Hydrological studies for dams in ungauged small catchments, 324-328

available live storage capacity of reservoir, 328
capacities in reservoirs after sediment deposition, 327
design flood by frequency analysis, 326
estimated values of design flood calculated by different methods, 325
introduction, 324
methods and materials, 325
peak discharges, 326
results and discussion, 325

I

IAA, 314, 317
ICMIT, 262
Incidence, 12
Influence of Influence of Thermal stratification on DO, 20-30
Influence of Thermal stratification on dissolved oxygen in Subhas Sarobar, Kolkata: introduction, 20
hypolimnetic dissolved oxygen, 24
map of Subhans Sarobar Lake, 22
materials and methods, 21-23
relationship of density and water temperature, 23, 24
– – water temperature and dissolved oxygen, 24-26
study area, 21
vertical distribution of temperature, 27-29
Institute of Microbial Technology (IMTECH), 237
Investigation of physico-chemical condition of major effluent points of Nambul river, Manipur, India: physio-chemical analysis of Nambul river, 291

J

Jalandhar, 330
JASCO-UVIDEC, 222
Jayanti, B., 257
Jayavarshni, G., 285
Jegadeeshkumar, D., 236
Johri, Tripati, 271
Joice, P. Esther, 118
Joice, P. Esther, 169
Joshi, M.V., 131
Juneja, Meenu, 95

K

Kalavathi, D., 169
Kale, A.D., 267
Kamble, D.K., 267
Kapoor, Baljeet S., 329
Kapurthala, 330
Karbi Anglong district, 253
Karbi tribe of Assam, India, 252-255
Karmakar, S., 347
Karuppasamy, K., 161
Kavitha, N., 72
Kharif, 332
Khullar, N.K., 324, 337
Kiran, B.R., 122, 226, 319
Kishan Sahkari Chini Mill Ltd., 47
Krishnapalai, 80
Kudali, 101
Kulkarni, K.M., 104, 206
Kumar, Anant, 157
Kumar, Arvind, 1
Kumar, Dinesh, 187
Kumar, K. Harish, 226, 319
Kumar, K.P. Ravindra, 122
Kumar, Pramod, 46
Kumar, Vijay, 271
Kumari, B. Lalitha, 198
Kumari, S. Binu, 118
Kumbakonam, 211
Kungumapriya, R., 236
KVIC, 300

L

LDH, 249
Limnological characteristics of Guruvayanakere Pond, 136-139
Literature levels, 97
Log-normal, 338

M

M. gypseum (Mg), 349
Magnesium in scalp hair and fingernails in relation to different parameters, 95-98
analysis, 97
introduction, 95
materials and methods, 96
materials and methods: wet acid digestion and preparation of water clear solution, 96
results and discussion, 97
variation of mean magnesium concentration in hair and nails in different age groups, 98
Mahesh, V., 152
Maili dam, 327
Male subject of Ajmer, 96
Malhotra, Manjeet, 141
Manjappa, S., 319
Mansonia, 162
Manufacture of sugar, 47
Matrix of life, 40
Mazumdar, A., 20
MC EME Mess, 33, 38
Medicine, 232
Mehra, Rita, 95
Metabolic activity, 208, 310
Metal levels, 96
Microbial contamination in drinking water: bacteriological standards, 181
cause, detection and remedy, 178-184
conventional method for detection of fecal coliforms, 182
coliforms, 181
fecal coliforms (thermo tolerant coliforms), 181
introduction, 178
microbiological analytical techniques, 180-181

some facts about water-related diseases, 180
untreated water maximum allowable limit, 181
disinfection, 184
filtration, 184
membrane filter technique (MF technique), 183
need for rapid technique, 183
sedimentation, 184
Standard Plat Count (SPC), 183
treatment for removal of organisms, 183
water disease outbreaks, 179
water entering the distribution system in piped supply after chlorination or disinfeciton, 181
Microorganisms found in gums and adhesives from post offices of Madurai city, 285-288
fungi identified from book binding shops, 288
introduction, 286
materials and methods, 286
partially characterization of bacteria obtained from post offices, 287
results and discussion, 286
Microsporum audounii, 347
Midya, Trilochan, 305
Miles, 290
Ministry of Non-Conventional Energy Sources, Govt. of India, New Delhi, 295
Mishra, Bhupesh Kumar, 50
Mishra, S.K., 84
Modern, 15
Modified Dairy Based Medium (MDBM), 237, 238
Mohr's salt, 174, 176, 177
Mukherji, S., 310
Mumtazuddin, S., 173
Mycelial growth, 349
Mysore paper mills, 101, 103
Mystus punctatus, 118, 119, 120
Mythili, P., 165

N

Nambul river, Manipur, India, 289-292
Narayana, J., 122, 226
Natarajan, G.M., 118, 169
Nawanshahar, 330
Neem bark, 274-276
Neem plant, 274
Nitrate pollution, 332
Nobel laureate, 40
Noise pollution level at Rewa town, 50-56
ambient noise parameter for various location at Rewa town winter season, 55
materials and methods, 51-52
noise parameter for different commercial zone, 54
– – – residential zone, 54
– – silence zone, 55
– – – traffic zone, 53
– – of industrial zone, 53
parameters studies, 52
pollution control standard for noise, 52
result and discussion, 52
Non-phumdi species, 7
North-eastern parts, 337
Notopterus notopterus, 121
NTU, 123
Nurpur Bedi block, 330
Nutrient agar (NA), 286
Nutrient status of Kanale tank, Sagara taluk with reference to diversity of aquatic macrophytes, 226-229
introduction, 226
macrophytic diversity in kanal tank, 228
methods, 227
morphometric and limnological features of Kanale tank, 227
nutrient status of Kanale tank, 229
results and discussion, 228
study area, 227

O

Objective, 330

P

P. granatum, 349
Palaeocurrent analysis from current beddings and pebbles of tura sandstone group of rocks occurring in and around Rangram area, Meghalaya, 278-284
introduction, 278
geology of the area, 279
method of study, 279
field investigation and interpretations, 280
table for vector calculation, 281
victor analysis of the maximum intercept of pebble, 282
– – – azimuth of the forest beds of current bedding, 283
discussion and conclusion, 283
Palanisamy, P., 118, 169
Pandey, Arvind K., 218
Parthi, N., 118
Particular area, 1
PAU, 295
Pawar, B.K., 267
Pawar, C.T., 131
Pawar, K., 141
Penicillium, 320
Periophthalmodon schlosseri, 121
Phormidium, 266
Physiological function, 173

Podaganari, 80
Pollen can function, 11
Polythene bottles, 188
Polyvinyl pyrrolidone (PVP), 84, 85
Polyvinyl pyrrolidone and dimethyl sulphoxide on ethyl cellulose transdermal patches, 84-90
 data showing the formulation of transdermal drug delivery system, 85
 evaluation of film, 86
 film characteristics, 86
 flatness test, 87
 method, 85
 preparation of matrix film, 85
 preparation of skin, 86
 results and discussion, 90
 scanning electron microscope study of prepared patches, 89
 SEM study, 88
  vitro skin permeation study, 88
 skin irritation test, 86
 WVT test, 87
Ponneelan, K.T.P.B., 165
Poonkothai, M., 165
Potamogeton, 2
Potato dextrose agar (PDA), 286
Potential difference across the cell of skin around the transmission tower, 271-273
Potential risks, 12
Potey, G.G., 218
PPO, 316, 317
Prakash, D.J., 127
Prakash, M.M., 141
Prasad, Mahavir, 187
Prathi, N., 169
Prediction of nitrate pollution of groundwater, 329-335
 introduction, 330
 map of observation wills (Bist Doab Tract, Punjab), 331
 materials and methods, 332
 nitrate N-content in groundwater, 333
 results and discussion, 332
Premna tomentosa, 79, 80
Primary significance, 97
Process of introgression, 11
Production and biochemical analysis of bacteriocin from immobilized and non-immobilized *Lactococcus lactis*, 236-240
 antibacterial activity of bacteriocin against various food pathogens, 239
 bacteriocin protein content of the cultures used in present study, 238
 biochemical characterization of food borne pathogens, 238
 discussion, 239
 introduction, 236-237
 materials and methods, 237
 results, 238
Pseudomonas aeruginosa, 165-168
Public debate, 12
Punjab Agricultural University, Ludhiana, 295
Purushothama, R., 122, 226, 319
Puttaiah, E.T., 100, 122, 319
PVP, 90

Q

Quenching of diphenylamine by benzoic acid and carbon tetrachloride in chloroform, 221-225
 energy, ionization potential, electron affinity, stoke shift an solvent parameter, 225
 experimental materials and methods, 222
 fluorescent intensity rations of dipheylamine of different concentration in chloroform, 222
 introduction, 221-222
 results and discussion, 222

R

Rabi, 332
Rahman, I., 347
Rajalakshmi, M., 285
Rajalakshmi, S., 72
Ramadas, K., 100, 136
Rani, Mary Esther, 285
Rani, Sobha, 32
Rao, Nirmala Babu, 32
Rathi, J. Matrtin, 79
Rattus norwegicus, 91-93
Regional Entomological Research Station at Virudhunagar, 162
Removal of heavy metals from electroplating industrial effluent using plants-phytoremediation, 152-156
Research officer (Botany), 80
Results and discussion, 233
Revathi, K., 236
Rhizopus, 320
River Arasalar, 211
River Bhadra, 102
River Cauvery, 211
River Nambul, 290
River Tunga, 102
River Tungabhadra, 102
Rohankar, P.H., 206
Role of soft computing and neural network in medical analysis, 261-263
Role of zinc in experimental myocardial infraction influence of abana: ayruvedic formulation, 248-251
 introduction, 248
 materials and methods, 249
 results and discussion, 249-251
 serum and heart of control and experimental animals, 249
Ropar district, 330
Rosee, P. Mourin, 285

Roy, D., 20
Rush, 257

S

Sabouraud Dextrose Agar (SDA), 232
Saha, Nimai Chandra, 305
Sahoo, S.K., 84
Sahu, D., 84
Salmonella antibodies, Triuchengode, Tamil Nadu, 72-76
antibodies against Salmonella in normal individuals reported by various workers, 76
antigenic structure of *salmonella*, 73
H antigens, 73
O antigens, 73
Vi antigens, 73
carriers in normal individuals in relation to sex, 76
classification of *Salmonella*, 73
introduction, 72-73
materials and methods, 73
precaution, 74
techniques, 74
results and discussion, 74-76
slide agglutination test, 74
tube agglutination test, 74
Salmonella antigens, 74
S. paratyphi, 72
S. typhi, 72, 74
Salvinia cuculata, 3
Samal, N.R., 20
Sandhya, B., 231
Santhi, R., 221
Sarkar, Nirmal Kumar, 114, 305
Saroor Nagar, 33
Sarotherodon mossambicus, 172
Sasikala, G., 118, 169
Sasikumar, C. Sheela, 152, 248
Sastry, K.V., 39
Scrotum, 93
Secunderabad Railway Station, 33
Sedimentation studies, 325
Sen, Supatra, 310
Serious problem for sugar, 12
Setaria italica, 14
Shanthi, B., 221
Shashikala, 257
Shivran, Hari Singh, 187
Shukla, Vineeta, 39
Shweta, 46
Singh, Bijay, 329
Singh, Indervir, 294
Singh, P.K., 241
Singh, R.V., 187
Singh, Sukhwinder, 300
Soil microorganisms, 200
Sonepat, 40
Soni, Ambica, 178
Sooch, Er. S.S., 337
Sooch, Er. Sarbjit Singh, 294, 300, 324, 329
Sooravan, T., 161
Sound level Ichalkaranji city, Maharashtra, 131-135
ambient air quality standards in respect of sound area, 132
database and methodology, 132
Ichalkaranji city
ambient sound level in different areas during night time, 134
growth of powerlooms and population, 133
monitoring location and ambient sound level in different areas, 133
introduction, 131-132
recommendations, 135
results and discussion, 132
Species distribution of the family verbenaceae in the valley district of Manipur, 241-247
introduction, 241
materials and methods, 242
results and discussion, 242-246
some commonly available verbenaceae species in the valley district of Manipur, 243
phytosociological studies of *Callicarpa macrophylla*, 244
phytosociological studies of *Clerodendron glandulosum*, 245
Spongy, 47
Srihari, V., 210
Standard and United Public Health Service (UPHS), 40
Standard Project Flood (SDF), 326
Staphylococcus aureus, 80, 81, 232
Status of drinking water quality awareness and its impact on student health
area, 194
causes of health and remedial measures, 197
drinking water quality awareness among school students of Buldana District, 196
findings and discussion, 195
hypotheses, 194
impact of water quality on health, 197
introduction, 193-194
limitations of the study, 195
method, 195
objective, 194
physico-chemical and microbiological analysis of drinking water samples from schools of Buldana District, 196
programmes organised by schools, 195
samples collected from schools of Buldana District of Maharashtra, 194
samples, 194

status of quality of drinking water availability at schools, 195
study of schools of Buldana District, 193-197
tools, 195
Suchitra, R., 165
Sujatha, A., 231
Sundar, I., 10
Suresh, J. Justin, 261
Suresha, G., 100, 136
Survey of Medicinal and Aromatic Plants Unit-Siddha, 80
Symploca
new report, 265, 265-266

T

T. mentagrophytes (Tm), 349
TAMAS, 51
TDS, 124, 166, 189
Temporary threshold shift (TTS), 51
Teron, Robindra, 252
Therohytes, 5, 7
Thombre, B.S., 193
Thrombocytopenic effect of buprenorphine in mice, 114-117
Tilapia mossambica, 306, 308
Tirunelveli District, Tamil Nadu, 80
Toxic effects, 348
Transgenic crops, 14
Trichophyton mentagrophytes, 347, 348
Trichophyton tonsurans (Tt), 348

U

Udhuwa Lake, Rajmahal, 1-7
Uma, S., 72
USPHS, 181
UV irradiation, 184

V

Vacuum filter, 47
Valuable toll, 128
Various authors, 222
Vats, R.P., 271
Vegetation ecology crisis of macrophytes of Udhuwa Lake, Rajmahal, 1-7
biological spectrum of the flora of Uduwa Lake, 6
discussion, 6
family-wise distribution of the macrophytes, 5
floristic composition in the study sites, 4
group-wise distribution of marcrophytes in the five study sites, 3
introduction, 1-2
life forms, 5
– from classification of macrophytic species in Udhuwa Lake, 6
materials and methods, 3
morphometry and bathymetric characters of Udhuwa Lake, 2
results, 3
study site, 2
Vibrio cholerae, 179
Vidyarani, W., 91
Vijayalakshmi, P., 236
Vishweswaraya iron and steel industries, 101
Vitex trifolia, 82

W

Water affected with bicycle manufacturing industrial wastes, 39-44
Water quality index, 40
Water supply in Indian, 179
WHO, 179, 231, 330, 348
WQI, 43, 44

X

Xiphophorus maculatus, 105

Y

Yadav, R.K., 157
Yashovarma, B., 136